Fusarium

Fusarium

Mycotoxins, Taxonomy and Pathogenicity

Editor

Łukasz Stępień

MDPI • Basel • Beijing • Wuhan • Barcelona • Belgrade • Manchester • Tokyo • Cluj • Tianjin

Editor
Łukasz Stpień
Department of Pathogen
Genetics and Plant Resistance,
Institute of Plant Genetics,
Polish Academy of Sciences
Poland

Editorial Office
MDPI
St. Alban-Anlage 66
4052 Basel, Switzerland

This is a reprint of articles from the Special Issue published online in the open access journal *Microorganisms* (ISSN 2076-2607) (available at: http://www.mdpi.com).

For citation purposes, cite each article independently as indicated on the article page online and as indicated below:

LastName, A.A.; LastName, B.B.; LastName, C.C. Article Title. *Journal Name* **Year**, *Article Number*, Page Range.

ISBN 978-3-03943-408-4 (Hbk)
ISBN 978-3-03943-409-1 (PDF)

Cover image courtesy of Lukasz Stepien.

Contents

About the Editor . . . . **vii**

Łukasz Stępień
Fusarium: Mycotoxins, Taxonomy, Pathogenicity
Reprinted from: *Microorganisms* **2020**, *8*, 1404, doi:10.3390/microorganisms8091404 . . . . **1**

Akos Mesterhazy, Andrea Gyorgy, Monika Varga and Beata Toth
Methodical Considerations and Resistance Evaluation against *F. graminearum* and *F. culmorum* Head Blight in Wheat. The Influence of Mixture of Isolates on Aggressiveness and Resistance Expression
Reprinted from: *Microorganisms* **2020**, *8*, 1036, doi:10.3390/microorganisms8071036 . . . . **5**

Maria E. Constantin, Babette V. Vlieger, Frank L. W. Takken and Martijn Rep
Diminished Pathogen and Enhanced Endophyte Colonization upon CoInoculation of Endophytic and Pathogenic *Fusarium* Strains
Reprinted from: *Microorganisms* **2020**, *8*, 544, doi:10.3390/microorganisms8040544 . . . . **33**

Wioleta Wojtasik, Aleksandra Boba, Marta Preisner, Kamil Kostyn, Jan Szopa and Anna Kulma
DNA Methylation Profile of *β-1,3-Glucanase* and *Chitinase* Genes in Flax Shows Specificity Towards *Fusarium Oxysporum* Strains Differing in Pathogenicity
Reprinted from: *Microorganisms* **2019**, *7*, 589, doi:10.3390/microorganisms7120589 . . . . **45**

Tomasz Góral, Aleksander Łukanowski, Elżbieta Małuszyńska, Kinga Stuper-Szablewska, Maciej Buśko and Juliusz Perkowski
Performance of Winter Wheat Cultivars Grown Organically and Conventionally with Focus on Fusarium Head Blight and *Fusarium* Trichothecene Toxins
Reprinted from: *Microorganisms* **2019**, *7*, 439, doi:10.3390/microorganisms7100439 . . . . **65**

Xavier Portell, Carol Verheecke-Vaessen, Rosa Torrelles-Ràfales, Angel Medina, Wilfred Otten, Naresh Magan and Esther García-Cela
Three-Dimensional Study of *F. graminearum* Colonisation of Stored Wheat: Post-Harvest Growth Patterns, Dry Matter Losses and Mycotoxin Contamination
Reprinted from: *Microorganisms* **2020**, *8*, 1170, doi:10.3390/microorganisms8081170 . . . . **87**

Andrea György, Beata Tóth, Monika Varga and Akos Mesterhazy
Methodical Considerations and Resistance Evaluation against *Fusarium graminearum* and *F. culmorum* Head Blight in Wheat. Part 3. Susceptibility Window and Resistance Expression
Reprinted from: *Microorganisms* **2020**, *8*, 627, doi:10.3390/microorganisms8050627 . . . . **105**

Valentina Spanic, Zorana Katanic, Michael Sulyok, Rudolf Krska, Katalin Puskas, Gyula Vida, Georg Drezner and Bojan Šarkanj
Multiple Fungal Metabolites Including Mycotoxins in Naturally Infected and *Fusarium*-Inoculated Wheat Samples
Reprinted from: *Microorganisms* **2020**, *8*, 578, doi:10.3390/microorganisms8040578 . . . . **123**

Jonas Vandicke, Katrien De Visschere, Siska Croubels, Sarah De Saeger, Kris Audenaert and Geert Haesaert
Mycotoxins in Flanders' Fields: Occurrence and Correlations with *Fusarium* Species in Whole-Plant Harvested Maize
Reprinted from: *Microorganisms* **2019**, *7*, 571, doi:10.3390/microorganisms7110571 . . . . **139**

Ying Tang, Pinkuan Zhu, Zhengyu Lu, Yao Qu, Li Huang, Ni Zheng, Yiwen Wang, Haozhen Nie, Yina Jiang and Ling Xu
The Photoreceptor Components FaWC1 and FaWC2 of *Fusarium asiaticum* Cooperatively Regulate Light Responses but Play Independent Roles in Virulence Expression
Reprinted from: *Microorganisms* **2020**, *8*, 365, doi:10.3390/microorganisms8030365 . . . . **161**

Molemi E. Rauwane, Udoka V. Ogugua, Chimdi M. Kalu, Lesiba K. Ledwaba, Adugna A. Woldesemayat and Khayalethu Ntushelo
Pathogenicity and Virulence Factors of *Fusarium graminearum* Including Factors Discovered Using Next Generation Sequencing Technologies and Proteomics
Reprinted from: *Microorganisms* **2020**, *8*, 305, doi:10.3390/microorganisms8020305 . . . . **179**

Giovanni Beccari, Łukasz Stępień, Andrea Onofri, Veronica M. T. Lattanzio, Biancamaria Ciasca, Sally I. Abd-El Fatah, Francesco Valente, Monika Urbaniak and Lorenzo Covarelli
In Vitro Fumonisin Biosynthesis and Genetic Structure of *Fusarium verticillioides* Strains from Five Mediterranean Countries
Reprinted from: *Microorganisms* **2020**, *8*, 241, doi:10.3390/microorganisms8020241 . . . . **209**

Lydia Woelflingseder, Nadia Gruber, Gerhard Adam and Doris Marko
Pro-Inflammatory Effects of NX-3 Toxin Are Comparable to Deoxynivalenol and not Modulated by the Co-Occurring Pro-Oxidant Aurofusarin
Reprinted from: *Microorganisms* **2020**, *8*, 603, doi:10.3390/microorganisms8040603 . . . . **227**

Thuluz Meza-Menchaca, Rupesh Kumar Singh, Jesús Quiroz-Chávez, Luz María García-Pérez, Norma Rodríguez-Mora, Manuel Soto-Luna, Guadalupe Gastélum-Contreras, Virginia Vanzzini-Zago, Lav Sharma and Francisco Roberto Quiroz-Figueroa
First Demonstration of Clinical *Fusarium* Strains Causing Cross-Kingdom Infections from Humans to Plants
Reprinted from: *Microorganisms* **2020**, *8*, 947, doi:10.3390/microorganisms8060947 . . . . **243**

About the Editor

Łukasz Stpień is a Full Professor of Agricultural Sciences at the Institute of Plant Genetics, Polish Academy of Sciences, Poznań (Poland). He earned his M.Sc. diploma in Plant Biotechnology (1999) at the University of Life Sciences in Poznań, Poland. He then completed his Ph.D. (2005), working on the identification of resistance genes in wheat, and his habilitation (2014), studying mycotoxin biosynthetic genes in *Fusarium* species (2014), both at the Institute of Plant Genetics, Polish Academy of Sciences, Poznań. Currently, Prof. Stpień is the Head of the Department of Pathogen Genetics and Plant Resistance at the IPG PAS in Poznań. He is also the leader of the Plant–Pathogen Interaction Team of the same Department. He has co-authored more than 60 research articles and reviews in JCR-listed journals, 4 patents and ca. 100 conference abstracts. He also supervised several Bachelor's, M.Sc. and Ph.D. theses. As a reviewer, he performed ca. 150 reviews of scientific articles and reviewed international Ph.D. theses and Horizon 2020 proposals as an Expert of the European Commission. Prof. Stpień has been a visiting researcher at Technical University in Munich (Germany); Institute of Epidemiology and Resistance, Aschersleben (Germany); Institute of Sciences of Food Production ISPA, CNR, Bari (Italy); and Technical University of Denmark, Lyngby (Denmark). His current research interests include plant–pathogen interaction studies using molecular, metabolomic and proteomic tools; using filamentous fungi for biotransformation of bioactive compounds; and looking for new tools for controlling feed and food contamination with fungal pathogens and mycotoxins.

microorganisms

Editorial

Fusarium: Mycotoxins, Taxonomy, Pathogenicity

Łukasz Stępień

Department of Pathogen Genetics and Plant Resistance, Institute of Plant Genetics, Polish Academy of Sciences, Strzeszyńska 34, 60-479 Poznań, Poland; lste@igr.poznan.pl; Tel.: +48-61-655-0286

Received: 7 September 2020; Accepted: 10 September 2020; Published: 12 September 2020

It has been over 200 years since *Fusarium* pathogens were described for the first time, and they are still in the spotlight of researchers worldwide, mostly due to their mycotoxigenic abilities and subsequent introduction of harmful metabolites into the food chain. The accelerating climatic changes result in pathogen populations and chemotype shifts all around the world, thus raising the demand for continuous studies of factors that affect virulence, disease severity and mycotoxin accumulation in plant tissues. This Special Issue summarizes recent advances in the field of *Fusarium* genetics, biology and toxicology.

An emphasis was bestowed upon trichothecene-producing species. *Fusarium graminearum* and *F. culmorum* are the prevailing deoxynivalenol (DON) producers. Inoculation of wheat with the mixture of isolates resulted in lower disease incidence than in the case of the single aggressive isolate, showing a considerable level of competition between the genotypes during the colonization of the plant [1]. A similar observation can be made when a pathogenic *Fusarium* strain is co-inoculated with the non-pathogenic endophytic strain. The endophyte decreases the efficiency of root infection by the pathogen, reducing the colonization and increasing the plant resistance [2]. The non-pathogenic strains may also alter the methylation patterns of the genes related to the pathogenesis response of the host plants, which results in an altered reaction to the pathogen encounter [3].

An organic farming system may also contribute to the increased incidence of *Fusarium* pathogens in the grain. The seed quality is lower, however, it is not clear if growing wheat without chemical protection results in the increased accumulation of mycotoxins when compared to the conventional farming system [4]. After the harvest, the stored grain may also be a source of mycotoxins, particularly when the storage conditions are favourable for fungal growth and proliferation. Water activity and spatial location of the inoculum are also important for grain colonization [5].

Environmental factors are known to play significant roles in disease progression. The precise time of inoculation in relation to the flowering is essential to define the "susceptibility window" for effective infection [6]. Moreover, it influences the levels of *Fusarium* metabolites and mycotoxins produced and accumulated in the grain as the infection proceeds [7]. The influence of weather on the contamination of plant material with *Fusarium* species and mycotoxins is even more obvious when the year-to-year variation is considered. A correlation is often found between weather conditions and the occurrence of the specific mycotoxin groups, e.g., fumonisins in maize [8]. Another factor influencing the virulence of *Fusarium* pathogens is light. Recent studies revealed the existence of a photo-sensor component which is not only responsible for the ecological adaptation of the pathogen, but also allows for the light regulation of the virulence expression [9]. The *Fusarium graminearum* virulence factors already discovered using modern New Generation Sequencing (NGS) and proteomic techniques were comprehensively reviewed [10], which will likely boost the research of other pathogenic species.

The mycotoxigenic abilities of the *Fusarium* populations have been widely studied for many years. In general, there is no correlation between the efficiency of the metabolite synthesis and geographical origin of the strains studied. Nevertheless, there are reports of significant differences in the toxin production that occur despite the genetic uniformity of the population over a large area [11]. The chemotypes emerging in unexplored geographic areas could be a serious threat. Constant monitoring of the main mycotoxin groups should be performed, as new analogues of well-known compounds can have similar or higher activities against plant, animal and human cells, as was discovered for the NX-3 trichothecene [12]. The ability of plant-pathogenic *Fusarium* species to infect humans has also been reported [13]. This confirms the enormous adaptability of the genus members in searching for new ecological niches.

In conclusion, constant progress in *Fusarium* research can be observed and is expected in the future in all areas of fungal biology, pathology and toxicology, especially with the aid of modern techniques, deployed to uncover the mechanisms of secondary metabolism regulation, interspecific molecular communication and virulence modulation by external and internal agents.

Funding: This research received no external funding.

Acknowledgments: I would like to thank all authors who contributed to this Special Issue, the reviewers who provided valuable and insightful comments, and all members of the Microorganisms Editorial Office for their professional assistance and constant support.

Conflicts of Interest: The author declares that no conflict of interest exists.

References

1. Mesterhazy, A.; Gyorgy, A.; Varga, M.; Toth, B. Methodical Considerations and Resistance Evaluation against F. graminearum and F. culmorum Head Blight in Wheat. The Influence of Mixture of Isolates on Aggressiveness and Resistance Expression. *Microorganisms* **2020**, *8*, 1036. [CrossRef]
2. Constantin, M.E.; Vlieger, B.V.; Takken, F.L.W.; Rep, M. Diminished Pathogen and Enhanced Endophyte Colonization upon CoInoculation of Endophytic and Pathogenic Fusarium Strains. *Microorganisms* **2020**, *8*, 544. [CrossRef] [PubMed]
3. Wojtasik, W.; Boba, A.; Preisner, M.; Kostyn, K.; Szopa, J.; Kulma, A. DNA Methylation Profile of β-1,3-Glucanase and Chitinase Genes in Flax Shows Specificity Towards Fusarium Oxysporum Strains Differing in Pathogenicity. *Microorganisms* **2019**, *7*, 589. [CrossRef] [PubMed]
4. Góral, T.; Łukanowski, A.; Małuszyńska, E.; Stuper-Szablewska, K.; Buśko, M.; Perkowski, J. Performance of Winter Wheat Cultivars Grown Organically and Conventionally with Focus on Fusarium Head Blight and Fusarium Trichothecene Toxins. *Microorganisms* **2019**, *7*, 439. [CrossRef] [PubMed]
5. Portell, X.; Verheecke-Vaessen, C.; Torrelles-Ràfales, R.; Medina, A.; Otten, W.; Magan, N.; García-Cela, E. Three-Dimensional Study of F. graminearum Colonisation of Stored Wheat: Post-Harvest Growth Patterns, Dry Matter Losses and Mycotoxin Contamination. *Microorganisms* **2020**, *8*, 1170. [CrossRef] [PubMed]
6. György, A.; Tóth, B.; Varga, M.; Mesterhazy, A. Methodical Considerations and Resistance Evaluation against Fusarium graminearum and F. culmorum Head Blight in Wheat. Part 3. Susceptibility Window and Resistance Expression. *Microorganisms* **2020**, *8*, 627. [CrossRef] [PubMed]
7. Spanic, V.; Katanic, Z.; Sulyok, M.; Krska, R.; Puskas, K.; Vida, G.; Drezner, G.; Šarkanj, B. Multiple Fungal Metabolites Including Mycotoxins in Naturally Infected and Fusarium-Inoculated Wheat Samples. *Microorganisms* **2020**, *8*, 578. [CrossRef] [PubMed]
8. Vandicke, J.; De Visschere, K.; Croubels, S.; De Saeger, S.; Audenaert, K.; Haesaert, G. Mycotoxins in Flanders' Fields: Occurrence and Correlations with Fusarium Species in Whole-Plant Harvested Maize. *Microorganisms* **2019**, *7*, 571. [CrossRef] [PubMed]
9. Tang, Y.; Zhu, P.; Lu, Z.; Qu, Y.; Huang, L.; Zheng, N.; Wang, Y.; Nie, H.; Jiang, Y.; Xu, L. The Photoreceptor Components FaWC1 and FaWC2 of Fusarium asiaticum Cooperatively Regulate Light Responses but Play Independent Roles in Virulence Expression. *Microorganisms* **2020**, *8*, 365. [CrossRef] [PubMed]
10. Rauwane, M.E.; Ogugua, U.V.; Kalu, C.M.; Ledwaba, L.K.; Woldesemayat, A.A.; Ntushelo, K. Pathogenicity and Virulence Factors of Fusarium graminearum Including Factors Discovered Using Next Generation Sequencing Technologies and Proteomics. *Microorganisms* **2020**, *8*, 305. [CrossRef] [PubMed]
11. Beccari, G.; Stępień, Ł.; Onofri, A.; Lattanzio, V.M.T.; Ciasca, B.; Abd-El Fatah, S.I.; Valente, F.; Urbaniak, M.; Covarelli, L. In Vitro Fumonisin Biosynthesis and Genetic Structure of Fusarium verticillioides Strains from Five Mediterranean Countries. *Microorganisms* **2020**, *8*, 241. [CrossRef] [PubMed]

12. Woelflingseder, L.; Gruber, N.; Adam, G.; Marko, D. Pro-Inflammatory Effects of NX-3 Toxin Are Comparable to Deoxynivalenol and not Modulated by the Co-Occurring Pro-Oxidant Aurofusarin. *Microorganisms* **2020**, *8*, 603. [CrossRef] [PubMed]
13. Meza-Menchaca, T.; Singh, R.K.; Quiroz-Chávez, J.; García-Pérez, L.M.; Rodríguez-Mora, N.; Soto-Luna, M.; Gastélum-Contreras, G.; Vanzzini-Zago, V.; Sharma, L.; Quiroz-Figueroa, F.R. First Demonstration of Clinical Fusarium Strains Causing Cross-Kingdom Infections from Humans to Plants. *Microorganisms* **2020**, *8*, 947. [CrossRef] [PubMed]

microorganisms

Article

Methodical Considerations and Resistance Evaluation against *F. graminearum* and *F. culmorum* Head Blight in Wheat. The Influence of Mixture of Isolates on Aggressiveness and Resistance Expression

Akos Mesterhazy [1,*], Andrea Gyorgy [2], Monika Varga [1,†] and Beata Toth [1,2]

[1] Cereal Research Non-Profit Ltd., 6726 Szeged, Hungary; varga.j.monika@gmail.com (M.V.); beata.toth@gabonakutato.hu (B.T.)
[2] NAIK Department of Field Crops Research, 6726 Szeged, Hungary; gyorgyandrea88@gmail.com
* Correspondence: akos.mesterhazy@gabonakutato.hu
† Present address: Department of Microbiology, University of Szeged, 6726 Szeged, Hungary.

Received: 27 May 2020; Accepted: 8 July 2020; Published: 13 July 2020

Abstract: In resistance tests to Fusarium head blight (FHB), the mixing of inocula before inoculation is normal, but no information about the background of mixing was given. Therefore, four experiments (2013–2015) were made with four independent isolates, their all-possible (11) mixtures and a control. Four cultivars with differing FHB resistance were used. Disease index (DI), Fusarium damaged kernels (FDK) and deoxynivalenol (DON) were evaluated. The isolates used were not stable in aggressiveness. Their mixtures did not also give a stable aggressiveness; it depended on the composition of mix. The three traits diverged in their responses. After the mixing, the aggressiveness was always less than that of the most pathogenic component was. However, in most cases it was significantly higher than the arithmetical mean of the participating isolates. A mixture was not better than a single isolate was. The prediction of the aggressiveness level is problematic even if the aggressiveness of the components was tested. Resistance expression is different in the mixing variants and in the three traits tested. Of them, DON is the most sensitive. More reliable resistance and toxin data can be received when instead of one more independent isolates are used. This is important when highly correct data are needed (genetic research or cultivar registration).

Keywords: disease index (DI); fusarium damaged kernels (FDK); deoxynivalenol (DON); host-pathogen relations; phenotyping FHB

1. Introduction

Mixing of isolates is a general methodical procedure used to produce inoculum for artificial inoculation. In most cases, no reason is given as to why it is used. It is known that the isolates of the *Fusarium* spp. have a strong variability in aggressiveness [1–3]. As mixing in seedling tests strongly influences aggressiveness [1], it is important to know what the influence of mixing on the disease-causing capacity is. It is clear now that *Fusarium graminearum* and *Fusarium culmorum* do not have vertical races and the resistance is race-non-specific [4–7]. Another important feature is that the resistance is also species-non-specific [8,9], meaning that the same quantitative traits locus (QTL) gives protection against all the *Fusarium* species tested. Highly significant differences were detected in aggressiveness within the *F. graminearum* and *F. culmorum* populations [4,10–12]. In addition, the aggressiveness does not seem to be stable [13], as proven by the many significant isolate/year interactions [14,15].

In this paper, and our previous publications, we used the aggressiveness term for the disease-causing capacity of the given inocula, as virulence is taken for the race-specific pathogens like rusts. The term pathogenicity is referred to the disease-causing capacity of the genus itself [4].

Table 1 shows a cross section of the literature working with mixtures. Research task, plant, media for increasing inoculum, conidium concentration, number of participating isolates in the inoculum and the data of visual symptoms, Fusarium damaged kernels (FDK) and deoxynivalenol (DON) were followed.

Table 1. Experimental data of Fusarium resistance and pathogenicity tests from papers using mixtures of isolates.

Author	Ref. No.	Plant	Application	Medium	Inoculation S or P	*Fusarium* spp.	No. of Isolates	Con. Conc.	Aggressiveness Visual	FHB Visual	Vis. Min.–Max.%	FDK	FDK Min.–Max.%	DON	DON Min.–Max. mg/kg
Andersen et al. 2015	[16]	wheat	Path	MBA	S	gram.	10 *v/o*	5×10^4	medium	DI *	20–50	no	no	high	7–75
Andersen et al. 2015	[16]	wheat	Path	MBA	P	gram.	10 *v/o*	5×10^4	medium/low	DI *	30–40	no	no	low/med	5–15
Garvin et al. 2009	[17]	wheat	QTL	n.g.	P	gram.	3 *v/o*	1×10^5	very high	DI	20–100	no	no	no	no
Busko et al.	[18]	wheat	Path	wheat seed	S	culm	3 *v/o*	5×10^5	no	no	no	no	no	no	no
Pirseyedi et al. 2019	[19]	durum	QTL	n.g.	P	gram.	3	5×10^5	very high	DI	0–100	no	no	no	no
Amarasighe 2010	[20]	wheat	Fung	CMC	S	gram.	7	5×10^4	medium	DI	0–28	medium	7.4–50	Medium	6.8–30.1
Bai and Sharen 1996	[5]	wheat	Res	MBM	P	gram.	n.g.	4×10^4	high	I	7–100	no	no	no	no
Bai et al. 1999	[21]	wheat	QTL	MBM	P	gram	10	4×10^4	low	DI	0.8–10.7	no	no	no	no
Basnet et al. 2012	[22]	wheat	QTL	n.g.	S	gram	n.g.	8×10^4	high	DI	1–90	high	30–90	no	no
Buerstmayr 2002	[23]	wheat	QTL	MBM	P	gram. + culm.	2	5×10^4	high	I	1–9	no	no	no	no
Buerstmayr 2011	[24]	wheat	QTL	MBM	S	gram.	1	2.5×10^4	medium	S (AUDPC)	114–946	no	no	no	no
Buerstmayr 2011	[24]	wheat	QTL	MBM	S	culm.	1	5×10^4	medium	S (AUDPC)	32–967	no	no	no	no
Dong et al. 2018	[25]	wheat	Gen	n.g.	P	gram.	30–39	$8–10 \times 10^4$	high	I	50–100	no	no	medium	0–30
Dong et al. 2018	[25]	wheat	Gen	n.g.	S	gram.	30–39	$8–10 \times 10^4$	high	S	20–80	no	no		
Chen et al. 2007	[26]	wheat	QTL	PDA	P	gram.	3	5×10^5	medium	DI	10–55	no	no	no	no
Chu et al. 2011	[27]	durum	QTL	n.g.	P	gram	3	5×10^4	very high	DI	14–75	high	1–100	low	0–38
Cowger et al. 2010.	[28]	wheat	Path	MBB	S	gram.	4 *v/o*	1×10^4, 1×10^5	no	no	no	low	3–18	low/medium	2–16.7
Li et al. 2011	[29]	wheat	QTL	MBM	P	gram.	10	1×10^4	low	DI	0–1	no	no	no	no
Cutberth et al. 2006.	[30]	wheat	QTL	n.g.	P	gram.	3	5×10^5	high	DI	0–100	no	no	no	no
D'Angelo et al. 2014	[31]	wheat	Fung	CMC	S	gram.	10	n.g.	low	DI	1.6–10.4	no	no	very low	0.3–8
Kollers et al. 2013	[32]	wheat	QTL	n.g.	S	gram.	nFg, nFc	5×10^4	medium	DI	0–34	no	no	no	no
Yang et al. 2005	[33]	wheat	QTL	PDA + CMC	S	gram.	4	5×10^5	high	DI	7–97.8	high	2.8–95.7	no	no
Gaertner et al. 2008	[34]	wheat	Res	oat grain	S	culm.	20	5×10^6	high	1–9	2–16	no	no	low/medium	2–16
Evans et al. 2012	[35]	barley	Path	MBA	S	gram.	3	2×10^5			n.a.			high	2–4/spike
Perlikowsi et al. 2017	[36]	trit.	Gen	n.g.	S	culm.	3	5×10^4	medium	DI	12–24	high	14–82	high	0.5–113
Yi et al. 2018	[37]	wheat	QTL	n.g.	P	gram.	4	1×10^5	low	DI	0.58–96	no	no	no	no
Yi et al. 2018	[37]	wheat	QTL		S	gram.	4	1×10^5	medium	DI	0.9–24.4	no	no	no	no
Gervais et al. 2003	[38]	wheat	QTL	barley seed	S	culm.	more	1×10^6	medium	1–9	2.5–8	no	no	no	no
Goral et al. 2002	[39]	trit.	Res	n.g.	S	culm.	10	1×10^5	low	S	7–20	no	no	high	7.5–118
Goral et al. 2015	[40]	wheat	Gen	n.g.	S	culm.	3 *v/o*	5×10^5	medium	DI	8–33	medium	15–37	low	2.5–7.6
Goral et al. 2015	[40]	wheat	Gen	n.g.	P	culm.	3 *v/o*	5×10^5	medium	DI	4.8–32	no	no	no	no
Hao et al. 2012	[41]	wheat	Gen	n.g.	P	gram.	more	5×10^4	n.a.	DI			n.g.		

Table 1. *Cont.*

Author	Ref. No.	Plant	Application	Medium	Inoculation S or P	*Fusarium* spp.	No. of Isolates	Con. Conc.	Aggressiveness Visual	FHB Visual	Vis. Min.–Max.%	FDK	FDK Min.–Max.%	DON	DON Min.–Max. mg/kg
He et al. 2013	[42]	wheat	Res	LBA	S	gram.	more	5×10^5	high	DI	0–89	no	no	low/med.	0.1–21.4
He et al. 2014.	[43]	wheat	Res	LBA	S	gram.	5	5×10^5	low	DI	0–15.9	low	5–42	low	0.2–7.05
Hilton et al. 1999	[44]	wheat	Res	PDA	S (?)	4 F. spp.	4 *v/v*	2.5×10^5	high	DI	20–78	no	no	no	no
Chen et al. 2005	[45]	wheat	Gen	n.g.	P	gram.	4	n.g.	n.g.	DI	S-MR	no	no	no	no
Klahr et al. 2007	[46]	wheat	Res	n.g.	S	culm.	more	1×10^6	medium/high	DI(AUDPC)	65–1403	no	no	no	no
Lin et al. 2006	[47]	wheat	QTL	n.g.	S	gram.	4	n.g.	low	I	0.05–0.82	no	no	no	no
Liu et al. 1997	[48]	wheat	Res	PDA	S	culm.	8	1×10^5	medium	DI	10–85	high	33–70	med./low	4–16
Forte et al. 2014	[49]	wheat	Gen	n.g.	P	gram.	more	5×10^4	high	DI	8–99	no	no	no	no
Malihipour et al. 2015	[50]	wheat	Gen	n.g.	S (field)	gram.	4	5×10^4	high	I, S	2–87	low	1–23	low	0.2–4.2
McCartney et al. 2015	[51]	wheat	QTL	n.g.	S	gram.	5	5×10^4	medium	DI	15–55	no	no	no	no
Muhovski et al. 2012	[52]	wheat	Gen	n.g.	P	gram.	more	1×10^5	n.g.	DI	n.g.	no	no	no	no
Osman et al. 2015	[53]	wheat	Res	rice grain	P	gram.	2 and 5	1×10^5	n.a.	DI	ng	no	no	no	no
Jones et al. 2018	[54]	wheat	Fung	oat grain	S	mix 5 F. spp.	20	1×10^5	medium	DI(AUDPC)	653	no	no	low	0–3.13
Liu et al. 2019	[55]	wheat	QTL	spawn	n.a.	gram.	20	n.a.	high	DI	6–83	no	no	no	no
Miedaner et al. 2017	[56]	wheat	QTL	n.g.	P	gram.	3	1×10^5	high	DI	20–100	no	no	no	no
Oliver et al. 2006	[57]	wheat	Gen	n.g.	P	gram.	3	1×10^4	medium/high	DI	5–57	no	no	no	no
Tamburic et al. 2017	[58]	wheat	QTL	Bilay	S	gram.	4	5.5×10^3	high	DI	1.6–67	no	no	high	0.5–47.2
Otto et al. 2002	[59]	durum	QTL	n.g.	P	gram.	3	n.g.	high	DI	15–62	no	no	no	no
Ding et al. 2011	[60]	wheat	Gen	n.g.	S	gram.	4	4×10^4	medium	DI	17–51	no	no	no	no
Klahr et al. 2011	[61]	wheat	QTL	n.g.	S	culm.	more	n.g.	n.g.	n.g.	n.g.	no	no	no	no
Zwart et al. 2008	[62]	wheat	Res	n.g.	S	Fg + Fc	2	1×10^5	medium	I,S,	DI	high	33–68	no	no
Oliver et al. 2006	[57]	wheat	QTL	n.g.	P	gram	3	n.g.	n.g.	DI	7.1–57.5	no	no	no	no
Miedaner et al. 2006	[63]	wheat	QTL	n.g.	S	culm.	2	5×10^5	medium	DI	6–35	no	no	high	3–60

The whole table: n.g. = not given, no = not tested, n.a. not applicable, Headings: Plants: trit. = triticale, Application: Path = pathology, Fung: fungicide research, Res = resistance research, tests, Gen = genetic aspects, QTL: quantitative traits locus Medium to increase fungi: LBA = lima bean medium MBA = mung bean agar, MBM mungo been medium liquid PDA = potato dextrose agar, CMC = carboxyl methyl cellulose, SNA = synthetic nutrient-poor agar, Inoculation of ears: S = spray, P = point, *Fusarium* spp.: gram. = graminearum, culm. = culmorum, Fg + Fc: mixture of the two species. No. of isolates: *v/v* volume/volume, when 4, each have 25% in the pooled inoculum, nFg, nFc, = mixture of the two species without giving the number of the isolates, therefor "n" before, Conidium concentration: 1×10^4, 1×10^5: = two different concentrations were used in the same paper, FHB (Fusarium head blight) Visual: * = 20 days after inoculations, I = Incidence, S = severity, VSS 1–9 = visual scale 1–9, DI = disease index, DAI = Days after inoculation.: AUDPC: area under disease progress curve, FDK = Fusarium damaged kernel, DON: deoxynivalenol.

Numerous authors used spray inoculation (n = 22) [18,20,22,24,28,31,39,46,62,63], and point inoculation was used in 20 cases [16,23,40,49,53]. The Chinese authors work mostly with point inoculation [21,26,27,29,37,41,45]. Many American sources also use this [5,25,26], partly with Chinese scientists working in the US, or from US–China collaboration. However, in increasing numbers, spray inoculations. In some cases, papers are found where both inoculation methods are used parallel [16,24,25,37,40]. Mixtures are made mostly from different isolates of the same *Fusarium* species; in several cases, the different chemotypes are mixed. However, without mixing, no tests were made, so nothing can be said about the effect of the mixing. We have an example that the inoculation was made separately with *F. graminearum* and *F. culmorum*, and then data were pooled for ANOVA [24]. The number of isolates in the mix varied from 2 to 39. In the eight cases, the participating inocula were adjusted before mixing to the given concentration, and then in three inocula, one-third of the amount was pooled to secure the same rate of the given inocula in the pooled version. For the others, we do not have such information, and in several cases no isolate number was given; this case is marked with "more" in the column no. of isolates in Table 1. Aggressiveness before the test was made only in one case [40]; for others, no test was performed. In several cases, the selection of the isolates was made based on experience of earlier years. The conidium concentration is very variable from 10,000 to one million. There is no explanation for this. This means that besides the mixing, the adjusting conidium concentration can also cause problems. There are two conclusions. There is no control of aggressiveness from side of the mixing and diluting. Therefore, only after finishing the test will be clear, whether the necessary aggressiveness could have been secured to achieve the necessary reliability of the experiment. The fifty papers were listed, but in four papers, two lines were used as the authors have applied different inoculation methods or different *Fusarium* spp. Thus, the total number of the cases is 54. The aggressiveness level was evaluated by the presented visual data in this paper. Nineteen cases were found in the high to —very high aggressiveness group, 18 were classified medium or medium/high, eight had low or medium/low level, and in ten cases, no data were printed (not tested or not given). From the 54 cases, only 17 proved good and acceptable, the others were of lower level with moderate differentiation power or even less. This shows, clearly, that securing the necessary aggressiveness could be secured at 36% of the cases. In many cases, disease index was found; in other cases, severity was mentioned, but looking at the data average, severity was indicated, so these were also considered as disease index. In older literature, this was normal. FDK severity was tested only in eleven cases; five cases were high, three cases low, and two medium severity. In 37 cases, we have no data. DON was measured in 17 cases, five cases had high numbers, two were medium, and 10 were low or low/medium and medium/low qualifications. It is important that, in several cases, high aggressiveness in visual symptoms resulted in low DON yield in grains [27,34,42,50]. However, in one case, one poor visual rate showed high DON contamination [39]. The data show that the response to visual symptoms, FDK and DON is not the same. The most important task is the reduction of the DON contamination. The problem is that the least research is done in this field, and only in five cases were the data suitable to analyze DON response; this is less than 10% of the cases.

In the cited literature, the number of isolates in the mixtures varied between 2 and 39. The conidium concentration was set to between 5500 and 5×10^6. This leads to the following question: is mixing and adjusting isolates not significant, or does it have a significant influence on inoculation results? From the papers, we did not get any information. The fact that everybody worked with the best thought conidium concentration and mixing—the published results do not support this probable conviction. However, we thought that the questions should be answered. Therefore, one should know what really happens when different isolates are mixed. After the test, we will know more, how the mixing is working and whether the aggressiveness of the composite inoculum could be.

An important thing should also be considered. Suppose that the aggressiveness problem can be solved for the one inoculum used normally (single isolate or mixture); the question remains whether the single inoculum can provide the reliability of the testing needed for scientific purposes in genetic analyses, variety registration trials, etc. Snijders [64,65] applied four *F. culmorum* isolates from Research

Institute for Plant Protection IPO-DLO, Wageningen, NL (IPO 39–01, IPO 329–01, IPO 348–01 and IPO 436–01). Ranking of isolates and the height of the infection were different and variety responses showed high variability. Further results also showed significant isolate-year interactions [66–71], e.g., changing ranks in different years. Besides the changing isolate ranking, the variety ranking differed, that also can be a problem in resistance classification. It seems [1] that the more aggressive isolates keep their aggressiveness much better (following dilution) than the less aggressive ones. It is supposed that the mixtures may have a similar picture.

Therefore, this study focused on three main objectives. First, making inoculations with four *Fusarium* isolates in every possible combination to observe the range of plant reactions as widely as possible. Second, to gain more reliable information about the response of cultivars with differing resistance levels, and the structure of resistance expression in order to understand the behavior of the isolates and their mixtures, depending on their aggressiveness level, and to study FDK and DON responses. Here, the changing variety ranks are especially important. Third, as the different traits (FHB, FDK and DON) often do not respond the same way, obtaining more information that would promote regulation of these traits at different aggressiveness levels.

2. Materials and Methods

2.1. Plant Material

Four winter wheat cultivars received from the Cereal Research Nonprofit Ltd., Szeged were tested (Table 2) with differing resistance levels. Their resistance or susceptibility have been verified, years ago, both under natural and artificially inoculated regimes.

Table 2. Winter wheat genotypes in the tables and figures, Szeged, 2013–2015.

Genotype	Resistance Class
GK Fény	MR
GK Garaboly	S
GK Csillag	MR
GK Futár	S

MR = moderately resistant, S = susceptible, GK is the abbreviation of the Hungarian name of Cereal Research Ltd., as breeding institute.

2.2. Field Conditions and Experimental Design

In the field tests, the recent basic methodologies were followed [70,71]. The tests lasted three seasons (2013, 2014 and 2015). As the mean for FDK was 66% in 2013 and 13.3% 2014, it was decided to continue the experiment. In 2015, two independent tests were performed with the same isolates, but different inocula, so four experiments were performed and evaluated as a unit. The plant material was sown and evaluated in the nursery of the Cereal Research Nonprofit Ltd. in Szeged, Hungary (46°14′24″ N, 20°5′39″ E) (Kecskes Experimental Station). The field experiments were conducted in four replicates in a randomized complete block design. The plot size was 1 × 5 m. For the 16 groups of heads, one plot was planned as a unit. Sowing was done in mid-October by using a Wintersteiger Plotseed TC planter (Wintersteiger GmbH, Ried, Austria). (Temperature data originate from the National Meteorological Station, Szeged, 1000 m from the nursery; precipitation was measured daily at 7.00 a.m. in the Kecskes Station, about 2–3 hundred m from the actual plots.) The weather data were similar in May and June (precipitation 2013 265 mm, 2014 142 mm). Concerning temperature, the monthly means for both years were 17.2 °C in May; in June, 19.9 °C and 20 °C were the corresponding data (2015 also showed very close data). The only difference is that the 2014 January–April had 110 mm rain and 2013 brought 224 mm rain. The driest year was 2015, with 112 mm winter, 68 mm May and 22 mm June precipitation.

2.3. Inoculum Production and Inoculation

F. graminearum and *F. culmorum* are the most important causal agents [72] and two isolates of each species were involved for testing. In the tests, four isolates were used, from *F. culmorum*, the Fc 12375 (1) that were isolated from wheat stalk inside space mycelium from a greenhouse test in the greenhouse of Cereal Research Inst. in 1977. The Fc 52.10 (2) and the two *F. graminearum* isolates, Fg 19.42 (3) and Fg 13.38 (4), originated from naturally contaminated wheat grains (2010). Their monosporic lines were used in the tests. To propagate inoculum the bubble-breeding method was used [1,3,10] on liquid Czapek-Dox medium. As aggressiveness is a variable trait [4,10,69,70], 50% more inocula were produced and the best ones were chosen for use. This way it was possible to put the aggressiveness under control. The aggressiveness of the isolates was done by the Petri dish method [1,3] (Figure 1). The inocula were stored until usage at 4 °C. Since the amount of material from the flowering plots was checked on the previous day, only that amount was separated following careful mixing from content of the 10 L balloon, as was necessary for that given day. The rate of the inocula in the mixtures was 50–50% with two components, one-third for three and one fourth with four components. They were made in the afternoon before inoculation. The suspension was fragmented by the Eta Mira household mixer machine (Czech Republic) with a 1 L volume-mixing unit.

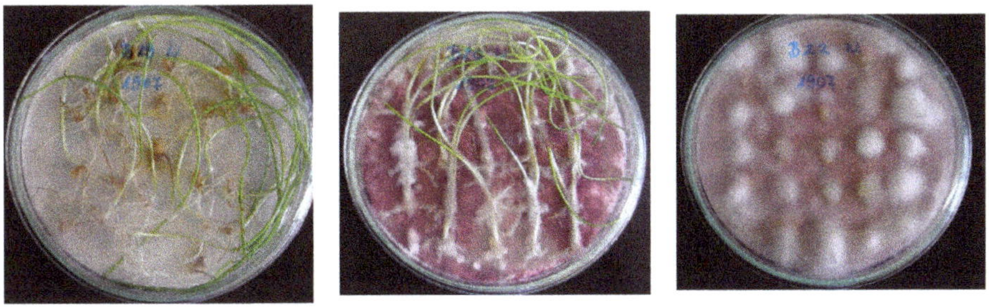

Figure 1. Aggressiveness test with different isolates of a moderately susceptible genotype No. 1907. Low (B19 **left**), medium (B20 **middle**) and highly aggressive (B22 **right**) isolates of *F. graminearum*. Original inocula, without dilution or mixing (the pictures are illustrations to show the aggressiveness differences within *F. graminearum*).

The inoculation was made at full flowering with spray inoculation. First, the control heads were covered at the end of the plot by a polyethylene bag without inoculation, with only sterilized water being sprayed. This was necessary to avoid cross inoculation from the suspension treated groups of heads. Then each plot was inoculated with 15 inocula (isolates 1, 2, 3, 4, 1 + 2, 1 + 3, 1 + 4, 2 + 3, 2 + 4, 3 + 4, 1 + 2+3, 1 + 2+4, 1 + 3+4, 2 + 3+4 and 1 + 2 + 3 + 4) on group of heads within a plot. As the mixtures were mixed *v/v* basis (at three components, one third was given from every component [16–18,40,44], in the counting of the effect of the mixing, the arithmetical mean was applied. This was applied earlier [2]. This was proportional with the volumes. When no interaction occurs between components, the arithmetical mean functions. If this is not the case, the arithmetical mean serves as control to compare the performance of the mixtures. In this case, as we had aggressiveness data of the participating components, the effect of the mixing could be measured. Each group of heads consisted of 15–20 heads and about 50 cm distance between each other to prevent cross inoculation between the isolates. The groups were positioned about 30 cm from the edge in two lines, in each, 8 and 7 groups of heads were spray inoculated (about 15 mL suspension each). Following inoculation, the sprayed heads were covered by polyethylene bags for 48 h [69]. As head size was different, larger heads needed more suspension to have the same coverage as smaller ones. After removing the bags, the groups of heads remained loosely bound at half height of the plants, not to disturb assimilation of the leaf system.

2.4. Evaluation of the Disease and Toxin Analysis

The evaluation of the visual symptoms was done on 10, 14, 18, 22 and 26 days after inoculation. In warmer years, the 22nd day was the last, because of the yellowing, the head symptoms could not be evaluated properly [4,14]. During evaluation, the percentage of the spikelets in the heads of groups were directly estimated as percentage value. Threshing was made carefully to not lose lighter infected grains (Seed Boy, Wintersteiger AG., Ried, Austria); fine cleaning was performed by an Ets Plaut-Aubry air separator (41290 Conan-Oucques, Conan-Oucques, France). In both cases, air speed was regulated so that light Fusarium infected grain remained. Then a visual evaluation of the FDK grains followed, expressed as a percentage value.

For toxin analyses, 6 g of the individual samples was separated for milling by a Perten Laboratory mill (Laboratory Mill 3310, Perten Instruments, 126 53 Hägersten, Sweden). Toxin extraction and DON toxin analysis was done according to Mesterházy et al. [14], where 1 g of fine milled wheat sample of the 6 g milled grain sample was extracted with 4 mL of acetonitrile/water (84/16, *v/v*) for 2.5 h in a vertical shaker. All chemicals and toxin standards were purchased from Sigma-Aldrich (1117 Budapest, 23. October Street 6–10). After centrifugation (10,000 rpm, 10 min), 2.5 mL of the extract was passed through an activated charcoal/neutral alumina solid phase extraction (SPE) column (Sigma-Aldrich Supelco, volume 5 mL, filled with 1 g mixture of 20 g Al_2O_3/Sigma/and 1 g activated carbon coal/Sigma/, prepared in the laboratory) at a flow rate of 1 mL/min. Then, 1.5 mL of the clear extract was transferred to a vial and evaporated to dryness at 40 °C under vacuum. The residue was dissolved in 500 µL of acetonitrile/water (20/80, *v/v*). Liquid chromatographic separation and quantification was made on an Agilent 1260 HPLC system (Agilent Technologies, Santa Clara, CA, USA) equipped with a membrane degasser, a binary pump, a standard autosampler, a thermostated column compartment and a diode array detector. DON was separated on a Zorbax SB-Aq (4.6 × 50 × 3.5 µm) column (Agilent) equipped with a Zorbax SB-Aq guard column (Agilent) (4.6 × 12.5 × 5 µm) thermostated at 40 °C. The mobile phase A was water, while mobile phase B was acetonitrile. Validation of DON was made by using the DON control toxin from Sigma with a regular dilution grade series. So all concentrations were within the scope of the validation line. The gradient elution was performed as follows: 0 min, 5% B; 5 min, 15%B; 8 min, 15%B; 10 min, 5% B; 12 min, 5% B. The flow rate was set to 1 mL/min. The injection volume was 5 µL. DON was monitored at 219 nm.

2.5. Statistical Analysis

The visual data for the 4–5 readings for a group of heads were averaged and they served as entries into the ANOVA method. For DON the yield of every group of heads was analyzed for DON. The four-way ANOVA was done according to the functions described in Sváb [73] and Weber [74] with the help of the built-in functions of Microsoft Excel. In the controls for visual evaluation, no visible infection of FHB and FDK was found. The data were not considered in the variance analysis, but the DON showed a low-level contamination in the control, so these data were included in the ANOVA analysis. To evaluate the significance of the two-ways and three-ways interactions, we followed the methodology suggested by Weber [74].

3. Results

3.1. Visual Data, Disease Index

Table 3 presents the data across experiments. The four isolates had very distinct aggressiveness across experiments (Table 3A). The general tendency is that the most aggressive isolate, Fg 19.42, had the highest value alone, all combinations produced less aggressiveness. On the other hand, the least aggressive isolate showed higher aggressiveness in mixture in all cases than when it was used alone. The two least aggressive isolates gave results closer to the more weakly aggressive Fc 52.10. All eleven mixtures showed lower aggressiveness than the most aggressive component. The difference between inoculum means is highly significant, the maximum is 50.7, the minimum is 5.08, the distance is 45.7, and the limit of significance LSD 5% for the 15 inocula is 2.06 (64 replicated behind each mean). As the mixing was made on *v/v* rate, it was anticipated that without specific mixing effects of the components, the arithmetical mean of the participating components will show the postulated performance. This was not the case. The measured aggressiveness across genotypes was seven cases higher than the mean of the components, and in four cases, the mean was lower than the arithmetical mean. The real data were between 74% and 173%. This means that the resulting aggressiveness is very variable and its aggressiveness cannot be forecasted. At present, we do not know the reason; further research should solve the problem. We can state, however, that the mixing will reduce the aggressiveness of the most aggressive component in each mixed inoculum, but balances much from the lower aggressive components in positive direction. On average, the mean shows a 28% increase in aggressiveness compared to the hypothetic arithmetical mean model. At the same time, the correlation coefficients, all are significant at $p = 0.001$ or higher at r-values between 0.976 and 0.985, indicating a similar response (Table 3B, $n = 4$)). This was not unexpected as we found it many times working with different isolates independently.

As the LSD 5% value 4.12% is valid for any difference among the data in Table 3A, the problem of significance can be identified without problem. When this is smaller than 4.12, no significant difference can be shown, when larger, it is proved. In many cases the behavior of genotypes (lines) or inocula (columns) is not so, and strongly varies. At low aggressiveness, no significant difference in resistance occurs for Isolates Fc52.10 and Isolate 4 Fg13.38 or the mixture 2 + 4. In other cases, such as Fc12375 3 + 4, the differences between genotypes are significant, with nearly 50% difference between them.

The variety reactions were compared for every inoculum ($n = 15$) so that the cultivar data were expressed at each cultivar to the mean of the four cultivars.

When the correlations are counted between the responses of the cultivars to different isolates, from the 105 correlations, only 13 were significant at $p = 5\%$ (Table 3C). Seven of the 11 mixed inocula contained Isolate 3. For the others, we have (altogether) four cases. The very variable correlations clearly show that the ranking of the genotypes at the different inocula (isolates and their mixtures) present a high diversity. In three inocula, the difference between genotypes is not significant (LSD 5% is smaller than 4.12%. Six of the genotypes have three cases without significant difference. Five cases were with no difference between two genotypes. Only one inoculum presented significant difference between all genotypes. This was the case also for the means of the four cultivars. For us the real problem is here—which inoculum is optimum to present differences in variety resistance? From the disease index, it seems that the mixing did not give to better differentiation of the genotypes. In this respect, the mixing is not the approach that would bring us closer to a more powerful methodology.

The four experiments (Table 4) had the same means for isolates and their combinations, but the means of the experiment differences are much larger, i.e., 44%, 8%, 26% and 28.5% (2013, 2014, 2015a, 2015b). The data proved that the differentiation between genotypes at low infection pressure is rather poor and not reliable compared to the data of the other years (Table 4B). All correlations ($n = 4$) where 2014 is a partner, gave correlations of r = 0.50, r = 0.66 and r = 0.69. For the rest, the correlations are between r = 0.85 and r = 0.99. Looking at the genotype correlations for the four years ($n = 15$) (Table 4C) from the 105 correlations, 52 were significant. Isolates 3 and 4 showed the least significant correlations with other inocula. It seems that aggressiveness level has a much higher importance in experimentation then mixing has. This is partly ecology-dependent, but is a result, also, of interaction between the aggressiveness level and differentiation of genotypes.

Table 3. Response of the wheat cultivars to the different isolates and their mixtures. Disease index data (%) across experiments, Szeged, 2013–2015.

Isolates	Cultivars, DI%				Mean	Counted	Mean/
Table 3A/Mixtures	GK Garaboly	GK Csillag	GK Fény	GK Futár			Counted%
1 Fc 12375	35.83	20.96	22.18	25.26	26.06	26.06	100.00
2 Fc 52.10	5.42	5.75	4.05	5.09	5.08	5.08	100.00
3 Fg 19.42	60.94	38.68	47.70	55.78	**50.78** [1]	50.78	100.00
4 Fg 13.38	12.30	11.82	10.43	11.42	11.49	11.49	100.00
1 + 2	26.53	20.17	26.47	25.74	24.73	15.57	158.84
1 + 3	46.24	31.55	32.24	37.22	**36.81**	38.42	95.83
1 + 4	25.60	20.46	21.54	24.13	22.93	25.39	90.32
2 + 3	49.37	36.98	39.10	43.39	**42.21**	27.93	151.16
2 + 4	6.85	6.55	5.48	5.95	6.21	8.28	74.93
3 + 4	45.38	32.24	38.52	40.27	**39.10**	31.13	125.60
1 + 2 + 3	40.62	32.58	31.15	39.04	**35.85**	27.30	131.30
1 + 2 + 4	27.37	21.95	21.33	22.63	23.32	14.21	164.11
1 + 3 + 4	34.64	31.04	29.13	34.31	**32.28**	29.44	109.64
2 + 3 + 4	45.99	35.45	37.09	36.89	**38.86**	22.45	173.09
1 + 2 + 3 + 4	35.83	32.36	29.15	32.18	**32.38**	23.35	138.66
Control	0.00	0.00	0.00	0.00	0.00	0.00	0.00
Mean	31.18	23.66	24.72	27.45	26.75	22.30	128.49 [2]
LSD 5% mix.					2.06		
LSD 5% between cultivars (cvs)					1.06		
LSD 5% between any data in the table (genotype x inoculum, AxC interaction): 4.12							

[1] Bold printed in Mean: Isolate 3 and its mixtures, [2] Mean for mixtures only

1, 2, 3, 4 = Isolates to be mixed per se. 1 + 2 = mix Is. 1 and 2, 1 + 2 + 3 = Is. 1 and 2 and 3, etc.

Table 3B/Correlations	GK Garaboly	GK Csillag	GK Fény	GK Futár
GK Csillag	0.9676 ***			
GK Fény	0.9825 ***	0.9730 ***		
GK Futár	0.9854 ***	0.9735 ***	0.9898 ***	

*** $p = 0.001$

Table 3C/Inocula	3	2 + 3	3 + 4	2 + 3 + 4	1 + 3	1 + 2 + 3	1 + 2 + 3 + 4	1 + 3 + 4	1	1 + 2	1 + 2 + 4	1 + 4	4	2 + 4
2 + 3	0.95													
3 + 4	0.98 *	0.94												
2 + 3 + 4	0.78	0.91	0.85											
1 + 3	0.89	0.99 *	0.89	0.95										
1 + 2 + 3	0.87	0.91	0.77	0.70	0.89									
1 + 2 + 3 + 4	0.55	0.77	0.53	0.78	0.85	0.81								
1 + 3 + 4	0.76	0.82	0.64	0.59	0.82	0.98 *	0.82							
1	0.86	0.97 *	0.89	0.98 *	0.99 *	0.83	0.84	0.74						
1 + 2	0.82	0.65	0.86	0.53	0.54	0.43	0.02	0.27	0.54					
1 + 2 + 4	0.75	0.92	0.79	0.97 *	0.97 *	0.80	0.91	0.74	0.98 *	0.37				
1 + 4	0.98 *	0.98 *	0.94	0.81	0.94	0.94	0.70	0.86	0.90	0.69	0.83			
4	0.31	0.56	0.26	0.58	0.67	0.68	0.96 *	0.74	0.65	−0.27	0.76	0.49		
2 + 4	0.18	0.47	0.18	0.59	0.60	0.52	0.92	0.58	0.61	−0.35	0.74	0.36	0.97 *	
2	−0.07	0.18	−0.16	0.18	0.29	0.42	0.74	0.55	0.26	−0.63	0.41	0.13	0.90	0.88

* significant at $p = 0.05$, limit r = 0.95

1, 2, 3, 4 = Isolates to be mixed per se. 1 + 2 = mix Is. 1 and 2, 1 + 2 + 3 = Is. 1 and 2 and 3, etc.

Table 4. Response during the experiments to the different isolates and their mixtures. Disease index data (%) across cultivars, Szeged, 2013–2015.

Table 4A/Isolates	Years				Mean
Mixtures	2013	2014	2015a	2015b	
1 Fc 12375	40.13	9.17	26.93	28.00	26.06
2 Fc 52.10	8.33	1.09	4.88	6.01	5.08
3 Fg 19.42	61.13	26.00	54.00	61.98	50.78
4 Fg 13.38	14.03	6.62	11.78	13.54	11.49
1 + 2	44.96	5.07	23.29	25.58	24.73
1 + 3	66.19	15.50	31.58	33.98	36.81
1 + 4	48.21	5.73	18.58	19.21	22.93
2 + 3	70.94	5.33	44.04	48.54	42.21
2 + 4	10.75	4.10	4.62	5.36	6.21
3 + 4	68.00	5.60	43.06	39.75	39.10
1 + 2 + 3	57.63	17.44	32.31	36.02	35.85
1 + 2 + 4	52.99	1.85	17.63	20.81	23.32
1 + 3 + 4	56.95	5.09	30.25	36.83	32.28
2 + 3 + 4	53.24	8.97	44.56	48.65	38.86
1 + 2 + 3 + 4	54.06	11.14	31.54	32.77	32.38
Mean	44.22	8.04	26.19	28.57	26.75
LSD 5%	4.12	4.12	4.12	4.12	2.06

1, 2, 3, 4 = Isolates to be mixed per se. 1 + 2 = mix Is. 1 and 2, 1 + 2 + 3 = Is. 1 and 2 and 3, etc.

Table 4B/Correlations	2013	2014	2015a	2015b
2014	0.5036			
2015a	0.8734 ***	0.6682 **		
2015b	0.8577 ***	0.6952 **	0.9903 ***	

*** $p = 0.001$, ** $p = 0.01$.

Table 4C/Inocula	1	2	3	4	1 + 2	1 + 3	1 + 4	2 + 3	2 + 4	3 + 4	1 + 2 + 3	1 + 2 + 4	1 + 3 + 4	2 + 3 + 4
2	0.99 *													
3	0.91	0.93												
4	0.94	0.96 *	0.99 *											
1 + 2	0.99 *	0.98 *	0.85	0.89										
1 + 3	0.94	0.93	0.73	0.79	0.98 *									
1 + 4	0.92	0.91	0.69	0.76	0.97 *	1.00 *								
2 + 3	1.00 *	1.00 *	0.93	0.96 *	0.98 *	0.92	0.90							
2 + 4	0.82	0.82	0.54	0.63	0.90	0.97 *	0.97 *	0.80						
3 + 4	1.00 *	0.98 *	0.88	0.92	0.99 *	0.94	0.93	0.99 *	0.83					
1 + 2 + 3	0.97 *	0.96 *	0.80	0.86	0.99 *	0.99 *	0.99 *	0.96 *	0.94	0.97 *				
1 + 2 + 4	0.94	0.93	0.73	0.79	0.98 *	1.00 *	1.00 *	0.92	0.97 *	0.94	0.99 *			
1 + 3 + 4	0.99 *	1.00 *	0.90	0.94	0.99 *	0.96 *	0.94	0.99 *	0.86	0.98 *	0.98 *	0.96 *		
2 + 3 + 4	0.94	0.95 *	0.99 *	0.99 *	0.89	0.78	0.75	0.96 *	0.60	0.93	0.84	0.78	0.92	
1 + 2 + 3 + 4	0.99 *	0.98 *	0.85	0.89	1.00 *	0.98 *	0.97 *	0.98 *	0.89	0.99 *	0.99 *	0.98 *	0.99 *	0.89

* $p = 5\%$, Limit: r = 0.95.

1,2,3,4 = Isolates to be mixed per se. 1 + 2 = mix Is. 1 and 2, 1 + 2 + 3 = Is. 1 and 2 and 3, etc.

3.2. *Fusarium Damaged Kernels (FDK)*

The values of the FDK data (Table 5) are much higher than the DI data; the mean was for DI 26.7% and for FDK 44.9. In the controls, no visual infection was recorded, so all infections originated from the artificial inoculation (Table 5A). The reduction of the aggressiveness through mixing is significant, but in extent, less than that of the FHB values. Here, the difference between the measured FDK and the counted is larger, 36% mean increase could be registered. The combinations having Isolate 3 (Fg 19.42) have a mean higher than 50% and in one case, higher than 60%. The mixing produced data compared to the arithmetical mean of the aggressiveness of the components between 91 and 184%. Actually, every mixture variant has more or less differing aggressiveness levels. The different compositions mean characteristic aggressiveness differences, which also influence the expression of resistance. The correlations between genotype means across years (Table 5B, $n = 15$) highly significant correlation above r = 0.90, indicating the similarity of the response of cultivars to different isolates. However, when we compare the aggressiveness of the mixture and the mean of participating isolates, we receive large deviations. From 91% to 184%, every possibility can occur. In Table 5C ($n = 4$) where the isolate reactions were compared for genotypes ($n = 4$), the variability in the correlations grow significantly, indicating different responses of the genotypes to the individual inocula.

Table 5. Response of the wheat cultivars to the different isolates and their mixtures. FDK data (%) across experiments, Szeged, 2013–2015.

Table 5A/Isolates	Cultivars				Mean	Counted	Mean
Mixtures	GK Garaboly	GK Futár	GK Fény	GK Csillag		Counted	Counted%
1 Fc 12375	56.88	43.44	36.69	38.06	43.77	43.77	100.00
2 Fc 52.10	9.39	5.51	6.38	5.05	6.58	6.58	100.00
3 Fg 19.42	82.63	73.00	68.75	53.13	**69.38** [1]	69.38	100.00
4 Fg 13.38	26.56	22.50	17.50	20.44	21.75	21.75	100.00
1 + 2	46.25	47.63	44.38	32.33	42.64	25.17	169.41
1 + 3	68.94	54.19	45.31	37.50	**51.48**	56.57	91.01
1 + 4	48.44	42.44	36.69	31.63	39.80	43.20	92.11
2 + 3	70.25	54.56	43.89	48.82	**54.38**	37.98	143.19
2 + 4	18.88	11.94	11.81	11.88	13.63	14.16	96.19
3 + 4	66.13	61.81	59.88	41.88	**57.42**	45.56	126.03
1 + 2 + 3	79.38	64.38	56.56	48.44	**62.19**	39.91	155.83
1 + 2 + 4	50.75	36.51	34.08	34.71	39.01	24.03	162.34
1 + 3 + 4	63.31	51.13	46.44	49.75	**52.66**	44.96	117.11
2 + 3 + 4	72.13	59.81	51.91	56.06	**59.98**	32.57	184.16
1 + 2 + 3 + 4	70.94	53.75	53.13	57.19	**58.75**	35.37	166.11
Mean	55.39	45.51	40.89	37.79	44.89	33.81	136.68 [2]
LSD 5% Mix.	7.65	7.65	7.65	7.65	7.65	3.82	
LSD 5% cv						1.97	

[1] Bold printed in Mean: Isolate 3 and its mixtures, [2] Mean only for mixtures,

1, 2, 3, 4 = Isolates to be mixed per se. 1 + 2 = mix Is. 1 and 2, 1 + 2 + 3 = Is. 1 and 2 and 3, etc.

Table 5B/Correlations	GK Garaboly	GK Futár	GK Fény
GK Futár	0.9735 ***		
GK Fény	0.9492 ***	0.9866 ***	
GK Csillag	0.9521 ***	0.9169 ***	0.9041 ***

*** $p = 0.001$.

Table 5C/inocula	1	2	3	4	1 + 2	1 + 3	1 + 4	2 + 3	2 + 4	3 + 4	1 + 2 + 3	1 + 2 + 4	1 + 3 + 4	2 + 3 + 4
2	0.88													
3	0.79	0.81												
4	0.95 *	0.70	0.64											
1 + 2	0.48	0.49	0.90	0.34										
1 + 3	0.94	0.87	0.95 *	0.85	0.74									
1 + 4	0.90	0.81	0.97 *	0.81	0.81	0.99 *								
2 + 3	0.99 *	0.82	0.73	0.98 *	0.40	0.91	0.87							
24	0.95 *	0.96 *	0.72	0.85	0.35	0.87	0.80	0.93						
3 + 4	0.62	0.69	0.97 *	0.45	0.97 *	0.85	0.89	0.54	0.54					
1 + 2 + 3	0.94	0.88	0.95 *	0.84	0.74	1.00 *	0.99 *	0.90	0.87	0.85				
1 + 2 + 4	0.98 *	0.94	0.75	0.90	0.40	0.90	0.85	0.97 *	0.99 *	0.58	0.91			
1 + 3 + 4	0.99 *	0.86	0.69	0.95 *	0.32	0.87	0.82	0.99	0.97 *	0.49	0.87	0.99 *		
2 + 3 + 4	0.99 *	0.82	0.71	0.98 *	0.38	0.90	0.85	1.00 *	0.93	0.52	0.89	0.97 *	0.99 *	
1 + 2 + 3 + 4	0.91	0.89	0.56	0.85	0.15	0.76	0.68	0.91	0.98 *	0.35	0.77	0.96 *	0.96 *	0.92

* $p = 5\%$, Limit: $r = 0.95$.

1, 2, 3, 4 = Isolates to be mixed per se. 1 + 2 = mix Is. 1 and 2, 1 + 2 + 3 = Is. 1 and 2 and 3, etc.

However, the cultivar responses often differ in the different isolates and mixture. The two more resistant genotypes are the same we found for FHB, GK Fény and GK Csillag. In eight cases, the response of the two cultivars does not significantly differ from each other. In five cases, GK Csillag has lower value, and only in one case GK Fény. GK Garaboly and GK Futár do not show significant difference in six cases; in all other cases, GK Futár has lower values. The larger differences are more stable, GK Csillag has better resistance in each inocula compared to GK Garaboly, but compared to GK Futár, in seven cases, no significant difference was found. From Table 5A, the correlation between the genotype reactions were also computed (Table 5C). Of the 105 possible correlations, 29 were significant. The non-significant correlations varied strongly. This was 29, more than double than was found in DI (13, Table 3C). Of the 29, twenty-five were found between inocula containing Isolate 3 in one or more partner inocula for the correlation test. This shows that FDK provides a closer correlation matrix. This would mean that a mixture automatically does not solve the problem and does not secure a stable level. The problem is as it was for the DI—that in a regular case we have only one test result, and not 15 as in this test. It is sure that a mixture does not provide the increased security of testing we hope from it. Except for the several low aggressive versions we know better to avoid, even the same aggressiveness does not always guarantee the same variety response. Comparing 2 + 3 + 4 and 1 + 2 + 3 + 4 at 59.98 and 58.75 mean aggressiveness in this trait, in the first case, Futár is significantly more susceptible, and in the other case, they have the same number. This means that we have to look for another solution.

The response of the experiment means (Table 6) strongly differs; 2013 gave the highest FDK values, 2014 was five times less and 2015a and 2015b showed similar results to 2013. The response of the different mixtures was very variable. At high infection pressure, there were rather small differences between FDK values at different isolates and mixtures, except isolate 2 and the combination of isolates 2 + 4 that gave significantly less FDK than the others. There is another feature that needs attention. In 2013, isolate 3 gave 84% FDK, in 2014 39.88%. However, at the same performance in 2013 the 2 + 3 gave only 6%, 3 + 4 gave 9.94% and 1 + 3 + 4 only 6.38%. Another example is 1 + 2 + 4 where 72.40% and 0.84% are the two corresponding data. However, in a less epidemic year, independent of the causing agents, the forecasting of the numbers for a heavy epidemic situation is hardly possible.

Table 6. Response of wheat to the different isolates and their mixtures. FDK data (%) of the experiments across wheat cultivars, Szeged, 2013–2015.

Isolates	Experiments				Mean
Table 6A/Mixtures	2013	2014	2015a	2015b	
1 Fc 12375	69.69	14.75	44.06	46.56	43.77
2 Fc 52.10	4.22	0.73	11.31	10.06	6.58
3 Fg 19.42	84.25	39.88	74.06	79.31	69.38
4 Fg 13.38	39.06	7.50	19.69	20.75	21.75
1 + 2	65.94	5.58	46.56	52.50	42.64
1 + 3	79.25	21.69	55.94	49.06	51.48
1 + 4	70.00	9.81	39.06	40.31	39.80
2 + 3	82.38	6.14	61.25	67.75	54.38
2 + 4	23.94	5.13	13.13	12.31	13.63
3 + 4	84.44	9.94	70.94	64.38	57.42
1 + 2 + 3	75.00	35.00	72.50	66.25	62.19
1 + 2 + 4	72.40	0.84	38.13	44.69	39.01
1 + 3 + 4	78.63	6.38	61.88	63.75	52.66
2 + 3 + 4	82.50	15.59	69.63	72.19	59.98
1 + 2 + 3 + 4	82.50	21.56	68.44	62.50	58.75
Mean	66.28	13.37	49.77	50.16	44.89
LSD 5%	7.65	7.65	7.65	7.65	3.82
LSD Experiment 5%					1.97

1, 2, 3, 4 = Isolates to be mixed per se. 1 + 2 = mix Is. 1 and 2, 1 + 2 + 3 = Is. 1 and 2 and 3, etc.

Table 6B/Correlations	2013	2014	2015a
2014	0.5344 *		
2015a	0.9278 ***	0.6786 **	
2015b	0.9392 ***	0.6262 **	0.9813 ***

*** $p = 0.001$, ** $p = 0.01$, * $p = 0.05$.

Table 6C/Inocula	1	2	3	4	1 + 2	1 + 3	1 + 4	2 + 3	2 + 4	3 + 4	1 + 2 + 3	1 + 2 + 4	1 + 3 + 4	2 + 3 + 4
2	0.36													
3	0.94	0.65												
4	0.98 *	0.15	0.84											
1 + 2	0.97 *	0.55	0.99 *	0.90										
1 + 3	0.99 *	0.34	0.90	0.97 *	0.94									
1 + 4	1.00 *	0.28	0.90	0.99 *	0.95 *	0.99 *								
2 + 3	0.96 *	0.59	1.00 *	0.88	1.00 *	0.93	0.93							
2 + 4	0.97	0.16	0.83	1.00 *	0.89	0.98 *	0.99 *	0.87						
3 + 4	0.95 *	0.62	0.98 *	0.87	0.98 *	0.95 *	0.93	0.99 *	0.87					
1 + 2 + 3	0.91	0.69	0.97 *	0.81	0.96 *	0.91	0.88	0.97 *	0.82	0.99 *				
1 + 2 + 4	1.00 *	0.36	0.94	0.97 *	0.98 *	0.98 *	0.99 *	0.97 *	0.96 *	0.95 *	0.90			
1 + 3 + 4	0.96 *	0.62	0.99 *	0.87	1.00 *	0.94	0.93	1.00 *	0.87	0.99 *	0.98 *	0.96 *		
2 + 3 + 4	0.94	0.66	1.00 *	0.84	0.99 *	0.92	0.91	1.00 *	0.84	0.99 *	0.98 *	0.94	1.00 *	
1 + 2 + 3 + 4	0.97 *	0.56	0.97 *	0.90	0.98 *	0.97 *	0.95 *	0.98 *	0.90	1.00 *	0.99 *	0.96 *	0.99 *	0.98 *

* $p = 5\%$. Limit: r = 0.95.

1,2,3,4 = Isolates to be mixed per se. 1 + 2 = mix Is. 1 and 2, 1 + 2 + 3 = Is. 1 and 2 and 3, etc.

It seems that the aggressiveness of the mixtures cannot be predicted based on the individual aggressiveness of the four basic isolates (Table 6B, $n = 15$). It depends also on the ecology, but the individual isolates behave differently in most of the mixtures. The correlations data support this, although the 2014 data do so only moderately. Significantly, the data of the higher epidemic situations correlate much better, above r = 0.90, when 2013 and the two 2015 experiments are compared. This agrees well with what we found for FHB data. The correlations between experiment reactions for every inocula were presented (Table 6C, $n = 4$). From the 105 possible correlations, 57 were significant compared to DI in Table 4 with 52 significant correlations. We have non-significant correlations where aggressiveness was very low, such as isolate 2 (Fc 52.10), where no other inocula gave significant correlation with this set.

3.3. DON Contamination

The variety specific data (Table 7) show a similar picture to the FDK data. The four isolates showed larger aggressiveness differences, and their combinations showed rather variable performance, depending on their combinations (Table 7A). It seems the mixing resulted in both a decreasing of the aggressiveness level of the most aggressive Fg 19.42 isolate and in a reduction of DON contamination; and this tendency was true not only for DON but also for DI and FDK. On the other hand, it is also true that the aggressiveness of the mixture was, in most cases, significantly higher than the arithmetical means of the mixtures measured by the performance of the DON contamination original of the participating inocula. The isolates and isolate combinations of the low pathogenic isolates showed normally low aggressiveness and low DON levels, indicating that the use of such isolates does not give suitable inoculum for inoculation. The correlations between cultivars against the 15 different inocula (Table 7B, n = 15) to the used inocula (individual and mixed), as well as mixed inocula, were highly significant between r = 0.81–0.95. Here, also, the correlations between the means across cultivars and individual performances were given. These were higher, between r = 0.90 and 0.98 than that of the numbers between cultivars. The conclusion is that the means are better resistance indicators than any of the isolates and their mixtures present.

Table 7. Response of the wheat cultivars to the different isolates and their mixtures. DON data (mg/kg) across experiments, Szeged, 2013–2015.

Isolates	**Cultivars, DON, mg/kg**				**Mean**	**Counted**	**Mean */**
Table 7A /Mixtures	**GK Garaboly**	**GK Futár**	**GK Fény**	**GK Csillag**	**Mean**		**Counted%**
1 Fc 12375	70.42	48.28	36.84	35.57	47.78	47.78	100.00
2 Fc 52.10	18.99	7.95	10.34	7.84	11.28	11.28	100.00
3 Fg 19.42	158.56	117.39	94.60	53.95	**106.13** [1]	106.13	100.00
4 Fg 13.38	3.78	4.40	4.26	3.67	4.03	4.03	100.00
1 + 2	58.56	63.62	41.00	28.50	47.92	29.53	162.28
1 + 3	126.33	66.39	33.17	31.40	**64.32**	76.95	83.59
1 + 4	45.66	33.92	28.79	20.02	32.10	25.90	123.91
2 + 3	116.63	66.29	51.69	60.90	**73.88**	58.70	125.85
2 + 4	10.29	4.50	5.39	5.43	6.40	7.66	83.61
3 + 4	124.46	70.36	78.36	40.47	**78.41**	55.08	142.37
1 + 2 + 3	120.70	85.78	44.61	38.35	**72.36**	55.06	131.42
1 + 2 + 4	61.18	29.92	26.34	27.53	36.24	28.58	126.82
1 + 3 + 4	97.25	64.18	41.66	54.87	**64.49**	52.64	122.50
2 + 3 + 4	125.21	89.21	54.30	68.41	**84.28**	40.48	208.21
1 + 2 + 3 + 4	106.08	59.19	56.08	58.50	**69.96**	42.30	165.38
Control	3.38	2.39	1.60	2.51	2.47	2.47	100.00
Mean	77.97	50.86	38.07	33.62	50.13	40.28	134.18 [2]
LSD 5%	18.71	18.71	18.71	18.71	9.35		
LSD 5% var.					4.67		

[1] Bold printed: Isolate 3 and its combinations, [2] Mean only for mixtures.

1,2,3,4 = Isolates to be mixed per se. 1 + 2 = mix Is. 1 and 2, 1 + 2 + 3 = Is. 1 and 2 and 3, etc.

Table 7B/Correlations	**GK Garaboly**	**GK Futár**	**GK Fény**	**GK Csillag**
GK Futár	0.9516 ***			
GK Fény	0.8997 ***	0.9106 ***		
GK Csillag	0.8740 ***	0.8467 ***	0.8097 ***	
Mean	0.9846 ***	0.9760 ***	0.9434 ***	0.9095 ***

*** p = 0.001.

Table 7C/Inocula	**Control**	**4**	**2 + 4**	**2**	**1 + 4**	**1 + 2 + 4**	**1**	**1 + 2**	**1 + 3**	**1 + 3 + 4**	**1 + 2 + 3 + 4**	**1 + 2 + 3**	**2 + 3**	**3 + 4**	**2 + 3 + 4**
4	−0.60														
2 + 4	0.79	−0.55													
2	0.69	−0.38	0.98 *												
1 + 4	0.63	0.08	0.76	0.85											
1 + 2 + 4	0.86	−0.43	0.97 *	0.96 *	0.87										
1	0.84	−0.21	0.87	0.88	0.95 *	0.96 *									
1 + 2	0.39	0.49	0.29	0.40	0.83	0.50	0.72								
1 + 3	0.85	−0.21	0.86	0.88	0.94	0.96 *	1.00 *	0.72							
1 + 3 + 4	0.95 *	−0.40	0.86	0.82	0.83	0.95 *	0.97 *	0.60	0.97 *						
1 + 2 + 3 + 4	0.86	−0.47	0.98 *	0.96 *	0.84	1.00 *	0.95 *	0.46	0.94	0.94					
1 + 2 + 3	0.78	−0.03	0.74	0.77	0.96 *	0.88	0.98 *	0.84	0.98 *	0.93	0.85				
2 + 3	0.92	−0.45	0.94	0.91	0.85	0.99 *	0.97 *	0.52	0.97 *	0.98 *	0.99 *	0.90			
3 + 4	0.54	−0.02	0.84	0.93	0.96 *	0.88	0.89	0.65	0.88	0.76	0.87	0.85	0.83		
2 + 3 + 4	0.93	−0.29	0.81	0.78	0.86	0.92	0.97 *	0.69	0.98 *	0.99 *	0.91	0.96 *	0.96 *	0.76	
3	0.58	0.15	0.71	0.81	1.00 *	0.82	0.92	0.86	0.92	0.80	0.80	0.95 *	0.81	0.94	0.84

* p = 0.05. Limit: r = 0.95

1,2,3,4 = Isolates to be mixed per se. 1 + 2 = mix Is. 1 and 2, 1 + 2 + 3 = Is. 1 and 2 and 3, etc.

The resistance expression differs in the 15 inocula. The genotypes that did not differ significantly to a given inoculum having lower differences than the LSD 5% is (Table 7A). In twelve cases, GK Garaboly is significantly more susceptible than GK Csillag and GK Fény. The more resistant GK Futár has higher DON contamination to GK Garaboly only in one case. GK Csillag shows lower DON contamination in seven cases than GK Fény, and in five cases, the GK Fény produces less DON. The difference for mean does not reach the significance. This is all independent from the inoculum, should it be pure isolate or mixture. It can be stated, also, that the large resistance differences have the highest chances to be significant, except in the three cases without significant deviation. At lower resistance differences even otherwise significant differences cannot be demonstrated as the data show at different inocula compared with the man cultivar reaction (Table 7A)

At two inocula, the cultivars did not show any significant difference. This is true, also, for the control that had natural infection. The DON contamination of the naturally infected heads was low compared to the artificially inoculated samples. On the other side, they are not without risk, as they were higher than the EU limit of 1.25 mg/kg. Four inocula showed three not differing genotypes. Six inocula could not differentiate between two genotypes, and in two cases, every genotype differed from the other. This was the case also for the means. Counting the correlations (Table 7C), being significant, 30 cases were found of the 105 total correlations between inocula. The control values correlated with inocula only in one case. In 24 cases, the Isolate 3 and its combinations showed significant correlations, for the other possibilities we had only six cases. The question is, again, what is the best inoculum? It seems that mixed inocula are not better than the single spore lines are. We have chances for good or very good differentiation at the highest aggressive inocula, but not in every case. For demonstrating no significant differences, we have a chance at low aggressiveness. One solid conclusion is clear. There is no proof that mixtures would be better than pure isolates. However, we should have more inocula than one to present more useful results for a QTL phenotyping or a registration trial. Since, for us, the DON reaction is the most important, we provide the data for it in Figure 2.

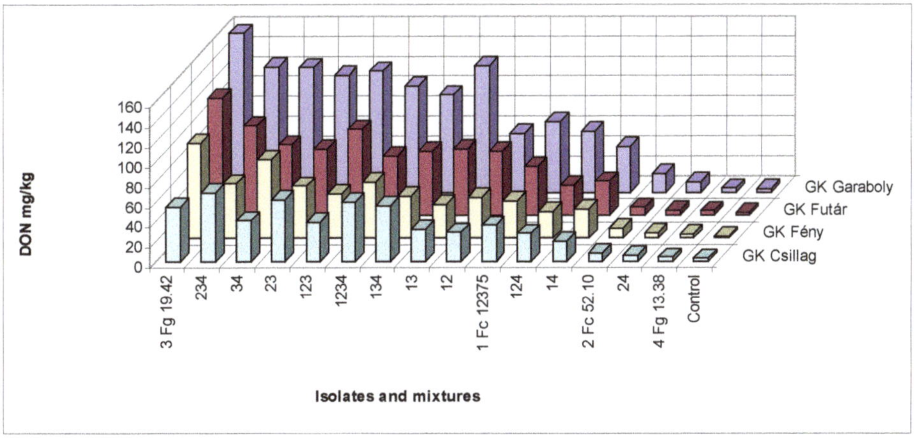

Figure 2. Influence of mixing isolates on the resistance expression of the four cultivars with differing FHB resistance. DON contamination across years and experiments. LSD 5% between any bars in the figure is 18.71.

There are large differences between experiments (Table 8). The epidemic conditions in 2015 significantly increased the DON contamination; even the FHB and FDK values did not predict these high DON data values (Table 8A). The correlations with the 2014 data and other experiments were moderate, whereas, the others were highly significant with a correlation of r = 0.90 and above (Table 8B). The data also prove that a low aggressiveness level does not allow a proper distinction between the effects of different inocula. As one year does not give well-balanced results, in spite of the higher number of isolates, the tests in different years have a high importance. The correlations between the four experiments were tested (Table 8C). Moreover, 89 cases from the 120 significant relations were registered. This shows a rather good agreement between responses of different experiments. It should be noted that the control values correlated at a high ratio with the artificial inoculation data. For DI 52 and FDK 57, significant correlations were found. It seems that the best correlation matrix was found for DON.

Table 8. Response of wheat to the different isolates and their mixtures. DON data (mg/kg) of the experiments across wheat cultivars, Szeged, 2013–2015.

Isolates	Experiments				Mean
Table 8A/Mixtures	**2013**	**2014**	**2015a**	**2015b**	
1 Fc 12375	32.98	5.74	72.85	79.55	47.78
2 Fc 52.10	1.38	0.70	20.65	22.39	11.28
3 Fg 19.42	52.36	20.82	166.73	184.60	106.13
4 Fg 13.38	1.86	1.05	6.59	6.63	4.03
1 + 2	24.39	3.03	76.96	87.29	47.92
1 + 3	46.62	9.37	111.46	89.86	64.32
1 + 4	29.58	1.59	50.65	46.58	32.10
2 + 3	49.85	2.44	102.20	141.03	73.88
2 + 4	2.81	1.75	11.59	9.46	6.40
3 + 4	51.53	4.48	142.16	115.50	78.41
1 + 2 + 3	35.26	15.14	127.27	111.77	72.36
1 + 2 + 4	29.98	0.61	58.86	55.51	36.24
1 + 3 + 4	36.11	3.41	115.20	103.24	64.49
2 + 3 + 4	52.11	5.91	132.52	146.59	84.28
1 + 2 + 3 + 4	43.16	7.91	129.64	99.14	69.96
Control	1.51	1.12	3.98	3.27	2.47
Mean Ecp.	30.72	5.32	83.08	81.40	50.13
LSD 5%	18.71	18.71	18.71	18.71	4.67

1, 2, 3, 4 = Isolates to be mixed per se. 1 + 2 = mix Is. 1 and 2, 1 + 2+3 = Is. 1 and 2 and 3, etc.

Table 8B/Correlations	GK Garaboly	GK Futár	GK Fény	GK Csillag
B/Correlations	**2013**	**2014**	**2015a**	**2015b**
2014	0.5552 *			
2015a	0.9333 ***	0.7251 **		
2015b	0.9171 ***	0.6934 **	0.9458 ***	
Mean	0.9490 ***	0.7217 **	0.9870 ***	0.9836 ***

*** $p = 0.001$, ** $p = 0.01$, * $p = 0.05$.

Table 8C/Inocula	1	2	3	4	1 + 2	1 + 3	1 + 4	2 + 3	2 + 4	3 + 4	1 + 2 + 3	1 + 2 + 4	1 + 3 + 4	2 + 3 + 4	1 + 2 + 3 + 4
2	0.95 *														
3	0.99 *	0.99 *													
4	0.97 *	0.99 *	1.00 *												
1 + 2	0.99 *	0.98 *	1.00 *	0.99 *											
1 + 3	0.96 *	0.92	0.94	0.95 *	0.95 *										
1 + 4	0.97 *	0.86	0.92	0.91	0.93	0.97									
2 + 3	0.98 *	0.93	0.97 *	0.94	0.98 *	0.90	0.92								
2 + 4	0.94	0.97 *	0.96 *	0.98 *	0.95 *	0.97 *	0.90	0.87							
3 + 4	0.97 *	0.93	0.95 *	0.96 *	0.96 *	1.00 *	0.97 *	0.90	0.97 *						
1 + 2 + 3	0.97 *	0.98 *	0.98 *	0.99 *	0.97 *	0.98 *	0.92	0.91	1.00 *	0.98 *					
1 + 2 + 4	0.98 *	0.90	0.95 *	0.94 *	0.96 *	0.98 *	1.00	0.94	0.93	0.98 *	0.95 *				
1 + 3 + 4	0.98 *	0.96 *	0.98 *	0.99 *	0.98 *	0.99 *	0.96 *	0.93	0.98 *	0.99 *	0.99 *	0.98 *			
2 + 3 + 4	1.00 *	0.97 *	0.99 *	0.98 *	1.00 *	0.96 *	0.96 *	0.98 *	0.95 *	0.97 *	0.97 *	0.98 *	0.98 *		
1 + 2 + 3 + 4	0.95 *	0.93	0.94	0.96 *	0.94	1.00 *	0.95 *	0.88	0.98 *	1.00 *	0.99 *	0.97 *	0.99 *	0.95 *	
Control	0.94	0.96 *	0.95 *	0.98*	0.95 *	0.98 *	0.91	0.87	1.00 *	0.98 *	0.99 *	0.93	0.98 *	0.94	0.99 *

* $p = 0.05$. Limit: r = 0.95.

1, 2, 3, 4 = Isolates to be mixed per se. 1 + 2 = mix Is. 1 and 2, 1 + 2+3 = Is. 1 and 2 and 3, etc.

The variance analyses (Table 9) revealed large similarities between the different traits (Table 9A). The three-way interactions were not significant, so an additional F test to the "Within" category was not necessary. All main effects were highly significant. It is more important that the interactions were much smaller than the individual main effects were (Table 9B). This is significant in all cases in differences between the main effect and interactions for FHB and FDK. The variety effect against A × B is not significant for DON indicating that the main effect is more impressionable than the FHB or FDK are. The genotype (A) effect is significant over A × C interaction, indicating a good stability of resistance.

Table 9. Response of wheat to the different isolates and their mixtures. ANOVAs for the three traits, FHB, FDK and DON, Szeged, 2013–2015.

Table 9A /Source of Variance	df	DI [a] MS	F	*p*	FDK [b] MS	F	*p*	df	DON [c] MS	F	*p*
Variety A	3	3076.8	86.8	***	14,160.0	116.2	***	3	44,671.8	61.3	***
Experiment B	3	59,962.0	1690.8	***	120,215.8	986.8	***	3	322,433.4	442.3	***
Inoculum C	14	11,287.9	318.3	***	21,562.8	177.0	***	15	53,877.0	73.9	***
A × B	9	110.6	3.1	***	919.7	7.5	***	9	33,981.2	46.6	***
A × C	42	156.4	4.4	***	387.8	3.2	***	45	8180.8	11.2	***
B × C	42	1235.1	34.8	***	1699.9	14.0	***	45	14,704.9	20.2	***
A × B × C	126	62.4	1.8	n.s.	203.2	1.7	n.s.	135	621.8	0.9	n.s.
Within	720	35.5			121.89			768	729.0		
Total	959							1023			

*** $p = 0.001$, n.s. = not significant. Analysis between two-way interactions and main effects.

Table 9B /Interactions, and df [d]	df 9A	F	*p*	F	*p*	df	F	*p*
A × B, df = 9	A, df 3	27.81	***	15.40	***	A, df 3	1.31	n.s.
	B, df 3	542.15	***	130.71	***	B, df 3	9.49	**
A × C, df 42	A, df 3	19.67	***	36.51	***	A, df 3	5.46	*
	C, df 14	72.17	***	54.80	***	C, df 15	6.59	***
BcC, df 42	B, df3	48.55	***	70.72	***	B × C,45; B, df 3	21.93	***
	C, df 14	9.14	***	12.68	***	B × C,45; B, df 15	2.31	*

*** $p = 0.001$, ** $p = 0.01$, * $p = 0.05$, n.s. = not significant. [a] = Disease Index, [b] = Fusarium damaged kernel, [c] = deoxynivalenol, [d] = degree of freedom.

3.4. Interrelations between Traits

Figure 3 shows the means of the three traits of the different tested inocula. The data correlate well, but we see that isolates 2 and 4 behave differently from the others. Isolate 4 causes relatively more FHB and FDK, but it is poor in DON production. Isolate 2 behaves oppositely, here the visual data are much lower, but the DON data are higher. Their mixture shows well the transitional result between components. The correlations are very close, they are between r = 0.95 and 0.97 (calculated without the control). The rates for DI/DON and FDK/DON were also calculated (Figure 4). For 1% FDK we had DON content between 1.53 to 0.19 mg/kg. Isolate 3 and its mixtures gave values between 1.25 and 1.53, the others between 0.19–1.12. This clearly means that the forecast of DON contamination via FDK is not possible. We draw also similar conclusions for the DON/FHB rate; however, the different behavior of the DI/DON rate is clear, following a decrease to 1 + 3 the rate increases again. It shows also that the FDK is more precise to describe the DON relation than DI is.

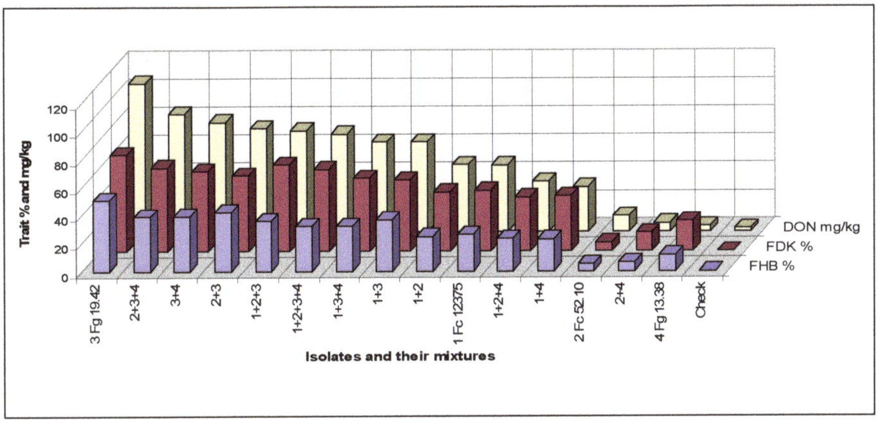

Figure 3. Mixtures and aggressiveness, means for the three traits, Szeged, 2013–2015. LSD 5% between any data for DI 2.06%, FDK 3.82% and DON 9.35 mg/kg.

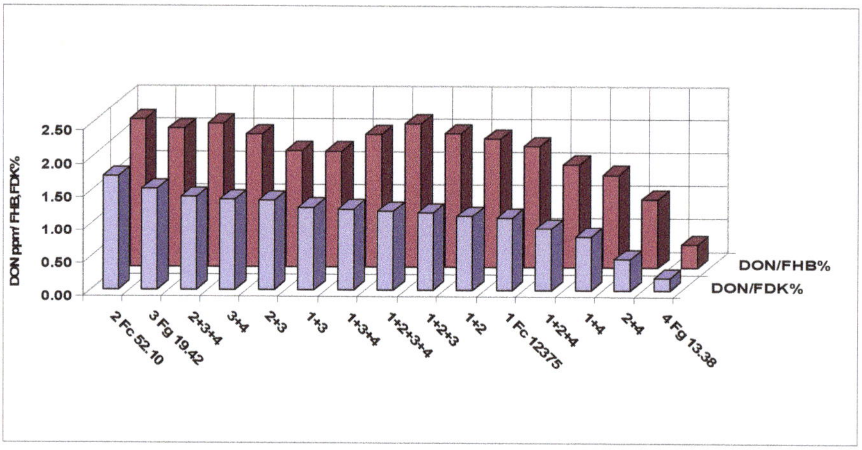

Figure 4. Influence of mixtures on FHB test results. DON production for one percent FHB and FDK infection, 2013–2015.

The cultivar reactions as means across inocula and years illustrate the well-known fact that visual head data are less informative for DON than the FDK data are (Figure 5). Based on the FHB values, no DON forecast is possible. GK Csillag is important as this cultivar performed well after a fungicide treatment than any other cultivars tested. Based on data from the milling industry, all staples were bought from this cultivar in 2010 when we had the national FHB epidemic. The correlation between FHB and DON is $r = 0.66$, which is not highly significant, whereas between FDK and DON the correlation is $r = 0.9968$, and is significant at $p = 0.001$. It means that the FDK signalizes the DON content much better than the visual scores.

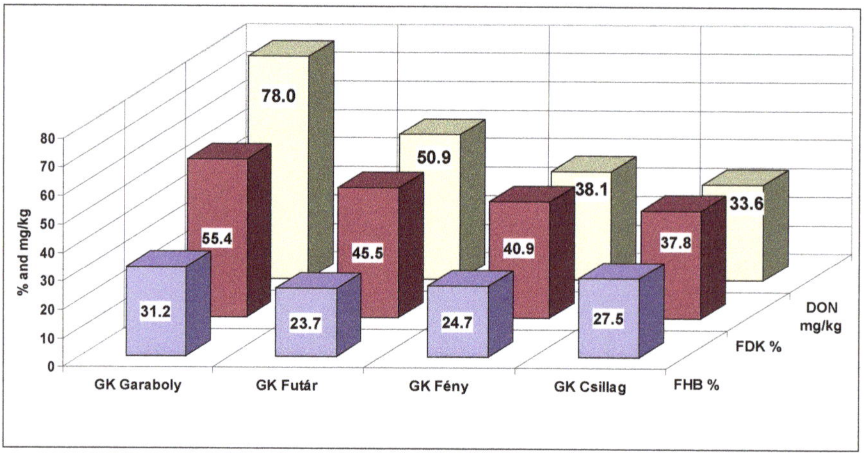

Figure 5. General performance of the cultivars across experiments and inocula for FHB (DI), FDK and DON contamination. LSD 5%: DI: 1.06, FDK: 1.97, DON: 4.68.

Correlations were counted between DI, FDK and DON for every cultivars tested to see whether there are different variety to the 15 inocula responses or not (Figure 6). The first conclusion is that the FDK/DON correlations are much closer than the DI/DON correlations, this agrees well with earlier experience. The DI/FDK data did not show any significant relations in the four cultivars, indicating that the same cultivar might have different responses at different inocula and mixtures of inocula.

The correlations between DI, FDK and DON were counted also for all isolates separately for the four cultivars (Table 10). In five inocula, all three correlations are highly significant and nearly no difference exist between them. Two of them were single isolates and three mixtures were found (1 + 3, 1 + 4 and 1 + 2 + 4). In eight cases, the FDK/DON correlations were closer than the DI/DON correlations. Thus, it seems that the isolates and their mixtures may have different profiles that can influence resistance expression. The only advantage of the mixture is that the very diverse behavior of 2 Fc 52.10 and 4 Fg 13.38 could be balanced by the mixing to some extent by increasing the aggressiveness. It is important that the DI/DON correlation across all inocula is much lower (r = 0.6668) than the FDK/DON relation (r = 0.9968). The DI/FDK is also low (r = 0.6170). This means that for forecasting DON the FDK is more precise means than visual scores are. This is an argument that FDK and DON contaminations should be treated more seriously than it normally happens.

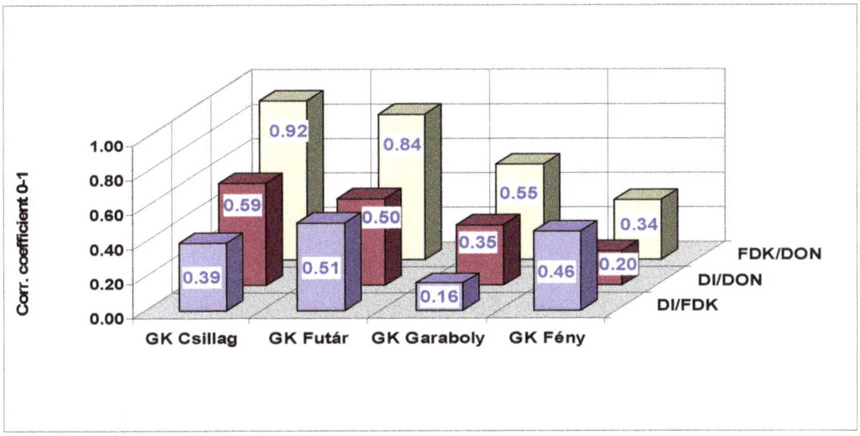

Figure 6. Correlations between genotype reactions to different FHB traits (data from Figures 2–4, for each isolate the performance of the cultivars was related to the mean of genotypes). Limit for $p = 0.05 = 0.51$, $p = 0.02 = 0.59$, $p = 0.001 = 0.77$.

Table 10. Correlations between DI, FDK and DON contamination the four cultivars ($n = 4$) for each isolates and mixtures and their means, Szeged, 2013–2015.

Inocula	Correlations between		
	DI/FDK	DI/DON	FDK/DON
1 Fc 12375	0.9894 *	0.9941 *	0.9947 *
2 Fc 52.10	0.0282	0.1054	0.9950 *
3 Fg 19.42	0.9790 *	0.9842 *	0.9870 *
4 Fg 13.38	0.8732	−0.6943	−0.3513
1 + 2	0.9577 *	0.7515	0.8971
1 + 3	0.9760 *	0.9996 *	0.9698 *
1 + 4	0.9907 *	0.9638	0.9904 *
2 + 3	0.9353	0.9084	0.9819 *
2 + 4	0.7020	0.7011	0.9846 *
3 + 4	0.9492	0.9552 *	0.8570
1 + 2 + 3	0.8668	0.9532	0.9754 *
1 + 2 + 4	0.9973 *	0.9943 *	0.9991 *
1 + 3 + 4	0.7703	0.8500	0.9906 *
2 + 3 + 4	0.9100	0.8746	0.9929 *
1 + 2 + 3 + 4	0.8984	0.8695	0.9804 *
Mean [a]	0.6170	0.6668	0.9968 *
Average of 15 inocula [b]	0.8549	0.7474	0.8830

* $p = 0.05$, [a] Means for all inocula. [b] averages of correlations (from inoculum 1 to mixture 1+2+3+4) across all inocula.

4. Discussion

4.1. Influence of Mixing

The mixing of isolates is a generally applied method. This is done, not only to regulate aggressiveness, but it can be a necessity when a larger amount of inoculum is needed than the present methods can produce. The cited literature (16–63) is not consequent. The number of components strongly varies. Moreover, many authors seemingly feel that they are more secure when more isolates are mixed, and they think that a mixture comes closer to the variability found in nature. However, we did not find anything pertaining to this problem in the literature. Thus, it was not clear what the result of the mixing considering the aggressiveness is. As nobody, except one [40], used the aggressiveness test, the mixing was made without an aggressiveness control. The lesson is that the different mixtures did not provide the same aggressiveness. For this reason, nobody can hope that a mixture automatically will have high aggressiveness. This was not an accident; the experimental results do not support such assumptions. Additionally, the reactions differed between DI, FDK and DON contamination. Large and significant differences were found between the different mixtures and pure isolates. There is another problem; that the conidium concentration was adjusted to different values. Of course, this could influence the effect of mixing. Most people used 5×10^4 or 5×10^5 conidia/mL, but (once) 10^6 conidia/mL was applied. Here, we remark only that to the uncontrolled mixing and uncontrolled adjusting of conidium concentration was a general rule. The effect of dilution was influenced by the isolate, the host variety through resistance, and the aggressiveness level, but other agents may also be present. In laboratory and greenhouse seedling tests [1], mixing of isolates did not give the mean of aggressiveness of the participating inocula aggressiveness in greenhouse experiments and the different mixtures reacted differently. However, these greenhouse tests could not provide any information about the behavior of the mixtures for field conditions. In the best case, it was mentioned in the papers that the isolates were aggressive in earlier tests and, therefore, the authors hoped in a successful infection. Therefore, the general experimental praxis and the earlier data pressed us to investigate this problem.

The result of the mixing depends most on the aggressiveness level of the participating inocula. From the low pathogenic isolate, no highly aggressive inocula can be produced, but from an aggressive isolate from the same test tube, inocula with differing aggressiveness could be produced [75]. When we see the most aggressive isolate (inoculum), the mixture decreases its aggressiveness (each case compared to the most aggressive component). When we compare the aggressiveness of the mixture to the arithmetical mean of the individual components, it becomes clear that in most cases the aggressiveness is higher, with even 50–100% difference in some cases than the mean aggressiveness level counted for the participating inocula. On the other side, the resulting aggressiveness was always lower than that of the most aggressive component. Therefore, the exact forecasting of the result of the mixing is hardly possible. This means that the most aggressive isolate can significantly overbalance the arithmetical mean, but still does not reach the level of the original aggressiveness of the most aggressive component. The system is much more complicated than it was thought earlier. The DI/DON and FDK/DON rates showed an interesting picture. The DI/DON rate was much more variable; this was found mostly when Fc. 12,375 isolate was a participant of the mixture or was alone. We do not know whether this, an instinct attribute, is for other variants of the DI/DON and FDK/DON rations, which were rather similar. It is also considerable that Fg 13.38 has a relatively high DON and FDK ratio and a very low DON content, the other three isolates perform similarly with lower DI, medium FDK and higher DON contamination with strong differences. This is supported also by the higher number of significant correlations from Tables 3C, 5C and 7C.

Uncontrolled mixing and diluting may result in low aggressiveness level and bad differentiation of the genotypes. The only advantage is that it stabilizes the variation in aggressiveness of the inocula compared to the individual components. However, it does not solve the problem of the use of one inoculum. As we do not have races at hand, the most important argument (the mixtures represent a better scope of the pathogenic population) is refuted. In this test series, this view could not be supported. The question is which of the 15 the best solution is. We cannot say each inoculum behaved differently. Having an inoculum the resistance can be underestimated (high aggressiveness) or underestimated low aggressiveness; therefore, several aggressiveness levels are needed to reach more reliable results. On the other hand, the different differentiation of the genotypes at different aggressiveness levels needs a correction mechanism, by the use of more isolates to produce more reliable results. This has a significance in QTL analysis where the large differences can be evaluated well, but the chance to identify for lower or medium effective QTL is not very convincing. The conclusion is that the mixing does not solve the problem to produce top quality inoculum for artificial inoculation. Even the aggressiveness of the participating inocula is known, an exact forecast of the aggressiveness of the mixture is not possible—especially when we would like to know FDK and DON reaction. In resistance response, the mixtures are not better, or worse, than a single aggressive inoculum.

There is another problem. As in this test, also, the case is, the mean performance of different independently used inocula produce a mean that describes much better the resistance level than any of the inocula alone. Earlier research [69–71] supports its usefulness. The use of more independent isolates is costly. Therefore, it should be used where high quality phenotyping is necessary for genetic research and other scientific purposes. It is needed to qualify new cultivars with higher or high resistance to FHB. The risk analysis needs careful work. The behavior of the most aggressive isolate resembles well to the combining ability of a given crossing partner in different crosses. In plant breeding, this is a well inheritable trait. It is supposed that such research would be important to make here, to be able to select the isolates that give the highest aggressiveness if possible.

An aggressiveness test is necessary to qualify inoculum before use. No mixing, no adjusting inoculum can secure the high aggressiveness needed for efficient work. When we want to have a mixture, an aggressiveness test can help choose the suitable inocula for mixing.

4.2. Resistance Expression

The correlations between cultivar reactions to the 15 isolates and mixtures for DI are between $r = 0.96$ and 0.98, $p = 0.001$. However, we have about tenfold difference in aggressiveness. From the 15 aggressiveness level the different inocula and their mixtures represent, only 1–2 cases were found where all genotypes significantly differed from each other. Normally two or three genotypes should have been marked without significant difference for the given inoculum. The data are clear that the DON contamination show closer relation with FDK than with DI. Seeing the three traits, the least reliable difference between genotypes is at disease index; larger is at FDK and the largest is at DON. However, this depends also on ecological conditions. It occurred that a higher DI was followed a poorer FDK and DON presentation. As normally only one data set is given, which is the set of the 15 possible cases shown, is a question. It seems to be clear that one inoculum surely will not describe medium or lower resistance differences, with more isolates, it can be done much better, but large resistance differences can be identified with an aggressive inoculum with a high probability. This is a confirmation of earlier results where resistance component kernel resistance and specific DON resistance were measured [2,4,14,15,70]. Resistance expression depends on the composition of the mixture, resistance level, conidium concentration and ecological conditions, so a prediction of the result of the mixture is hardly possible.

4.3. Genetic Aspects

There is a new development in the genetic mapping work. QTLs were identified with different role [76]. QTLs can regulate only visual symptoms, visual symptoms and FDK, FDK and DON and QTLs were identified influencing all three traits. Similar results were published by recent authors [77–81]. This will have the consequence that mapping should be extended to FDK and DON that is generally not done. For the crossing partners, this information is inevitable. From the generally used DI mapping [17,19,21–25,29,30,32,33,37,56,61], this aspect falls out and has only reduced significance of the QTL identified.

4.4. Breeding Aspects

Large-scale mass selection can be done well by using a mixture with high aggressiveness. In this case, an efficient negative selection should only be done. This should also be measured otherwise, so much aggressiveness can be lost that the necessary high selection pressure cannot be provided.

In an advanced stage of breeding, it is necessary to determine the amount of resistance in the given variety. As responses to DI, FDK and DON may differ, we should test them with more isolates (inocula) with different levels of aggressiveness. This provides a better answer in phenotyping for genetic studies, cultivar registration tests or any other tasks that need high preciosity.

It is not thought that one should be anxious about high aggressiveness to diminish resistance differences. The results prove the contrary; the resistance differences are much higher at high aggressiveness. The papers using high on very high aggressiveness prove this clearly [17,19,22,27,42,58]. Therefore, to identify DON overproduction or DON resistance at least medium or high aggressiveness is needed [69]. A test [15] with four isolates against FHB on 26 genotypes with variable aggressiveness showed that highly resistant lines and cultivars like Sumai 3 and the Szeged line Sgv/NB/MM/Sum3 yielded 0.9 mg/kg DON, the most susceptible line gave between 110 and 433 mg/kg. It seems that in spite of the high aggressiveness the differentiation of the genotypes remained good, even much better than at use of lower aggressiveness. When flowering differences are 2–3 weeks between cultivars, because of the changing environmental conditions their correct comparison is often problematic. By using control genotypes with known resistance levels, the management of the resistance tests can be done. However, in a genetic analysis, with the same flowering time differences we have problems, because we should treat the whole population as a unit and therefore data are often not comparable and the QTLs are often artifacts. There are also good examples. In the framework of the USWBSI (United States Wheat and Barley Scab Initiative (www:scabusa.org), multilocation ring tests are made every year to test wheat and Barley genotypes where beside DI, also FDK and DON response, is evaluated [82]. We follow this working way since more than 30 years. This is the only way that can secure the lower DON contamination of the next variety generations. We hope only that scientific research will recognize the need.

5. Conclusions

The Fusarium resistance testing is a very sensitive experimental system. The pathologically uncontrolled mixing and conidium concentration adjusting cannot secure the stable high aggressiveness needed to receive high differentiation of the genotypes [16–63]. Mixing and conidium concentration adjusting, which are applied without controlling by plant pathological methods, cannot secure the stable high aggressiveness needed to receive a correct data on differentiation of the genotypes. It also became clear that FDK or DON specific behavior could not be verified exactly by single inocula. In the mixing variants with the same level of DI, very diverse results for FDK and DON can be recorded. It is also sure that one isolate, or mixture, should it be selected as good as possible, provides only one pathogenic level. Without controlling for aggressiveness of the dilution, concentration or mixing can lead to artifacts. Again, without using more isolates independently, will not have the chance to present the phenotypic data that are necessary to receive reliable data for the identification and validation in medium and low effects of QTL. There are strong arguments to change the resistance testing methodology followed until now.

Author Contributions: Conceptualization, A.M.; methodology, A.M., A.G.; M.V., and B.T.; investigation, A.G., B.T., M.V.; data curation, A.G. and A.M.; writing original draft preparation, A.M. and B.T.; writing—review and editing, A.M., supervision, A.M.; project administration, B.T. and A.M. All authors have read and agreed to the published version of the manuscript.

Funding: The authors are thankful for the financial support of the MycoRed FP7 (KBBE-2007-2-5-05) and GOP-1.1.1-11-2012-0159 projects and GINOP 2.2.1-15-2016-00021 (supported by Hungarian and EU sources) and the national Hungarian TUDFO/51757/2019/ITM project, 2019–2023.

Acknowledgments: The authors acknowledge the kind help of Matyas Cserhati in correcting the English of the manuscript.

Conflicts of Interest: The authors declare no conflict of interest.

Abbreviations

ANOVA	analysis of variance
AUDPC	area under the disease progress curve
DF	degree of freedom
DI	Disease index
DON	deoxynivalenol
FHB	Fusarium damaged kernel
GK	before variety names: Hungarian abbreviation for Cereal Research institution
LSD	limit of significant difference
QTL	quantitative trait locus
SPE	solid phase extraction column

References

1. Mesterhazy, A. The effect of inoculation method on the expression of symptoms caused by *Fusarium graminearum* Schwabe on wheat in seedling stage. In *Symposium on Current Topics in Plant Pathology*; Akadémiai Kiadó, Ed.; Acad. Publ. House of the Hung. Acad. Sci.: Budapest, Hungary, 1975; issued 1977; pp. 223–232.
2. Mesterhazy, A. The role of aggressiveness of *Fusarium graminearum* isolates in the inoculation tests on wheat in seedling state. *Acta Phytopathol. Acad. Sci. Hung.* **1981**, *16*, 281–292.
3. Mesterhazy, A. Effect of seed production area on the seedling resistance of wheat to *Fusarium* seedling blight. *Agronomie* **1985**, *5*, 491–497. [CrossRef]
4. Mesterhazy, A. Types and components of resistance against *Fusarium* head blight of wheat. *Plant Breed.* **1995**, *114*, 377–386. [CrossRef]
5. Bai, G.H.; Shaner, G. Variation in *Fusarium graminearum* and cultivar resistance to wheat scab. *Plant Dis.* **1996**, *80*, 975–979. [CrossRef]
6. Dill-Macky, R. Inoculation methods and evaluation of *Fusarium* head blight resistance in wheat. In *Fusarium Head Blight in Wheat and Barley*; Leonard, K.J., Bushnell, W.R., Eds.; APS Press: St. Paul, MN, USA, 2003; Chapter eight; pp. 184–210, ISBN 0-89054-302-x.
7. van Eeuwijk, F.A.; Mesterhazy, A.; Kling, C.I.; Ruckenbauer, P.; Saur, L.; Buerstmayr, H.; Lemmens, M.; Maurin, M.; Snijders, C.H.A. Assessing non-specificity of resistance in wheat to head blight caused by inoculation with European strains of *Fusarium culmorum, F. graminearum* and *F. nivale*, using a multiplicative model for interaction. *Theor. Appl. Genet.* **1995**, *90*, 221–228. [CrossRef]
8. Mesterhazy, A. Role of deoxynivalenol in aggressiveness of *Fusarium graminearum* and *F. culmorum* and in resistance to *Fusarium* head blight. *Eur. J. Plant Pathol.* **2002**, *108*, 675–684. [CrossRef]
9. Mesterhazy, A.; Buerstmayr, H.; Tóth, B.; Lehoczki-Krsjak, S.; Szabó-Hevér, Á.; Lemmens, M. An improved strategy for breeding FHB resistant wheat must include Type I resistance. In Proceedings of the 5th Canadian Workshop on Fusarium Head Blight, Winnipeg, WB, Canada, 27–30 November 2007; Clear, R., Ed.; pp. 51–66.
10. Mesterhazy, Á.; Tóth, B.; Varga, M.; Bartók, T.; Szabó-Hevér, Á.; Farády, L.; Lehoczki-Krsjak, S. Role of fungicides, of nozzle types, and the resistance level of wheat varieties in the control of *Fusarium* head blight and deoxynivalenol. *Toxins* **2011**, *3*, 1453–1483. [CrossRef] [PubMed]
11. Miedaner, T.; Gang, G.; Geiger, H.H. Quantitative-genetic basis of aggressiveness of 42 isolates of *F. culmorum* for winter rye head blight. *Plant Dis.* **1996**, *80*, 500–504. [CrossRef]

12. Talas, F.; Würschum, T.; Reif, J.C.; Parzies, H.K.; Miedaner, T. Association of single nucleotide polymorphic sites in candidate genes with aggressiveness and deoxynivalenol production in *Fusarium graminearum* causing wheat head blight. *BMC Genet.* **2012**, *13*, 1–10. Available online: http://www.biomedcentral.com/1471-2156/13/14 (accessed on 28 August 2018). [CrossRef]
13. Gale, L.R. Population Biology of *Fusarium* species causing head blight of grain crops. In *Fusarium Head Blight of Wheat and Barley*; Leonard, K.J., Bushnell, W.R., Eds.; APS Press: Minnesota, MN, USA, 2003; pp. 120–132, ISBN 0-89054-302-X.
14. Mesterhazy, A.; Bartók, T.; Mirocha, C.M.; Komoróczy, R. Nature of resistance of wheat to *Fusarium* head blight and deoxynivalenol contamination and their consequences for breeding. *Plant Breed.* **1999**, *118*, 97–110. [CrossRef]
15. Mesterhazy, A.; Bartók, T.; Kászonyi, G.; Varga, M.; Tóth, B.; Varga, J. Common resistance to different *Fusarium* spp. causing Fusarium head blight in wheat. *Eur. J. Plant Path.* **2005**, *112*, 267–281. [CrossRef]
16. Andersen, K.F.; Madden, V.; Paul, P.A. Fusarium head blight development and deoxynivalenol accumulation in wheat as influenced by post-anthesis moisture patterns. *Phytopathology* **2015**, *105*, 210–219. [CrossRef] [PubMed]
17. Garvin, D.F.; Stack, R.W.; Hansen, J.M. Quantitative trait locus mapping of increased Fusarium head blight susceptibility associated with a wild emmer wheat chromosome. *Phytopathology* **2009**, *99*, 447–452. [CrossRef] [PubMed]
18. Buśko, M.; Góral, T.; Ostrowska, A.; Matysiak, A.; Walentyn-Góral, D.; Perkowski, J. The effect of *Fusarium* inoculation and fungicide application on concentrations of flavonoids (apigenin, kaempferol, luteolin, naringenin, quercetin, rutin, vitexin) in winter wheat cultivars. *Am. J. Plant Sci.* **2014**, *5*, 3727–3736. [CrossRef]
19. Pirseyedi, S.M.; Kumar, A.; Ghavami, F.; Hegstad, J.B.; Mergoum, M.; Mazaheri, M.; Kianian, S.F.; Elias, E.M. Mapping QTL for *Fusarium* Head Blight resistance in a Tunisian-derived durum wheat population. *Cereal Res. Commun.* **2019**, *47*, 78–87. [CrossRef]
20. Amarasinghe, C.C. Fusarium Head Blight of Wheat: Evaluation of the Efficacies of Fungicides towards *Fusarium Graminearum* 3-ADON and 15-ADON Isolates in Spring Wheat and Assess the Genetic Differences between 3-ADON Isolates from Canada and China. Master's Thesis, University of Manitoba, Winnipeg, MB, Canada, 2010.
21. Bai, G.H.; Kolb, F.L.; Shaner, G.; Domier, L.L. Amplified fragment length polymorphism markers linked to a major quantitative trait locus controlling scab resistance in wheat. *Phytopathology* **1999**, *89*, 343–348. [CrossRef]
22. Basnet Bhoja, R.; Glover, K.D.; Ibrahim, A.M.H.; Yen, Y.; Chao, S. A QTL on chromosome 2DS of 'Sumai 3' increases susceptibility to Fusarium head blight in wheat. *Euphytica* **2012**, *186*, 91–101. [CrossRef]
23. Buerstmayr, H.; Lemmens, M.; Hartl, L.; Doldi, L.; Steiner, B.; Stierschneider, M.; Ruckenbauer, P. Molecular mapping of QTLs for Fusarium head blight resistance in spring wheat. I. Resistance to fungal spread (Type II resistance). *Theor. Appl. Genet.* **2002**, *104*, 84–91. [CrossRef]
24. Buerstmayr, M.; Lemmens, M.; Steiner, B.; Buerstmayr, H. Advanced backcross QTL mapping of resistance to Fusarium head blight and plant morphological traits in a *Triticum macha* 3 *T. aestivum* population. *Theor. Appl. Genet.* **2011**, *123*, 293–306. [CrossRef]
25. Dong, H.; Wang, R.; Yuan, Y.; Anderson, J.; Pumphrey, M.; Zhang, Z.; Chen, J. Evaluation of the Potential for Genomic Selection to Improve Spring Wheat Resistance to Fusarium Head Blight in the Pacific Northwest. *Front. Plant Sci.* **2018**, *9*, 911. [CrossRef]
26. Chen, X.F.; Faris, J.D.; Hu, J.; Stack, R.W.; Adhikari, T.; Elias, E.M.; Kianian, S.F.; Cai, X. Saturation and comparative mapping of a major Fusarium head blight resistance QTL in tetraploid wheat. *Mol. Breed.* **2007**, *19*, 113–124. [CrossRef]
27. Chu, C.; Niu, Z.X.; Zhong, S.; Chao, S.; Friesen, T.L.; Halley, S.; Elias, M.E.; Dong, Y.; Faris, J.D.; Xu, S.S. Identification and molecular mapping of two QTLs with major effects for resistance to Fusarium head blight in wheat. *Theor. Appl. Genet.* **2011**, *123*, 1107–1119. [CrossRef] [PubMed]
28. Cowger, C.; Arrellano, C. Plump kernels with high deoxynivalenol linked to late *Gibberella zeae* infection and marginal disease conditions in winter wheat. *Phytopathology* **2010**, *100*, 719–728. [CrossRef]
29. Li, T.; Bai, G.H.; Wu, S.; Gu, S. Quantitative trait loci for resistance to fusarium head blight in a Chinese wheat landrace Haiyanzhong. *Theor. Appl. Genet.* **2011**, *122*, 1497–1502. [CrossRef] [PubMed]

30. Cuthbert, P.A.; Somers, D.J.; Julian, T.; Cloutier, S.; Brule'-Babel, A. Fine mapping *Fhb1*, a major gene controlling fusarium head blight resistance in bread wheat (*Triticum aestivum* L.). *Theor. Appl. Genet.* **2006**, *112*, 1465–1472. [CrossRef] [PubMed]
31. D'Angelo, D.L.; Bradley, C.A.; Ames, K.A.; Willyerd, K.T.; Madden, L.V.; Paul, P.A. Efficacy of fungicide applications during and after anthesis against Fusarium head blight and deoxynivalenol in soft red winter wheat. *Plant Dis.* **2014**, *98*, 1387–1397. [CrossRef]
32. Kollers, S.; Rodemann, B.; Ling, J.; Korzun, V.; Ebmeyer, E.; Argillier, O.; Hinze, M.; Plieske, J.; Kulosa, D.; Ganal, M.W.; et al. Whole Genome Association Mapping of Fusarium Head Blight Resistance in European Winter Wheat (*Triticum aestivum* L.). *PLoS ONE* **2013**, *8*, e57500. [CrossRef]
33. Yang, Z.; Gilbert, J.; Fedak, G.; Somers, D.J. Genetic characterization of QTL associated with resistance to Fusarium head blight in a doubledhaploid spring wheat population. *Genome* **2005**, *48*, 187–196. [CrossRef]
34. Gärtner, B.H.; Munich, M.; Kleijer, G.; Mascher, F. Characterisation of kernel resistance against *Fusarium* infection in spring wheat by baking quality and mycotoxin assessments. *Eur. J. Plant Pathol.* **2008**, *120*, 61–68. [CrossRef]
35. Evans, C.K.; Xie, W.; Dill-Macky, R.; Mirocha, C.J. Biosynthesis of Deoxynivalenol in Spikelets of Barley Inoculated with Macroconidia of *Fusarium graminearum*. *Plant Dis.* **2000**, *84*, 654–660. [CrossRef]
36. Perlikowski, D.; Wisniewska, H.; Kaczmarek, J.; Góral, T.; Ochodzki, P.; Kwiatek, M.; Majka, M.; Augustyniak, A.; Kosmala, A. Alterations in Kernel Proteome after Infection with *Fusarium culmorum* in two triticale cultivars with contrasting resistance to *Fusarium* head blight. *Front. Plant Sci.* **2016**, *7*, 1217. [CrossRef]
37. Yi, X.; Cheng, J.; Jiang, Z.; Hu, W.; Bie, T.; Gao, D.; Li, D.; Wu, R.; Li, Y.; Chen, S.; et al. Genetic analysis of Fusarium head blight resistance in CIMMYT bread wheat line c615 using traditional and conditional QTL mapping. *Front. Plant Sci.* **2018**, *9*, 573. [CrossRef]
38. Gervais, L.; Dedryver, F.; Morlais, J.Y.; Bodusseau, V.; Negre, S.; Bilous, M.; Groos, C.; Trottet, M. Mapping of quantitative trait loci for field resistance to Fusarium head blight in an European winter wheat. *Theor. Appl. Genet.* **2003**, *106*, 961–970. [CrossRef] [PubMed]
39. Góral, T.; Busko, M.; Ochy, H.; Jackowiak, H.; Perkowski, J. Resistance of triticale lines and cultivars to Fusarium head blight and deoxynivalenol accumulation in kernels. *J. Appl. Genet.* **2002**, *43A*, 237–248.
40. Góral, T.; Stuper-Szablewska, K.; Buśko, M.; Boczkowska, M.; Walentyn-Góral, D.; Wiśniewska, H.; Perkowski, J. Relationships Between Genetic Diversity and *Fusarium* Toxin Profiles of Winter Wheat Cultivars. *Plant Pathol. J.* **2015**, *31*, 226–244. [CrossRef]
41. Hao, C.; Wang, Y.; Hou, J.; Feuillet, C.; Balfourier, F.; Zhang, X. Association Mapping and Haplotype Analysis of a 3.1-Mb Genomic Region Involved in Fusarium Head Blight Resistance on Wheat Chromosome 3BS. *PLoS ONE* **2012**, *7*, e46444. [CrossRef] [PubMed]
42. He, X.; Singh, P.K.; Duveiller, E.; Schlang, N.; Dreisigacker, S.; Singh, R.P. Identification and characterization of international Fusarium head blight screening nurseries of wheat at CIMMYT, Mexico. *Eur. J. Plant Pathol.* **2013**, *136*, 123–134. [CrossRef]
43. He, X.; Singh, P.K.; Duveiller, E.; Schlang, N.; Dreisigacker, S.; Singh, R.P.; Payne, T.; He, Z. Characterization of Chinese wheat germplasm for resistance to Fusarium head blight at CIMMYT, Mexico. *Euphytica* **2014**, *195*, 383–395. [CrossRef]
44. Hilton, A.J.; Jenkinson, P.; Hollins, T.W.; Parry, D.W. Relationship between cultivar height and severity of Fusarium ear blight in wheat. *Plant Pathol.* **1999**, *48*, 202–208. [CrossRef]
45. Chen, P.; Liu, W.; Yuan, J.; Wang, X.; Zhou, B.; Wang, S.; Zhang, S.; Feng, Y.; Yang, B.; Liu, G.; et al. Development and characterization of wheat—*Leymus racemosus* translocation lines with resistance to Fusarium Head Blight. *Theor. Appl. Genet.* **2005**, *111*, 941–948. [CrossRef]
46. Klahr, A.; Zimmermann, G.; Wenzel, G.; Mohler, V. Effects of environment, disease progress, plant height and heading date on the detection of QTLs for resistance to *Fusarium* head blight in a European winter wheat cross. *Euphytica* **2007**, *154*, 17–28. [CrossRef]
47. Lin, F.; Xue, S.L.; Zhang, Z.Z.; Zhang, C.Q.; Kong, Z.X.; Yao, G.Q.; Tian, D.G.; Zhu, H.L.; Li, C.Y.; Wei, J.B.; et al. Mapping QTL associated with resistance to *Fusarium* head blight in the Nanda2419 Wangshuibai population. II: Type I resistance. *Theor. Appl. Genet.* **2006**, *112*, 528–553. [CrossRef]

48. Liu, W.; Langseth, W.; Skinnes, H.; Elen, O.N.; Sundheim, L. Comparison of visual head blight ratings, seed infection levels, and deoxynivalenol production for assessment of resistance in cereals inoculated with *Fusarium culmorum*. *Eur. J. Plant Path.* **1997**, *103*, 589–595. [CrossRef]
49. Forte, P.; Virili, M.E.; Kuzmanovic, L.; Moscetti, I.; Gennaro, A.; D'Ovidio, R.; Ceoloni, C. A novel assembly of *Thinopyrum ponticum* genes into the durum wheat genome: Pyramiding *Fusarium* head blight resistance onto recombinant lines previously engineered for other beneficial traits from the same alien species. *Mol. Breed.* **2014**, *34*, 1701–1716. [CrossRef]
50. Malihipour, A.; Gilbert, J.; Fedak, G.; Brûlé-Babel, A.L.; Cao, W. Characterization of agronomic traits in a population of wheat derived from *Triticum timopheevii* and their association with *Fusarium* head blight. *Eur. J. Plant Pathol.* **2016**, *144*, 31–43. [CrossRef]
51. McCartney, C.A.; Brûlé-Babel, A.L.; Fedak, G.; Martin, R.A.; McCallum, B.D.; Gilbert, J.; Hiebert, C.W.; Pozniak, C.J. *Fusarium* Head Blight Resistance QTL in the Spring Wheat Cross Kenyon/86ISMN 2137. *Front. Microbiol.* **2016**, *7*, 1542. [CrossRef] [PubMed]
52. Muhovski, Y.; Batoko, H.; Jacquemi, J.M. Identification, characterization and mapping of differentially expressed genes in a winter wheat cultivar (Centenaire) resistant to *Fusarium graminearum* infection. *Mol. Biol. Rep.* **2012**, *39*, 9583–9600. [CrossRef]
53. Osman, M.; He, X.; Singh, R.P.; Duveiller, E.; Lillemo, M.; Pereyra, S.A.; Westerdijk-Hoks, I.; Kurushima, M.; Yau, S.K.; Benedettelli, S.; et al. Phenotypic and genotypic characterization of CIMMYT's 15th international *Fusarium* head blight screening nursery of wheat. *Euphytica* **2015**, *205*, 521–537. [CrossRef]
54. Jones, S.; Farooqi, A.; Foulkes, J.; Sparkes, D.L.; Linforth, R.; Ray, R.V. Canopy and ear traits associated with avoidance of *Fusarium* head blight in wheat. *Front. Plant Sci.* **2018**, *9*, 1021. [CrossRef]
55. Liu, Y.; Salsman, E.; Fiedler, J.D.; Hegstad, J.B.; Green, A.; Mergoum, M.; Zhong, S.; Li, X. Genetic mapping and prediction analysis of FHB resistance in a hard red spring wheat breeding population. *Front. Plant Sci.* **2019**, *10*, 1007. [CrossRef]
56. Miedaner, T.; Gwiazdowska, D.; Waskiewicz, A. *Management of Fusarium Species and Their Mycotoxins in Cereal Food and Feed*; Frontiers Media: Lausanne, Switzerland, 2017; pp. 134–144. [CrossRef]
57. Oliver, R.E.; Xu, S.S.; Stack, R.W.; Friesen, T.L.; Jin, Y.; Cai, X. Molecular cytogenetic characterization of four partial wheat-*Thinopyrum ponticum* amphiploids and their reactions to *Fusarium* head blight, tan spot, and *Stagonospora nodorum* blotch. *Theor. Appl. Genet.* **2006**, *112*, 1473–1479. [CrossRef] [PubMed]
58. Tamburic-Ilincic, L.; Barcellos, R.S. Alleles on the two dwarfing loci on 4B and 4D are main drivers of FHB-related traits in the Canadian winter wheat population 'Vienna' x '25R47'. *Plant Breed.* **2017**, *136*, 799–808. [CrossRef]
59. Otto, C.D.; Kianian, S.F.; Elias, E.M.; Stack, R.W.; Joppa, L.R. Genetic dissection of a major *Fusarium* head blight QTL in tetraploid wheat. *Plant Mol. Biol.* **2002**, *48*, 625–632. [CrossRef] [PubMed]
60. Ding, L.; Xu, H.; Yi, H.; Yang, L.; Kong, Z.; Zhang, L.; Xue, S.; Jia, H.; Ma, Z.-Q. Resistance to Hemi-Biotrophic Infection Is Associated with Coordinated and Ordered Expression of Diverse Defense Signaling Pathways. *PLoS ONE* **2011**, *6*, e19008. [CrossRef] [PubMed]
61. Klahr, A.; Mohler, V.; Herz, M.; Wenzel, G.; Schwarz, G. Enhanced power of QTL detection for *Fusarium* head blight resistance in wheat by means of codominant scoring of hemizygous molecular markers. *Mol. Breed.* **2004**, *13*, 289–300. [CrossRef]
62. Zwart, R.S.; Muylle, H.; van Bockstaele, E.; Roldán-Ruiz, I. Evaluation of genetic diversity of *Fusarium* head blight resistance in European winter wheat. *Theor. Appl. Genet.* **2008**, *117*, 813–828. [CrossRef]
63. Miedaner, T.; Wilde, F.; Steiner, B.; Buerstmayr, H.; Korzun, V.; Ebmeyer, E. Stacking quantitative trait loci (QTL) for Fusarium head blight resistance from non-adapted sources in an European elite spring wheat background and assessing their effects on deoxynivalenol (DON) content and disease severity. *Theor. Appl. Genet.* **2006**, *112*, 562–569. [CrossRef] [PubMed]
64. Snijders, C.H.A.; van Eeuwijk, E.A. Genotype x strain interactions for resistance to *Fusarium* head blight caused by *Fusarium culmorum* in winter wheat. *Theor. Appl. Genet.* **1991**, *81*, 239–244. [CrossRef] [PubMed]
65. Snijders, C.H.A. Diallel analysis of resistance to head blight caused by *Fusarium culmorum* in winter wheat. *Euphytica* **1990**, *50*, 1–9. [CrossRef]
66. Mesterházy, Á. Selection of head blight resistant wheat through improved seedling resistance. *Plant Breed.* **1987**, *98*, 25–36. [CrossRef]

67. Mesterhazy, Á. A laboratory method to predict pathogenicity of *Fusarium graminearum* in field and resistance to scab. *Acta Phytopathol. Acad. Sci. Hung.* **1984**, *19*, 205–218.
68. Mesterházy, Á. Expression of resistance to *Fusarium graminearum* and *F. culmorum* under various experimental conditions. *J. Phytopathol.* **1988**, *133*, 304–310. [CrossRef]
69. Mesterhazy, A.; Lehoczki-Krsjak, S.; Varga, M.; Szabó-Hevér, Á.; Tóth, B.; Lemmens, M. Breeding for fhb resistance via *Fusarium* damaged kernels and deoxynivalenol accumulation as well as inoculation methods in winter wheat. *Agric. Sci.* **2015**, *6*, 970–1002. [CrossRef]
70. Mesterhazy, Á.; Varga, M.; György, A.; Lehoczki-Krsjak, S.; Tóth, B. The role of adapted and non-adapted resistance sources in breeding resistance of winter wheat to *Fusarium* head blight and deoxynivalenol contamination. *World Mycotoxin J.* **2018**, *11*, 539–557. [CrossRef]
71. György, A.; Tóth, B.; Varga, M.; Mesterhazy, A. Methodical considerations and resistance evaluation against *Fusarium graminearum* and *F. culmorum* head blight in wheat. Part 3. Susceptibility window and resistance expression. *Microorganisms* **2020**, *8*, 627. [CrossRef] [PubMed]
72. Mesterhazy, Á. *Fusarium* species of wheat in South Hungary, 1970–1983. *Cereal Res. Commun.* **1984**, *12*, 167–170.
73. Sváb, J. *Biometriai Módszerek a Kutatásban (Methods for Biometrics in Research)*, 3rd ed.; Mezőgazdasági Kiadó (Agr. Publ. House): Budapest, Hungary, 1981; pp. 1–557.
74. Weber, E. *Grundriss der Biologischen Statistik (Fundaments of the Biological Statistics)*; VEB Fisher Verlag: Jena, Germany, 1967.
75. Mesterhazy, Á. Breeding for resistance against FHB in wheat. In *Mycotoxin Reduction in Grain Chains: A Practical Guide*; Logrieco, A.F., Visconti, A., Eds.; Blackwell-Wiley: Ames, IA, USA; Chichester, UK; Oxford, UK, 2014; pp. 189–208, ISBN 978-0-8138-2083-5.
76. Szabo-Hever, Á.; Lehoczki-Krsjak, S.; Varga, M.; Purnhauser, L.; Pauk, J.; Lantos, C.; Mesterhazy, Á. Differential influence of QTL linked to *Fusarium* head blight, *Fusarium*-damaged kernel, deoxynivalenol content and associated morphological traits in a Frontana-derived wheat population. *Euphytica* **2014**, *200*, 9–26. [CrossRef]
77. Liu, S.; Griffey, C.A.; Hall, M.D.; McKendry, A.L.; Chen, J.; Brooks, W.S.; Brown-Guedira, G.; van Sanford, D.; Schmale, D.G. Molecular characterization of field resistance to *Fusarium* head blight in two US soft red winter wheat cultivars. *Theor. Appl. Genet.* **2013**, *126*, 2485–2498. [CrossRef]
78. Petersen, S. Advancing Marker-Assisted Selection for Resistance to Powdery Mildew and Fusarium Head Blight in Wheat. Ph.D. Thesis, North Carolina State University, North Carolina, NC, USA, 2015; p. 378.
79. Tessmann, E.W.; Dong, Y.-H.; Van Sanford, D.A. GWAS for Fusarium Head Blight Traits in a Soft Red Winter wheat Mapping Panel. *Crop Sci.* **2019**, *59*, 1823–1837. [CrossRef]
80. He, X.; Dreisigacker, S.; Ravi, P.; Singh, R.P.; Singh, P.K. Genetics for low correlation between *Fusarium* head blight disease and deoxynivalenol (DON) content in a bread wheat mapping population. *Theor. Appl. Genet.* **2019**, *132*, 2401–2411. [CrossRef]
81. Wu, L.; Zhang, Y.; He, Y.; Jiang, P.; Zhang, X.; Ma, H.-X. Genome-Wide Association Mapping of Resistance to Fusarium Head Blight Spread and Deoxynivalenol Accumulation in Chinese Elite Wheat Germplasm. 2019. Available online: https://doi.org/10.1094/PHYTO-12-18-0484-R (accessed on 11 June 2019).
82. Murphy, J.P.; Lyerly, J.H.; Acharya, R.; Page, J.; Ward, B.; Brown-Guedira, G. Southern uniform Winter Wheat Scab Nursery. 2019 Nursery Report. 2019, p. 31. Available online: www.scabusa.org (accessed on 1 June 2020).

microorganisms

MDPI

Article

Diminished Pathogen and Enhanced Endophyte Colonization upon CoInoculation of Endophytic and Pathogenic *Fusarium* Strains

Maria E. Constantin, Babette V. Vlieger, Frank L. W. Takken and Martijn Rep *

Molecular Plant Pathology, Faculty of Science, Swammerdam Institute for Life Sciences, University of Amsterdam, 1098 XH Amsterdam, The Netherlands; m.e.constantin@uva.nl (M.E.C.); babette.vlieger@student.uva.nl (B.V.V.); f.l.w.takken@uva.nl (F.L.W.T.)

* Correspondence: m.rep@uva.nl; Tel.: +31-6248-43359

Received: 12 March 2020; Accepted: 7 April 2020; Published: 9 April 2020

Abstract: Root colonization by *Fusarium oxysporum* (Fo) endophytes reduces wilt disease symptoms caused by pathogenic Fo strains. The endophytic strain Fo47, isolated from wilt suppressive soils, reduces *Fusarium* wilt in various crop species such as tomato, flax, and asparagus. How endophyte-mediated resistance (EMR) against *Fusarium* wilt is achieved is unclear. Here, nonpathogenic colonization by Fo47 and pathogenic colonization by Fo f.sp. *lycopersici* (Fol) strains were assessed in tomato roots and stems when inoculated separately or coinoculated. It is shown that Fo47 reduces Fol colonization in stems of both noncultivated and cultivated tomato species. Conversely, Fo47 colonization of coinoculated tomato stems was increased compared to single inoculated plants. Quantitative PCR of fungal colonization of roots (co)inoculated with Fo47 and/or Fol showed that pathogen colonization was drastically reduced when coinoculated with Fo47, compared with single inoculated roots. Endophytic colonization of tomato roots remained unchanged upon coinoculation with Fol. In conclusion, EMR against *Fusarium* wilt is correlated with a reduction of root and stem colonization by the pathogen. In addition, the endophyte may take advantage of the pathogen-induced suppression of plant defences as it colonizes tomato stems more extensively.

Keywords: colonization; *Fusarium*; endophyte; Fo47; wilt disease

---

1. Introduction

Among the most common inhabitants of soils are fungal isolates belonging to the *Fusarium oxysporum* (Fo) species complex [1–3]. Their presence is not limited to the soil—*Fusarium* hyphae can also colonize plant roots superficially and internally. The typically asymptomatic root colonization by Fo shows that Fo is mostly an endophyte [2,4]. Endophytic colonization is often restricted to the root cortex and endodermis, and fungal hyphae do not commonly reach xylem vessels. In resistant plants, limited spread of pathogenic isolates in the xylem correlates with the production of gums and tyloses, apparently halting the fungus at the early stages of infection [5]. In susceptible plants, this response appears to be induced too late, and the multiple occlusions that eventually block the xylem vessels affect the water transport of the plant, leading to wilting and death [6].

Pathogenic Fo isolates are currently grouped into 106 *formae speciales* (ff. ssp.) [7]. One of these is Fo f.sp. *lycoperisici* (Fol), which causes wilt disease in tomato plants. Disease symptoms caused by Fol can be strongly reduced upon pre or coinoculation with endophytic Fo strains [8,9]. One of the first *Fusarium* endophytes shown to reduce *Fusarium* wilt disease is Fo47, which was isolated from wilt-suppressive soils [10]. Since its discovery, Fo47 has been shown to reduce *Fusarium* wilt in a variety of plant species including tomato, asparagus, flax, chickpea, and cotton [8,9,11–14]. Endophyte-mediated resistance (EMR) against *Fusarium* wilts is not unique to Fo47, as other Fo isolates also have this capacity [15].

Moreover, even isolates pathogenic on another host can confer resistance, such as Fo f.sp. *dianthi* against Fol in tomato plants [16]. Additionally, other *Fusarium* species seem to be able to reduce disease symptoms induced by Fo [15,17]. For example, the *Fusarium solani*-K isolate, which colonizes tomato roots, reduces susceptibly against Fo f.sp. *radicis-lycopersici* [17]. These observations suggest that roots colonized by endophytic *Fusarium* are less susceptible to *Fusarium* wilt.

Currently, it remains unclear how *Fusarium*-mediated resistance against *Fusarium* wilt is achieved. A recent report showed that tomato lines deficient in salicylic acid accumulation, jasmonic acid biosynthesis, or ethylene production and sensing could still trigger EMR against *Fusarium* wilt disease [9]. To better understand the extent to which EMR may be host genotype-specific, a range of tomato lines and species were inoculated. Moreover, the contribution of the endophyte genotype in triggering EMR by using various *Fusarium* species was determined. Colonization of tomato roots by endophytic and pathogenic strains was assessed, as was the migration into the stem upon single and coinoculation. Our results showed that various *Fusarium* species can behave as endophytes in tomato roots and trigger similar resistance levels across different tomato species. Roots and stems of tomato plants coinoculated with Fo47 and Fol were found to be colonized to a lesser extent by Fol than when inoculated alone, but surprisingly Fo47 colonization of stems was enhanced in the presence of the pathogen.

2. Materials and Methods

2.1. Plant Lines and Fungal Strains

Bioassays were performed either on the tomato (*Solanum lycopersicum*) line C32 or Money Maker, susceptible to *Fusarium* wilt, or on the wild tomato relatives *S. pimpinellifolium* (accession LA1578) and *S. chmielewskii* (accessions LA1840, LA2663 and LA2695) in a climate-controlled greenhouse with a day-night temperature of 25°C, 16 h light/ 8 h dark, and a relative humidity of 65%. Fol029 (race 3, sFP2381, carrying phleomycin resistance [18]) and Fo47 (sFP1544, carrying hygromycin resistance, [19]) were inoculated on the aforementioned tomato lines. Bioassays using various *Fusarium* species were performed on the tomato line C32 using the wild-type pathogen Fol4287 (sFP801) [20], and endophyte Fo47 (sFP730) [10], *F. hostae* (sFP2236), *F. proliferatum* (sFP2240), *F. redolens* (sFP4856) [21], and *F. solani* (sFP895). To facilitate fungal reisolation from tomato stems transgenic fungi caring different resistance markers were used. Therefore, the endophytic strain Fo47 containing hygromycin resistance (sFP1544) and the pathogenic strain Fol4287 (race2, SFP3858) [22] or Fol029 (race 3, sFP2381) were used for the reisolation experiments. Strains used in bioassays are described in the supplementary material are the following: Fo47 (sFP1544, [19]), Fol4287 (sFP3059, [23]), Fo47 (sFP730), Fol4287 (sFP801), and Fol017 (sFP17, [24]).

2.2. Fusarium Inoculation and Disease Scoring

Fusarium strains were grown on potato dextrose agar (PDA) plates for seven to ten days at 25°C in the dark. An agar plug from these plates was transferred to 100 mL of NO3 minimal media (1% KNO3, 0.17% Yeast Nitrogen Base without amino acids and ammonia and 3% sucrose) and incubated at 150 rpm for 3-5 days at 25°C. Spores were filtered over one layer of miracloth filter (Millipore), centrifuged at 2000 rpm, washed with sterile water, and finally resuspended in sterile MiliQ-water. Ten-day-old or 13-days-old tomato seedlings were uprooted, and their roots were trimmed to facilitate *Fusarium* infection. Subsequently, tomato roots were dipped for five minutes in a suspension of 107 spores/mL or 107 spores/mL: 107 spores/mL (ratio 1:1) in the case of the coinoculation treatments if not stated differently. After inoculation, the seedlings were potted, and three weeks afterwards fresh weight and disease index were assessed as described [9]. In short, the disease score was based on the number of brown vessels and external growth wilting symptoms, where 0= healthy plant with no brown vessels; 1= brown vessel(s) only at the basal level; 2= one or two brown vessels at the cotyledon level, the plant still looks healthy; 3= at least three brown vessels and the plant shows clear external witling symptoms;

4= all vessels are brown and the plant shows clear size reduction; and 5= plant is dead. Fresh weight was measured by determining weight of each tomato plant cut at cotyledon level.

2.3. Fungal Recovery Assay

Tomato stems were collected three weeks after inoculation and surface sterilized with 70% ethanol as described [9]. In short, the ethanol was removed by pouring into a collection tube, and stems were rinsed twice with sterile water to remove the ethanol. Subsequently, the extremities of the stem were trimmed and one piece at the cotyledon and one at crown level were cut and placed on PDA plates containing 200 mg/L streptomycin and 100 mg/l penicillin to prevent bacterial growth. When transgenic fungal strains were used, the PDA plates also contained 100 mg/L zeocin or 100 mg/L hygromycin for ensuring selection of the right fungal strain. Plates were incubated for four days at 25°C in the dark, after which Fo outgrowth was assessed as follows: 0= no fungal outgrowth, 1= fungal outgrowth of stem pieces isolated from either crown or cotyledon level, and 2= fungal outgrowth from stems isolated at both crown and cotyledon level.

2.4. Fungal DNA Isolation and Sequencing

To confirm that the mycelium emerging from tomato stems corresponded to the originally inoculated strain, the mycelium was scraped from the PDA plate and used for gDNA isolation and PCR as described by the authors of [25]. In short, mycelium scrapes were placed in a 2 mL tube containing 400 µL TE buffer (1 mM EDTA pH = 8, 10 mM Tris pH= 8), 200-300 µL glass beads, and 300 µL phenol:chloroform (1:1), followed by disruption using a TissueLyser (Qiagen; Venlo, Netherlands) for 2 min at 30 Hz. Afterwards, the samples were centrifuged for 10 min. The upper phase was transferred to a new 2 mL tube and was subjected to another round of phenol:choloroform extraction. PCR amplification of *EF1-alpha* gene (see Table S1 for primers) was performed in 20 µL reaction consisting of of 0.4 µL dNTPs (10 mM), 4 µL SuperTaq buffer (10x), 0.1 µL SuperTaq (5 U/µl), 1 µL of DNA template, and 1 µL primers (5 pmol/µL). The cycling program for PCR amplification was 95°C for 5 min, 35 cycles of 30s at 95°C, 30s at 55°C, and 1 min at 72°C, with a final elongation step of 3 min at 72°C. The PCR amplicon was sequenced and compared with the *EF1-alpha* sequence of the strain inoculated originally.

2.5. Analysis of Fungal Colonization by Quantitative PCR

Tomato roots were harvested three weeks after inoculation, washed, snap-frozen in liquid nitrogen, and freeze-dried overnight. Samples were ground in a mortar cooled with liquid nitrogen using a pestle and approximately 100 mg of the resulting powder was using for gDNA isolation. gDNA isolation and purification were performed using GeneJET plant Genomic purification Kit (Thermo Scientific; Walthamm MA, USA). DNA concentration was estimated by spectroscopy using Nanodrop (Thermo Scientific, Walthamm MA, USA), and the quality of the gDNA was assessed by agarose gel electrophoresis. The 10 µL qPCR mixture contained 10 ng of gDNA, 10 pM of each primer, and 2 µL of HOT FirePolEvaGreen qPCR Mix Plus (Solis BioDyne; Tartu, Estonia) were performed in QuantStudioTM3 (Thermo Scientific; Walthamm MA, USA). The cycling program was set to 15 min at 95°C, 40 cycles of 15s at 95°C, 1 min at 60°C, and 30s at 72°C. The melting curve analysis was performed afterwards as follows: 15s at 95°C, 1 min at 60°C, and 15s at 95°C. The sequences of the primers used are summarized in Table S1 [26,27]. For InterGenic Spacer (IGS) primers two standard curves (four- or ten- times dilution) were performed with a starting concentration of 10 ng, resulting with a primer efficiently of 110 and 95%, respectively. Three technical replicates were used per biological sample, and data was normalized to plant tubulin gene level, using qbase+3.1 (Biogazelle; Ghent, Belgium).

2.6. Statistical Analyses

Data collected from bioassays (fresh weight, disease index) were analyzed using PRISM 7.0 (GraphPad) by performing a Mann–Whitney U- test. The data obtained from qPCR were analyzed with ordinary one-way ANOVA with Tukey's multiple comparisons test in the case of IGS and with an unpaired Student's t-test for *SIX8* and *SCAR*.

3. Results

3.1. Endophyte-Mediated Resistance Occurred at a 1:1 Ratio and Required Live Endophyte Spores

The endophytic strain Fo47 has been reported to trigger EMR when inoculated in plants at a concentration 10-100 times higher than that of the pathogen [28]. To determine the optimal concentration for a reliable EMR assay, tomato plants were coinoculated with Fo47 at the same ratio as the pathogen (1:1), or in 10- or 100-times excess (Figure S1a). Decreasing the pathogen concentration lowered the severity of disease symptoms (Figure S1a) but did not influence the level of EMR triggered by Fo47, since it was already quite strong at a 1:1 endophyte: pathogen ratio (Figure S1b). Based on this, all subsequent bioassays were carried out using 10^7 spores/mL for single inoculations and ratio 1:1 for coinoculation treatment. To determine whether a living endophyte is required to induce resistance, tomato seedlings were coinoculated with heat-killed spores of either Fo47 or Fol4287, together with living Fol4287 (Figure S1b). Since plants coinoculated with heat-killed spores and Fol4287 became diseased, it can be concluded that a living endophyte is required for triggering EMR (Figure S1b).

Fo strains within the same vegetative compatibility group (VCG) can form stable heterokaryons, while those belonging to different VCGs undergo cell death upon heterokaryon formation [29]. To test whether VCG incompatibility can result in a reduced viable concentration of Fol4287, and thereby reduce disease symptoms, two Fol strains of different VCG (Fol4287 (VCG030) and Fol017 (VCG031), [26]) were coinoculated. This resulted in disease symptoms that were indistinguishable from those observed upon single inoculation (Figure S1c). This observation implies that VCG incompatibility is unlikely to be an explanation for the suppression of disease symptoms by Fo47. Finally, to determine whether there is direct antagonism between Fo47 and Fol4287, two agar plugs with seven days old mycelium were placed on a PDA plate at a fixed distance from of each other. No visible inhibition halo or reduction of mycelial growth was observed (Figure S1d). Taken together, living Fo47 spores can efficiently confer resistance against Fol4287 when applied in a 1:1 ratio, without exhibiting obvious direct antagonism.

3.2. Fo47 Also Conferred Resistance against Fol in Wild Tomato Species

To test whether Fo47 can also trigger EMR in wild tomato species, 13-days-old uprooted tomato seedlings were inoculated with either water (mock), a spore solution of either Fo47 or Fol029, or both Fo47 and Fol029 in a 1:1 ratio. Based on the severity of disease symptoms, each plant received a disease index (DI) score ranging from 0 (healthy) to 5 (dead). Fo47 treatment caused no visible disease symptoms in any plant line, and fresh weight was indistinguishable from the mock (Figure 1b,c). Conversely, inoculation with Fol029 caused a visible growth reduction in all plant lines (Figure 1a). The cultivated tomato line *S. lycopersicum* C32 showed the most drastic weight reduction among the three-tomato species tested (Figure 1a,b) and displayed the most consistent disease symptoms (Figure 1c). *S. pimpinellifolium* and *S. chmielewskii* exhibited more varying disease symptoms (DI = 2 to 5) (Figure 1b and Figure S2a,b) upon Fol029 inoculation. Coinoculation of Fo47 with Fol029 resulted in increased fresh weight compared with solely Fol029 inoculated tomato lines (Figure 1a,b), but this difference was only statistically significant for *S. lycopersicum* and *S. chmielewskii* line LA2663. Similarly, tomato lines coinoculated with Fo47 and Fol029 showed reduced disease symptoms compared with plants single inoculated with Fol029 (Figure 1c, Figure S2a,b). In conclusion, Fo47 reduced disease symptoms caused by Fol029 in both cultivated and wild tomato species.

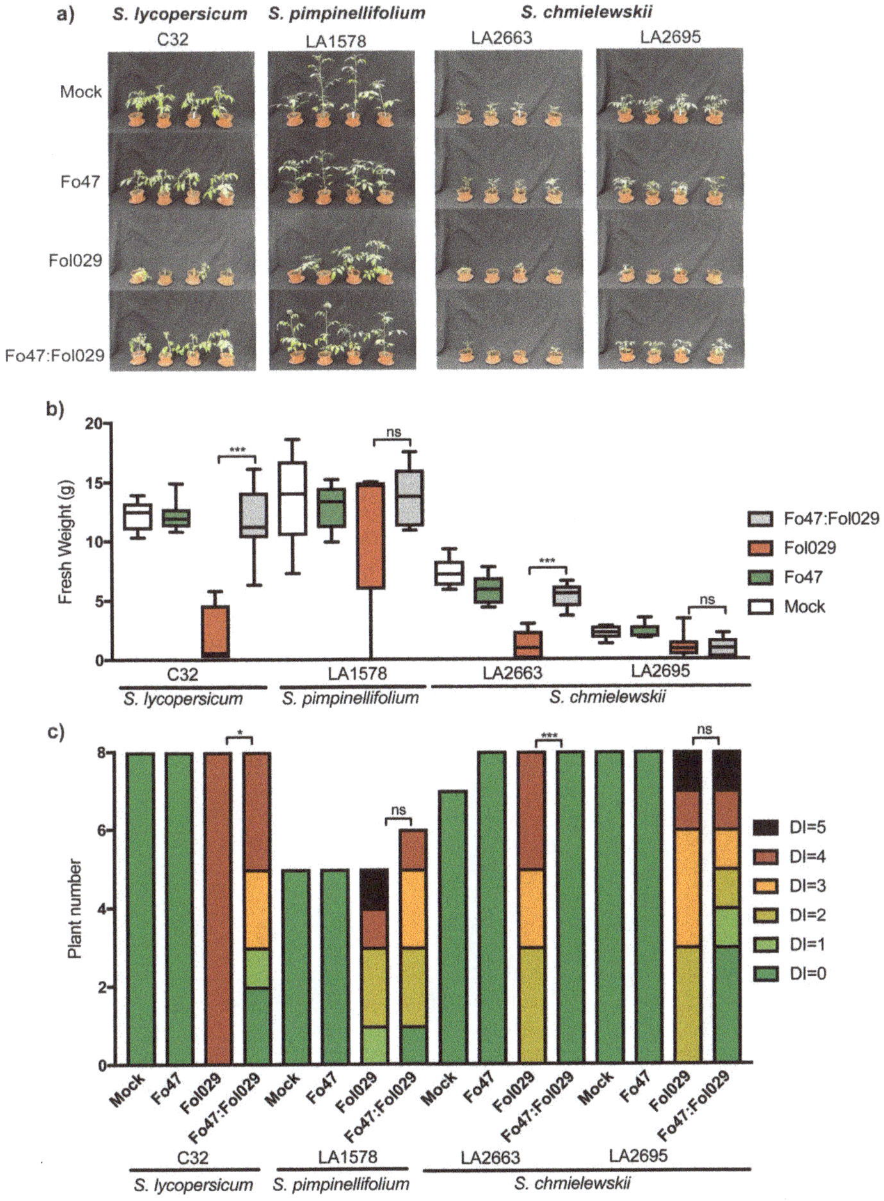

Figure 1. Fo47 can trigger resistance against Fol029 in (**a**) *Solanum lycopersicum* (C-32) and *S. chmielewski* (LA2663, LA2695). Thirteen-day-old tomato seedlings were inoculated with water (mock), Fo47, or Fol029, or coinoculated with Fo47 and Fol029. (**b**) Fresh weight and (**c**) disease index (DI) were assessed three weeks after inoculation where DI = 0 no brown vessels; DI = 1 brown vessel(s) only at basal level; DI = 2 one or two brown vessels at cotyledon level; DI = 3 three brown vessels at cotyledon level; DI = 4 all vessels are brown; DI = 5 the plant is dead. Data were analysed by a nonparametric Mann–Whitney U-test (^{ns} P > 0.05; * $p < 0.05$; *** $p < 0.001$).

3.3. Different Fusarium Species Behaved as Endophytes in Tomato Plants and Triggered Resistance against Fol

To test if the ability of conferring EMR against *Fusarium* wilt is more broadly present in the genus *Fusarium* tomato, bioassays with *Fusarium* spp. other than Fo were performed. Roots of tomato seedlings were inoculated in water or spore suspension of Fo47, *Fusarium redolens* (Fr), *Fusarium solani* (Fs), *Fusarium hostae* (Fh), or *Fusarium proliferatum* (Fp), together with Fol4287 in a 1:1 ratio. Inoculation of Fo47 or other *Fusarium* spp. alone did not result in visible differences compared to the mock inoculation control (Figure 2a) or in differences in fresh weight (Figure 2b). Therefore, the Fr, Fs, Fh, and Fp strains used here are not pathogenic on tomato plants (Figure 2c). Tomato plants inoculated with Fol4287 showed a reduction in fresh weight (Figure 2b) compared to the mock treatment and developed disease symptoms three weeks after inoculation (Figure 2c). In this experiment, these disease symptoms were not as severe as previously observed (Figure S1). Tomato plants coinoculated with Fol4287 and other *Fusarium* isolates (such as Fo47, Fr, Fs, Fh, and Fp) were taller than plants inoculated with Fol4287 alone (Figure 2a). Differences in fresh weight were significant for coinoculation with Fo47, Fr, and Fs (Figure 2b). In line with this, coinoculation treatment of Fol4287 with a nonpathogenic *Fusarium* strain resulted in reduced disease symptoms compared with Fol4287 alone (Figure 2c, Figure S3c). Fh and Fp did not consistently reduce *Fusarium* disease symptoms, suggesting that they may have a weaker effect than the other isolates tested.

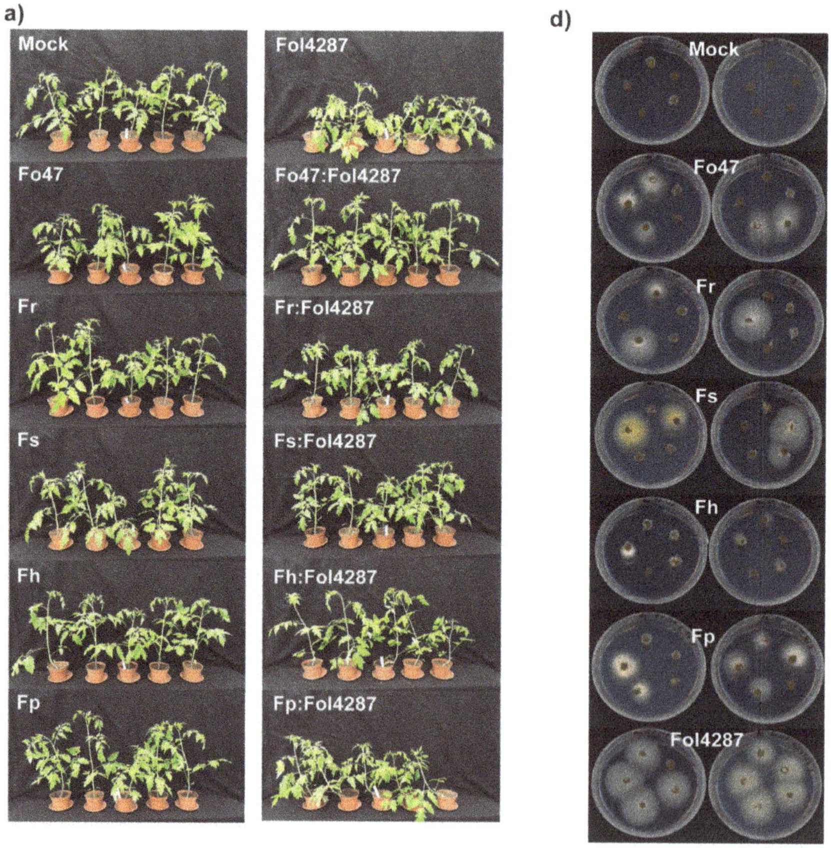

Figure 2. *Cont.*

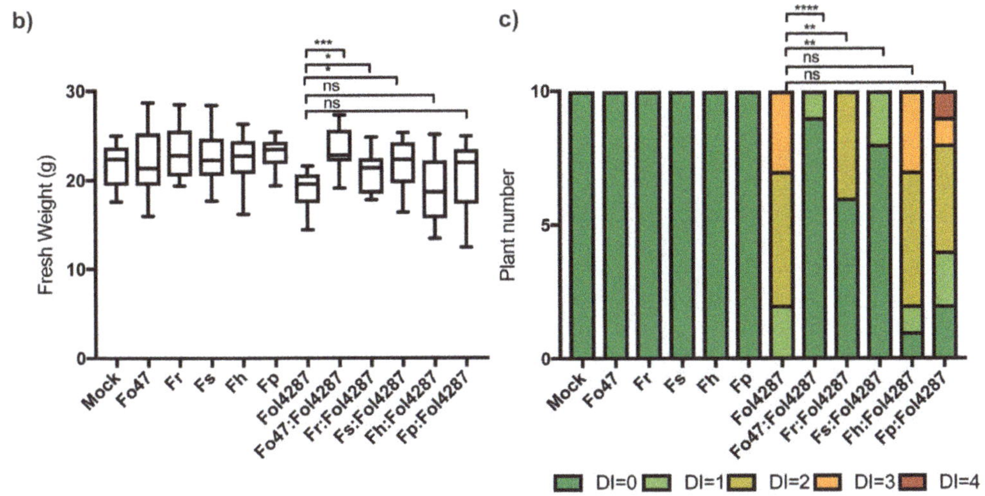

Figure 2. Fo47, *Fusarium redolens* (Fr), *Fusarium solani* (Fs), *Fusarium hostae* (Fh), and *Fusarium proliferatum* (Fp) can reduce *Fusarium* wilt disease symptoms in tomato. (**a**) Representative tomato plants three weeks after inoculation; (**b**) fresh weight and (**c**) disease index (DI) of tomato plants three weeks after inoculation. DI = 0 no brown vessels; DI = 1 brown vessel(s) only at basal level; DI = 2 one or two brown vessels at cotyledon level; DI = 3 three brown vessels at cotyledon level, DI = 4 all vessels are brown, DI = 5 the plant is dead. Data were analysed by a nonparametric Mann–Whitney U-test (^{ns} $P > 0.05$; * $p < 0.05$, ** $p < 0.01$; *** $p < 0.001$); (**d**) Ten tomato stems pieces from crown level showing *Fusarium* outgrowth on PDA plates after being incubated for four days in dark at 25 °C.

To examine whether the different *Fusarium* spp. are endophytes, tomato stems were harvested three weeks post inoculation, surface sterilized, and a piece of the stem at crown level was harvested. These were placed on PDA plates with antibiotics and incubated at 25°C as schematically depicted in Figure S4. After four days, mycelia emerged from stem pieces, which was used for gDNA isolation and *EF1-alpha* PCR followed by sequencing of the PCR product. This analysis revealed that the mycelium originated from the *Fusarium* strain used to inoculate the plants (Figure 2d). Following mock treatment, either no fungal mycelia emerged from the stems (Figure 2d), or mycelia emerged that did not correspond to *Fusarium* spp. (Figure S3d). The most frequently reisolated fungus from tomato stems was Fol4287, followed by Fo47, Fp and Fr, and Fs (Figure 2d, Figure S3d). The least frequently reisolated species from tomato stems was Fs (Figure 2, Figure S3); however, its presence was confirmed in one tomato stem. Taken together, Fr, Fs, Fh, and Fp can behave as endophytes, colonizing tomato stems and triggering EMR against Fo wilt in tomato.

3.4. Coinoculation of Fol with Fo47 Limited Colonization of tomato stems by Fol4287 while Fo47 Colonization was Increased

To determine the extent of tomato stem colonization of Fo47 and Fol4287 upon coinoculation, these strains were reisolated from tomato stems three weeks after inoculation. To discriminate between the strains, transgenic fungi carrying either hygromycin (Fo47) or phleomycin (Fol4287, Fol029) resistance were used in this experiment. As reported [9], Fo47 colonization was usually restricted to the crown level, and the fungus was rarely observed at cotyledon level (Figure 3a,b). In contrast to Fo47, Fol4287 was reisolated from both crown and cotyledon level in every case upon single inoculation (Figure 3a,b). Coinoculation of Fo47 and Fol4287 strongly reduced the extent of colonization by Fol4287 in tomato stems, and in few instances Fol4287 could not be reisolated from either crown or cotyledon level (Figure 3a,b). Surprisingly, coinoculation of Fo47 and Fol4287 led to more frequent reisolation of Fo47 from tomato stems at crown level and even cotyledon level (Figure 3a,b). This reduced extent of stem colonization by the pathogen and increased migration into the stem by Fo47 upon coinoculation was also observed in wild tomato species (Figure S5). It appears therefore that Fo47 can limit the spread of the pathogen in stems of various tomato species, while Fo47 colonization is enhanced in tomato stems when coinoculated with a pathogenic strain.

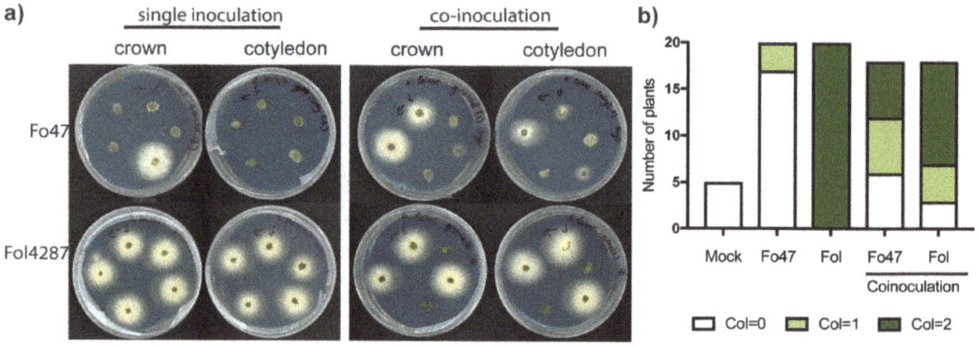

Figure 3. Fo47 colonizes tomato cotyledons upon coinoculation with Fol4287, while Fol4287 colonization is reduced. (**a**) Representative potato dextrose agar (PDA) plates with tomato pieces taken from crown and cotyledon level four days after incubation in the dark at 25°C; (**b**) plates were scored as following: No Fo outgrowth is represented as white (Col = 0), Fo outgrowth at either crown or cotyledon level is represented in light green (Col = 1) an outgrowth at both crown, and cotyledon level (Col = 2) is represented in dark green. This experiment was performed three times with similar results.

3.5. Fo47 Limited Fol Colonization in Tomato Roots

Next, the level of root colonization by Fo47 or Fol4287 was determined to assess whether it changes upon coinoculation. To do so, fungal biomass in tomato roots was measured by quantitative PCR (qPCR). Tomato roots were harvested three weeks after inoculation, and qPCR was performed using either Fo primers designed for InterGenic Spacer (IGS) region or primers for Fo47 (*SCAR* primers described [27]) and for Fol4287 (SIX8 primers) relataive to the plan tubulin. In line with the stem reisolation experiment, Fol4287 was found to colonize tomato roots approximately 20-fold more than Fo47 (Figure 4a). Upon coinoculation, total fungal biomass in tomato roots was similar to tomato roots only inoculated with Fo47 but reduced compared with roots inoculated only with Fol4287 ($p = 0.0349$, Figure 4a). To distinguish between Fo47 and Fol4287 colonization in coinoculated roots, Fo47- and Fol4287-specific primers were used (Figure 4b,c). Quantification by qPCR revealed that Fo47 colonization of tomato roots is similar in single inoculated or coinoculated roots (Figure 4b). In contrast, Fol4287 colonization is about 9-fold reduced in tomato roots coinoculated with Fo47 compared to when inoculated alone (Figure 4c). Overall, our data show that Fo47 is a poor root colonizer compared with Fol4287, but can drastically reduce the amount of Fol4287 in tomato roots upon coinoculation.

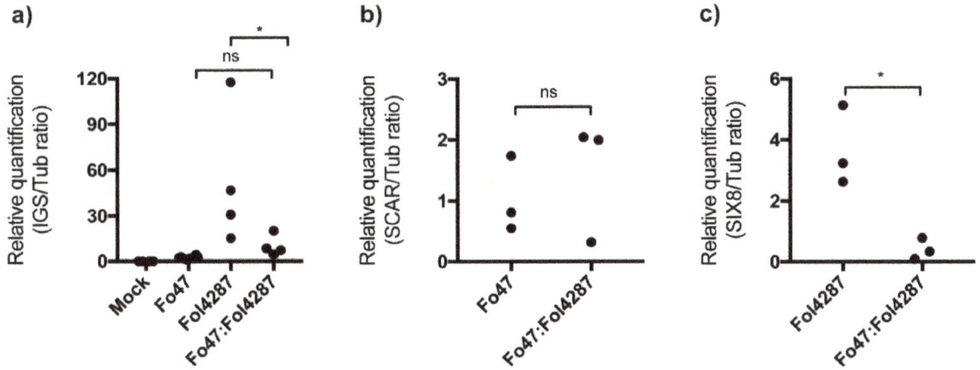

Figure 4. Coinoculated tomato roots are less extensively colonized by pathogenic strain Fol4287. Colonization was assessed by the qPCR using (**a**) Fo InterGenic Spacer (IGS) primers, (**b**) Fo47 marker *SCAR*, or (**c**) Fol marker *SIX8*. Data obtained by quantifying IGS were analysed by an ANOVA with a Tukey multiple comparison test, and in the case of data for *SCAR* and *SIX8* with a Student's t-test. (^{ns} $P > 0.05$; * $p < 0.05$; ** $p < 0.01$).

4. Discussion

Fo47 triggered resistance against Fol029 in two wild tomato species to a similar extent as in cultivated tomato. Various studies have reported that Fo47 can trigger resistance in a variety of plants: asparagus, flax, cucumber, eucalyptus, pepper, banana, and chickpeas [30]. This suggests that EMR is a conserved host trait. Coinoculation of Fo47 with Fol4287 reduced proliferation of the pathogen in both stems and roots, while colonization by the endophyte was enhanced only in stems. Therefore, EMR is likely achieved by limiting pathogen colonization of tomato roots and stems.

Additionally, like Fo, *F. redolens*, *F. solani*, *F. hostae*, and *F. proliferatum* behaved as endophytes and conferred protection against *Fusarium* wilt disease in tomato plants. Previous studies in tomato plants found that the endophytic strain *Fusarium solani*-K could confer resistance against Fo f.sp. *radicis-lycopersici* (Forl) [17]. Moreover, 205 Fo strains, 81 *F. proliferatum* strains, and other 16 *Fusarium* spp. have been shown to confer resistance against Forl [15]. This suggests that many *Fusarium* spp. share the capability of being endophytes and triggering EMR against *Fusarium* wilt.

Fungal quantification of infected tomato roots revealed that Fo47 drastically reduces root colonization by Fol4287. This finding is in line with an earlier study in which the biomass of pathogenic strain Fol8 was significantly reduced in tomato roots in presence of Fo47 [8]. Additionally, a reduction of Fol4287 and Fol029 colonization of tomato stems when coinoculated with Fo47 was observed. The lower amount of pathogen in roots and stems upon coinoculation correlates with the reduced disease symptoms observed. Similarly, the endophytic *Verticillium* strain Vt305 was observed to reduce the amount of the pathogenic species *Verticillium longisporum* in cauliflower roots, hypocotyl, and stem. This suggests that endophytes can trigger resistance by reducing the proliferation of a pathogen in the plant [31].

Strain Fo47 colonized tomato stems to a greater extent when coinoculated with Fol4287, compared to single inoculation, despite not colonizing the tomato roots more extensively. One possible explanation for this phenomenon is that Fo47 takes advantage of effector-triggered susceptibility facilitated by Fol. Fol is known to secrete effector proteins, such as Avr2, which facilitate Fol colonization of tomato stems [32]. Transgenic tomato stems over-expressing AVR2 without signal peptide are hyper-colonized by Fo47, compared to wild-type plants [33]. Therefore, it seems plausible that Fo47 benefits from Fol effector-triggered immune suppression to more extensively colonize tomato stems. There are other examples of such a "boost" in endophytic stem colonization facilitated by pathogens. The *Verticillium* endophyte Vt305 was found to accumulate to higher levels at 70 dpi in tomato stems but not the roots when coinoculated in a 1:1 ratio with the pathogen *Verticillium longisporum* [31]. In another case, infection of *Arabidopsis* plants with the white rust pathogen *Albugo* facilitated infection by *Phytophthora infestans*, which otherwise does not infect *Arabidopsis* [34].

Several questions about EMR remain unanswered. One is how Fo47 reduces root colonization by Fol. One possibility is that Fo47 limits the spread of Fol via direct antagonism, such as antibiosis or competition. Another possibility is indirect antagonism by inducing host resistance. These two possibilities are not mutually exclusive. No antibiosis was observed under in vitro conditions, and in our bioassay set-up it is unlikely that there is competition for growth on roots or for infection sites, since *Fusarium* was inoculated by exposing damaged roots to spores before transplanting them to the soil. Still, in planta competition between strains cannot be excluded, for instance, Fo47 might consume nutrients limiting Fol development. This hypothesis is difficult to test, and it does not easily fit with the observation that endophytic Fo can trigger resistance in a split root system against pathogenic Fo [13,16]. Additionally, the low colonization level of Fo47 in tomato roots seems unlikely to have a drastic impact on the level of nutrients. It is more likely, therefore, that Fo47 induces resistance responses in the host that result in reduced Fol colonization of roots and subsequent spread in the stems. The underlying mechanism of this is elusive but appears to be independent of the major defense-related hormones ethylene, salicylic acid, and jasmonic acid. Whether physical barriers such as papillae or lignified cell walls, induced by Fo47 colonization, are involved remains a question for future study [9].

In conclusion, *Fusarium*-mediated resistance against *Fusarium* wilt disease of tomato is common within the genus *Fusarium* and consists of limiting pathogen colonization inside roots and stem, while the extent of endophytic colonization in tomato stems is increased. Understanding how endophytic *Fusarium* species can suppress *Fusarium* wilt disease may help to improve the design of strategies to better control this soil-borne disease.

Supplementary Materials: The following are available online at http://www.mdpi.com/2076-2607/8/4/544/s1, Figure S1: Endophyte-mediated resistance set-up, Figure S2: Fo47 can trigger resistance against Fol029 in *Solanum lycopersicum* (C-32), **a**) *S. pimpinellifolium* (LA1578) and *S. chmielewski* (LA2663, LA2695) and in **b**) *S. chmielewski* (LA1840), Figure S3: Fo47, *Fusarium redolens* (Fr), *Fusarium solani* (Fs), *Fusarium hostae* (Fh), and *Fusarium proliferatum* (Fp) can suppress *Fusarium* wilt disease in tomato, Figure S4: Schematic representation of a tomato cotyledon harvested three-weeks after inoculation, surfaced sterilized with 70% ethanol, washed with sterile water twice, Figure S5: Fo47 reaches cotyledon levels upon coinoculation with Fol029, while pathogen colonization is reduced in *Solanum lycopersicum* (C-32) **a**) *S. pimpinellifolium* (LA1578) and *S. chmielewski* (LA2663, LA2695) **b**) *S. chmielewski* (LA1840), Table S1: Primer sequences used for (q)PCR analysis.

Author Contributions: Conceptualization, M.E.C., B.V.V., F.L.W.T., and M.R.; methodology, M.E.C. B.V.V.; formal analysis, M.E.C. B.V.V., investigation, M.E.C. B.V.V.; writing—original draft preparation, M.E.C.; writing—review and editing B.V.V., F.L.W.T., M.R., visualization, M.E.C.; B.V.V.; supervision, F.L.W.T., M.R.; funding acquisition, F.L.W.T., M.R. All authors have read and agreed to the published version of the manuscript.

Funding: This research was funded by the European Union's Horizon 2020 Research and Innovation Programme under the Marie Skłodowska-Curie, grant agreement No. 676480. FT obtains support from the NWO-Earth and Life Sciences funded VICI Project No. 865.14.003.

Acknowledgments: The authors thank Ludek Tikovsky and Harold Lemereis for their help in the greenhouse. We are thankful to Petra Bleeker at University of Amsterdam, for providing us the wild tomato lines, and to Francisco J. de Lamo, Jiming Li, and Kees Ketting for help with the bioassays.

Conflicts of Interest: The authors declare no conflict of interest.

References

1. Gordon, T.R.; Okamoto, D. Population-Structure and the Relationship between Pathogenic and Nonpathogenic Strains of Fusarium oxysporum. *Phytopathology* **1992**, *82*, 73–77. [CrossRef]
2. Demers, J.E.; Gugino, B.K.; Jimenez-Gasco Mdel, M. Highly Diverse Endophytic and Soil Fusarium oxysporum Populations Associated with Field-Grown Tomato Plants. *Appl. Environ. Microbiol.* **2015**, *81*, 81–90. [CrossRef]

3. Rocha, L.O.; Laurence, M.H.; Ludowici, V.A.; Puno, V.I.; Lim, C.C.; Tesoriero, L.A.; Summerell, B.A.; Liew, E.C.Y. Putative Effector Genes Detected in Fusarium oxysporum from Natural Ecosystems of Australia. *Plant Pathol.* **2016**, *65*, 914–929. [CrossRef]
4. Pereira, E.; de Aldana, B.R.V.; San Emeterio, L.; Zabalgogeazcoa, I. A Survey of Culturable Fungal Endophytes from Festuca rubra subsp. pruinosa, a Grass from Marine Cliffs, Reveals a Core Microbiome. *Front. Microbiol.* **2019**, *9*. [CrossRef] [PubMed]
5. Beckman, C.H.; Elgersma, D.M.; MacHardy, W.E. The Localization of Fusarial Infections in the Vascular Tissue of Single-Dominant-Gene Resistant Tomatoes. *Pytopathology* **1972**, *62*, 1256–1260. [CrossRef]
6. Yadeta, K.; Thomma, B.P.H.J. The Xylem as Battleground for Plant Hosts and Vascular Wilt Pathogens. *Front. Plant Sci.* **2013**, *4*, 97. [CrossRef] [PubMed]
7. Edel-Hermann, V.; Lecomte, C. Current Status of Fusarium oxysporum Formae Speciales and Races. *Phytopathology* **2019**, *109*, 512–530. [CrossRef]
8. Aime, S.; Alabouvette, C.; Steinberg, C.; Olivain, C. The Endophytic Strain Fusarium oxysporum Fo47: A Good Candidate for Priming the Defense Responses in Tomato Roots. *Mol. Plant-Microbe Interact.* **2013**, *26*, 918–926. [CrossRef]
9. Constantin, M.E.; de Lamo, F.J.; Vlieger, B.V.; Rep, M.; Takken, F.L.W. Endophyte-Mediated Resistance in Tomato to Fusarium oxysporum Is Independent of Et, Ja, and Sa. *Front. Plant Sci.* **2019**, *10*, 979. [CrossRef]
10. Alabouvette, C.; De La Broise, D.; Lemanceau, P.; Couteaudier, Y.; Louvet, J. Utilisation De Souches Non Pathogènes De Fusarium Pour Lutter Contre Les Fusarioses: Situation Actuelle Dans La Pratique1. *EPPO Bull.* **1987**, *17*, 665–667. [CrossRef]
11. He, C.Y.; Wolyn, D.J. Potential Role for Salicylic Acid in Induced Resistance of Asparagus Roots to Fusarium oxysporum f.sp. asparagi. *Plant Pathol.* **2005**, *54*, 227–232. [CrossRef]
12. Trouvelot, S.; Olivain, C.; Recorbet, G.; Migheli, Q.; Alabouvette, C. Recovery of Fusarium oxysporum Fo47 Mutants Affected in Their Biocontrol Activity after Transposition of the Fot1 Element. *Phytopathology* **2002**, *92*, 936–945. [CrossRef] [PubMed]
13. Kaur, R.; Singh, R.S. Study of Induced Systemic Resistance in Cicer Arietinum, L. Due to Nonpathogenic Fusarium oxysporum Using a Modified Split Root Technique. *J. Phytopathol.* **2007**, *155*, 694–698. [CrossRef]
14. Zhang, J.; Chen, J.; Jia, R.M.; Ma, Q.; Zong, Z.F.; Wang, Y. Suppression of Plant Wilt Diseases by Nonpathogenic Fusarium oxysporum Fo47 Combined with Actinomycete Strains. *Biocontrol Sci. Technol.* **2018**, *28*, 562–573. [CrossRef]
15. Bao, J.; Fravel, D.; Lazarovits, G.; Chellemi, D.; van Berkum, P.; O'Neill, N. Biocontrol Genotypes of Fusarium oxysporum from Tomato Fields in Florida. *Phytoparasitica* **2004**, *32*, 9–20.
16. Kroon, B.A.M.; Scheffer, R.J.; Elgersma, D.M. Induced Resistance in Tomato Plants against Fusarium-Wilt Invoked by Fusarium oxysporum f. sp. dianthi. *Neth. J. Plant Pathol.* **1991**, *97*, 401–408. [CrossRef]
17. Kavroulakis, N.; Ntougias, S.; Zervakis, G.I.; Ehaliotis, C.; Haralampidis, K.; Papadopoulou, K.K. Role of Ethylene in the Protection of Tomato Plants against Soil-Borne Fungal Pathogens Conferred by an Endophytic Fusarium solani Strain. *J. Exp. Bot.* **2007**, *58*, 3853–3864. [CrossRef]
18. Does, H.; Constantin, M.; Houterman, P.; Takken, F.; Cornelissen, B.; Haring, M.; Burg, H.; Rep, M. Fusarium oxysporum Colonizes the Stem of Resistant Tomato Plants, the Extent Varying with the R-Gene Present. *Eur. J. Plant Pathol.* **2018**, *154*, 55–65. [CrossRef]
19. Ma, L.J.; Does, H.; Borkovich, K.; Coleman, J.; Daboussi, M.; Di Pietro, A.; Dufresne, M.; Freitag, M.; Grabherr, M.; Henrissat, B.; et al. Comparative Genomics Reveals Mobile Pathogenicity Chromosomes in Fusarium. *Nature* **2010**, *464*, 367–373. [CrossRef]
20. Pietro, A.; Madrid, M.P.; Caracuel, Z.; Delgado-Jarana, J.; Roncero, M.I.G. Fusarium oxysporum: Exploring the Molecular Arsenal of a Vascular Wilt Fungus. *Mol. Plant Pathol.* **2003**, *4*, 315–325. [CrossRef]
21. Baayen, R.P.; vanDreven, F.; Krijger, M.C.; Waalwijk, C. Genetic Diversity in Fusarium oxysporum f. sp. dianthi and Fusarium redolens f. sp. dianthi. *Eur. J. Plant Pathol.* **1997**, *103*, 395–408. [CrossRef]
22. Vlaardingerbroek, I.; Beerens, B.; Schmidt, S.M.; Cornelissen, B.J.; Rep, M. Dispensable Chromosomes in Fusarium oxysporum f. sp. lycopersici. *Mol. Plant Pathol.* **2016**, *17*, 1455–1466. [CrossRef] [PubMed]
23. Vlaardingerbroek, I.; Beerens, B.; Rose, L.; Fokkens, L.; Cornelissen, B.J.C.; Rep, M. Exchange of Core Chromosomes and Horizontal Transfer of Lineage-Specific Chromosomes in Fusarium oxysporum. *Environ. Microbiol.* **2016**, *18*, 3702–3713. [CrossRef] [PubMed]

24. Mes, J.J.; Weststeijn, E.A.; Herlaar, F.; Lambalk, J.J.M.; Wijbrandi, J.; Haring, M.A.; Cornelissen, B.J.C. Biological and Molecular Characterization of Fusarium oxysporum f. sp. lycopersici Divides Race 1 Isolates into Separate Virulence Groups. *Phytopathology* **1999**, *89*, 156–160. [CrossRef]
25. van Dam, P.; de Sain, M.; ter Horst, A.; Gragt, M.; Rep, M. Use of Comparative Genomics-Based Markers for Discrimination of Host Specificity in Fusarium oxysporum. *Appl. Environ. Microbiol.* **2018**, *84*. [CrossRef]
26. Biju, V.C.; Fokkens, L.; Houterman, P.M.; Rep, M.; Cornelissen, B.J.C. Multiple Evolutionary Trajectories Have Led to the Emergence of Races in Fusarium oxysporum f. sp. lycopersici. *Appl. Environ. Microbiol.* **2017**, 83. [CrossRef]
27. Edel-Hermann, V.; Aime, S.; Cordier, C.; Olivain, C.; Steinberg, C.; Alabouvette, C. Development of a Strain-Specific Real-Time Pcr Assay for the Detection and Quantification of the Biological Control Agent Fo47 in Root Tissues. *FEMS Microbiol. Lett.* **2011**, *322*, 34–40. [CrossRef]
28. Fravel, D.; Olivain, C.; Alabouvette, C. Fusarium oxysporum and Its Biocontrol. *New Phytol.* **2003**, *157*, 493–502. [CrossRef]
29. Leslie, J.F. Fungal Vegetative Compatibility. *Annu. Rev. Phytopathol.* **1993**, *31*, 127–150. [CrossRef]
30. de Lamo, F.J.; Takken, F.L.W. Biocontrol by Fusarium oxysporum Using Endophyte-Mediated Resistance. *Front. Plant Sci.* **2020**, *11*. [CrossRef]
31. Tyvaert, L.; Franca, S.C.; Debode, J.; Hofte, M. The Endophyte Verticillium Vt305 Protects Cauliflower against Verticillium Wilt. *J. Appl. Microbiol.* **2014**, *116*, 1563–1571. [CrossRef] [PubMed]
32. Di, X.T.; Gomila, J.; Takken, F.L.W. Involvement of Salicylic Acid, Ethylene and Jasmonic Acid Signalling Pathways in the Susceptibility of Tomato to Fusarium oxysporum. *Mol. Plant Pathol.* **2017**, *18*, 1024–1035. [CrossRef] [PubMed]
33. de Lamo, F.J.; Simkovicova, M.; Fresno, D.H.; de Groot, T.; Tintor, N.; Rep, M.; Takken, F.L.W. Pattern-Triggered Immunity restricts host colonisation by endophytic Fusaria, but does not affect Endophyte-Mediated Resistance. (unpublished; manuscript submitted).
34. Prince, D.C.; Rallapalli, G.; Xu, D.Y.; Schoonbeek, H.J.; Cevik, V.S.; Asai, E.; Kemen, N.; Cruz-Mireles, A.; Kemen, K.; Belhaj, S.; et al. Albugo-Imposed Changes to Tryptophanderived Antimicrobial Metabolite Biosynthesis May Contribute to Suppression of Non Host Resistance to Phytophthora infestans in Arabidopsis thaliana. *BMC Biol.* **2017**, *15*, 20. [CrossRef] [PubMed]

microorganisms

MDPI

Article

DNA Methylation Profile of *β-1,3-Glucanase* and *Chitinase* Genes in Flax Shows Specificity Towards *Fusarium Oxysporum* Strains Differing in Pathogenicity

Wioleta Wojtasik [1,*], Aleksandra Boba [1], Marta Preisner [1,2], Kamil Kostyn [1,2], Jan Szopa [1,2] and Anna Kulma [1]

1 Department of Genetic Biochemistry, Faculty of Biotechnology, University of Wroclaw, Przybyszewskiego 63, 51-148 Wroclaw, Poland; aleksandra.boba@uwr.edu.pl (A.B.); marta.preisner@upwr.edu.pl (M.P.); kamil.kostyn@upwr.edu.pl (K.K.); szopa@ibmb.uni.wroc.pl (J.S.); anna.kulma@uwr.edu.pl (A.K.)

2 Department of Genetics, Plant Breeding and Seed Production, Faculty of Life Sciences and Technology, Wroclaw University of Environmental and Plant Sciences, pl. Grunwaldzki 24A, 50-363 Wroclaw, Poland

* Correspondence: wioleta.wojtasik@uwr.edu.pl

Received: 2 September 2019; Accepted: 18 November 2019; Published: 20 November 2019

Abstract: Most losses in flax (*Linum usitatissimum* L.) crops are caused by fungal infections. The new epigenetic approach to improve plant resistance requires broadening the knowledge about the influence of pathogenic and non-pathogenic *Fusarium oxysporum* strains on changes in the profile of DNA methylation. Two contrasting effects on the levels of methylation in flax have been detected for both types of *Fusarium* strain infection: Genome-wide hypermethylation and hypomethylation of resistance-related genes (*β-1,3-glucanase* and *chitinase*). Despite the differences in methylation profile, the expression of these genes increased. Plants pretreated with the non-pathogenic strain memorize the hypomethylation pattern and then react more efficiently upon pathogen infection. The peak of demethylation correlates with the alteration in gene expression induced by the non-pathogenic strain. In the case of pathogen infection, the expression peak lags behind the gene demethylation. Dynamic changes in tetramer methylation induced by both pathogenic and non-pathogenic *Fusarium* strains are dependent on the ratio between the level of methyltransferase and demethylase gene expression. Infection with both *Fusarium* strains suppressed methyltransferase expression and increased the demethylase (*demeter*) transcript level. The obtained results provide important new information about changes in methylation profile and thus expression regulation of pathogenesis-related genes in the flax plant response to stressors.

Keywords: flax; *Fusarium oxysporum*; pathogenic and non-pathogenic strains; sensitization; DNA methylation; PR genes

1. Introduction

Currently, the greatest losses in flax crops are caused by fungal infections. Fungal diseases are the cause of about 20% of losses of flax cultivation. They result in the reduction of crop yield, both the seed and fiber and deterioration of their quality, as well as feed and food obtained from them [1]. Among the genus *Fusarium*, saprophytic *Fusarium oxysporum* is the major flax pathogen. The strains of *Fusarium oxysporum* can be divided depending on their virulence into pathogenic and non-pathogenic strains. The first group infects various plant species causing wilt or root rot and finally leads to death of the plant. By contrast, the latter do not invade the vascular system of plants and do not kill the host. Moreover, non-pathogenic strains of *F. oxysporum* colonizing plants without symptoms of fusariosis can protect the host from infection by pathogenic strains [2,3]. Therefore, some selected, non-pathogenic strains are perceived as potential biological control agents because their resistance-inducing activity correlates with an increase in the activity of pathogenesis-related (PR) proteins: β-1,3-glucanase and chitinase [4,5].

PR proteins play a very important role in the plant response to pathogenic infections. They are locally accumulated in infection sites and surrounding tissues, as well as in uninfected tissues, providing resistance to plants against subsequent infection [6]. β-1,3-glucanase is responsible for the degradation of the fungal cell wall by the hydrolysis of β-1,3-glycosyl bonds in β-1,3-glucan, thereby releasing the fungal cell wall fragments (oligosaccharides), which act as elicitors (damage-associated molecular patterns, DAMPs) in a defensive reaction stimulating the production of other PR proteins and antifungal low molecular weight components (phytoalexins) [7]. Chitinases catalyze the cleavage of bonds between C1 and C4, two residues of N-acetyl-D-glucosamine in chitin, inhibiting the growth of fungal hyphae during the invasion of the intercellular space, and releasing fungal elicitors (chitooligosaccharides or chitin oligomers) that induce a plant's defense response by biosynthesis of Chitinases and other PR proteins [8].

There is a significant increase in the expression of both *β-1,3-glucanase* and *chitinase* genes during the infection, as confirmed in many plants, e.g., tomato, tobacco, soybeans, wheat, and peas [9,10]. At present, genetic engineering tools are frequently and routinely used to improve the plant defense system and thus plants' resistance against pathogens. The results of such an approach are genetically modified organisms (GMOs). One possibility to strengthen the plant defense system making it pathogen-resistant is upregulation or overexpression of pathogen-related genes or genes encoding enzymes involved in the synthesis of secondary metabolites [11–13].

A similar effect, intense synthesis of pathogen-related proteins, was observed in various plant species after plant treatment with a non-pathogenic strain of *F. oxysporum* when compared to the control plants [3,14]. This was followed by an increase in the resistance against pathogenic strains of *F. oxysporum* in cucumber and peas [3]. In contrast, a study conducted by Aime and co-workers revealed that colonization of tomato by a non-pathogenic strain resulted in a decreased mRNA level of *β-1,3-glucanases* and *chitinases* genes. Furthermore, the expression of this group of genes was observed only when tomato plants subsequently underwent infection with a pathogenic strain of *F. oxysporum*. Such response, called priming, implies the protective role of non-pathogenic strains by accelerating and strengthening the plant defense mechanisms only upon contact with the pathogen [15].

After exposure to non-pathogenic fungal strains, the plants develop acquired immunity to pathogenic infections. Moreover, in this way the immune memory associated with the process of preparing cells (priming) may be transferred to the next generation. Priming results from the accumulation in the cell of inactive mitogen-activated protein kinases (MPK3, MPK6), and nonexpressor of pathogenesis-related genes 1 (NPR1) (both mRNA and protein) that are activated after infection [16]. Priming also affects chromatin modifications: Mostly methylation and acetylation of histones and DNA, associated with activation of genes involved in plant resistance. A local pathogenic infection can change the status of methylation and acetylation of histones and in particular can affect the promoter sequences of genes in systemic tissues. Such epigenetic changes provide plants with a long-term immune memory inherited from generation to generation [17].

Among eukaryotes, the plant genome bears the highest levels of DNA methylation, with up to 50% in some species. DNA methylation and modification of associated chromatin proteins determines the epigenetic status of the genome. Depending on the sequence context of the cytosine residue to be methylated, we can define CG, CNG, or CNN (C = cytosine; G = guanine; N = nucleotide other than G) specific methylation types. Two types of maintenance methyltransferase exist in plants: DNA methyltransferase (MET1), which predominantly methylates CG sites, and plant-specific chromomethylase 3 (CMT3) that methylates CNG (N = A, C, or T) sites. The third type methylates CNN and is controlled by domains of rearranged methyltransferase 1 (DRM2), a homologue of mammalian de novo DNA methyltransferase DNMT3 [18]. De novo DNA methylation is conducted by the pathway that includes specific transcripts that are copied into dsRNA by RNA-dependent RNA polymerase 2 (RDR2). This siRNA associates with argonaute 4 (AGO4) in a complex that mediates DNA methylation [19].

Our preliminary data indicate that single base pair mapping (bisulfite-seq) of methylated cytosine in flax seedlings reveals that over 8% of CG, 4% of CNG, and 7.3% of CNN is methylated. For comparison, the methylation state of maize genome (mainly composed of repetitive elements) ranges around 75% and 62% of CG and CNG sequence contexts, respectively [20]. Genome-wide mapping of methylated cytosine in *Arabidopsis thaliana* (*A. thaliana*) revealed levels of 24% CG, 6.7% CNG, and 1.7% CNN methylation [21].

Many literature reports have described changes in DNA at the genome-wide level caused by environmental stress, which can be inherited [22]. Pathogenic infection and abiotic stress can lead to two contrasting effects at the level of methylation in plants: Hypermethylation at the genome-wide level and hypomethylation of resistance-related genes, which can correlate with regulation of the expression of the defense-related genes in plants [23].

Based on the current state of knowledge about the mechanisms of epigenetic inheritance, it is anticipated that changes in the flax epigenome introduced by infection with a non-pathogenic *Fusarium oxysporum* strain would be directional and transmitted to the offspring. The aim of the study is to accurately understand and compare the mechanisms of infection of pathogenic and non-pathogenic strains as well as the sensitizing effect of the non-pathogenic strain, and in particular to examine the DNA methylation not only throughout the genome but also in specific genes. PR proteins are essential to plant resistance to biotic stress, which makes them the main focus of this research. The findings of our research can help to improve crop resistance in the future.

2. Materials and Methods

2.1. Plant and Fungal Material

A fibrous flax variety (*Linum usitatissimum* L. cv. Nike) was cultured under strictly defined conditions: Medium: MS (Murashige and Skoog, 1962) (Sigma-Aldrich, Saint Louis, Missouri, USA) with 0.8% agar (Sigma-Aldrich, Saint Louis, Missouri, USA) and 1% sucrose (Chempur, Piekary Slaskie, Poland); humidity: 50%; temperature: 21 °C/16 °C; illuminance: 23 mmol/s/m^3; day/night: 16/8 h. The pathogenic strain *Fusarium oxysporum* f. sp. *linii* (Bolley) Snyder et Hansen (ATCC number: MYA-1201, Foln3) and the non-pathogenic strain *Fusarium oxysporum* (ATCC number: MYA-1198, Fo47) were bought from ATCC (Manassas, Virginia, USA) cultured on PDA (potato dextrose agar) medium (Sigma-Aldrich, Saint Louis, Missouri, USA) at 28 °C in darkness.

Molecular analysis (Figure S1): Seed germination, and growth of seedlings were carried out on Petri dishes under controlled conditions in a phytotron. Seven days after sowing, the seedlings were transferred (together with medium) to the PDA medium with either a pathogenic or non-pathogenic strain of *Fusarium oxysporum*, or to a PDA medium (control). Fungal strains were earlier grown on PDA medium for 3 and 5 days, respectively. Seedlings were harvested at the 6th, 12th, 24th, 36th, and 48th hour of incubation (100 for each stage), frozen in liquid nitrogen, stored at -70 °C, and powdered before use. Treated flax seedlings were prepared in three biological repetitions. Each treatment had its own control. In order to sensitize the flax plants with the strain Fo47, first flax seedlings were incubated with Fo47 for 12 h. Twelve hours was chosen as the optimal sensitization time because after this period we observed the largest increase in the expression levels of *β-1,3-glucanase 1* and *chitinase* in preliminary studies. Subsequently, plants were transferred and incubated with pathogenic Foln3. For controls, plants were grown for 12 h on PDA medium without fungi then transferred to new PDA medium also without fungi. After 6, 12, 24, 36, and 48 h, seedlings were collected separately.

Phenotypic analysis: Four-week-old flax plants (from in vitro culture) were transferred onto a PDA medium with non-pathogenic strains of *F. oxysporum*, pathogenic strains of *F. oxysporum*, and without fungi (control plants) for two days. Before use, *F. oxysporum* strain was grown on PDA medium for 5 days. Next, flax plants incubated with non-pathogenic strain of *F. oxysporum* were transferred onto pathogenic *F. oxysporum* or non-pathogenic strain of *F. oxysporum*, and control plants were transferred to pathogenic strain of *F. oxysporum* or a PDA medium without fungi. Flax plants incubated with pathogenic strain of *F. oxysporum* for two days were transferred onto pathogenic *F. oxysporum*. Phenotypic changes were observed after four and six days.

2.2. Identification of DNA Sequences

In order to find the complete coding sequences of the genes, we searched the flax genome (Acc. No. AFSQ00000000.1) using fragments of DNA/cDNA obtained previously in our laboratory, via homology search and PCR, to obtain clones containing the genomic sequence of the GOI (gene of interest). Flax clones were scanned using FGENESH, SoftBerry (Mount Kisco, NY, USA). The resulting coding sequence was compared to the sequences from different plant species available in NCBI using BLAST software.

2.3. DNA Isolation

Genomic DNA was isolated using DNeasy Plant Mini Kit (QIAGEN, Hilden, Germany) following the manufacturer's protocol. The DNA integrity was examined by gel electrophoresis on 1.0% (w/v) agarose and the DNA amount was checked by spectrophotometry method (Spectrophotometer Implen NanoPhotometer Pearl, München, Germany) at 260 nm.

2.4. Determination of Methylation Patterns of Genes

The methylation pattern of DNA was determined using the combination of restricted cleavage of the gene sequence at methylation sites (CCGG islands) using the restriction enzymes *MspI* and *HpaII* (New England Biolabs, Ipswich, Massachusetts, USA) and real-time PCR using primers flanking the methylation islands (cleavage sites). The difference between these two isoschizomers is that *HpaII* can cleave unmethylated CCGG, but *MspI* can cleave unmethylated CCGG as well as the sequence when the internal C residue is methylated CCmGG. Both of these enzymes do not cleave the sequence CCGG when the external and internal C is methylated (CmCmGG).

The reaction (in final volume 50 μL) of restriction cleavage by *HpaII* (250 units) and *MspI* (250 units) was run at 37 °C overnight on the template DNA (0.75μg), which later served as a template for real-time PCR reactions. Restrictions enzymes were inactivated by incubation at 80 °C for 20 min. The program used for real-time PCR was 95 °C for 10 min, and 40 cycles of denaturation for 10 s at 95 °C, annealing for 20 s at 57 °C, and extension for 25 s at 72 °C, and reaction mixtures contain (in final volume 25 μL): Master Mix 2x, forward and reverse primers in final concentration 0.5 μM and 37.5 ng DNA (final concentration 1.5 ng/μL).

The obtained results of digested and undigested DNA were obtained as x-fold of the relative quantification to *actin* (the reference gene). The pattern of gene methylation sites was calculated according to the equations: CCGG (non-methylated cytosines) = total DNA – DNA left after digest by *HpaII*; CCmGG (internal methylated cytosine) = DNA left after digest by *HpaII* - DNA left after digest by *MspI*; CmCmGG (external and internal methylated cytosine) = DNA left after digest by *MspI*. The changes in the different methylated cytosine in DNA (CCGG, CCmGG, and CmCmGG) were presented as x-fold relatively to the control.

Seven CCGG sites were analyzed (three in the promoter (P) and three in the exon (E) and one in the intron (I)) in the *β-1,3-glucanase 1* gene, twelve sites (eleven in the exon and one in the intron) in the *β-1,3-glucanase 2* gene, and eight sites in the exon in the *chitinase* gene. The *β-1,3-glucanase 1* gene sites were characterized by the following positions starting from ATG in DNA: P1 -1204-1201; P2 -1147-1144; P3 -344-341; E1 + 440-443; E2 + 1479-1482; E3 + 1763-1766; I1 + 2341-2344; the *β-1,3-glucanase 2* gene: E1 +50-53; E2 +432-435; E3 +690-693; E4 +883-886; E5 +891-894; E6 +1009-1012; E7 +1024-1027; E8 +1034-1037; E9 +1097-1100; E10 +1178-1181; E11 +1421-1424; I1 +1874-1877; and the *chitinase* gene: E1 +118-121; E2 +138-141; E3 +163 – 167; E4 +227-230; E5 +401-404; E6 +467-470; E7 +957-960;

E8 +1105-1108. A schematic diagram of the position of all putative sites and the relative position of the primers that were used to perform qPCR to estimate methylation levels is presented in Figure S2.

2.5. Determination of Total DNA Methylation

The global level of DNA methylation of flax infected by *Fusarium* strains was determined colorimetrically by measuring the level of 5-methylcytosine using the dedicated MethylFlash Methylated DNA Quantification Kit (EpiGentek Group Inc., Farmingdale, NY, USA) according to the manufacturer's instructions.

2.6. Gene Expression Analysis

The level of transcript for PR genes (*β-1,3-glucanase 1, β-1,3-glucanase 2, chitinase*), genes involved in chromatin modification (*chromomethylase 1, chromomethylase 3, repressor of silencing 1, demeter, argonaute 1, argonaute 4, RNA-dependent RNA polymerase 2*), and *actin* gene was measured using the quantitative PCR technique.

The total RNA was isolated from plant tissues by treatment with Trizol reagent (Life Technologies, Carlsbad, California, USA) according to the manufacturer's manual. Its quality was verified by gel electrophoresis on 1.5% (w/v) agarose containing 15% (v/v) formaldehyde and quantitated spectrophotometrically (Spectrophotometer Implen NanoPhotometer Pearl, München, Germany) at $\lambda = 260$ nm. In order to remove the residual genomic DNA, the samples were subjected to digestion with DNase I (Thermo Fisher Scientific, Waltham, Massachusetts, USA). Then, purified RNA served as a template for reverse transcription PCR to obtain cDNA. The reaction was designed using a High Capacity cDNA Reverse Transcription Kit (Applied Biosystems, Foster City, California, USA).

The quantitative PCR reactions were run using a SYBR dye-based set of reagents (DyNAmo HS SYBR Green qPCR Kit, Thermo Fisher Scientific, Waltham, Massachusetts, USA) on an Applied Biosystems StepOnePlus Real-Time PCR System (Life Technologies, Carlsbad, California, USA). Primers were designed to selectively amplify only plant sequences (not fungal, see Table S1) using LightCycler Probe Design Software 2 (Roche, Basel, Switzerland). Their specificity at annealing temperature was verified by analyzing the melting curves of the obtained reaction products. The reaction setup was designed according to the kit manufacturer's protocol. The program used for real-time PCR was 95 °C for 10 min, and 40 cycles of denaturation for 15 s at 95 °C, annealing for 20 s at 57 °C, and extension for 30 s at 72 °C, and reaction mixtures contain (in final volume 25 μL): Master Mix 2x, forward and reverse primers in final concentration 0.5 μM and 125 ng cDNA (final concentration 5 ng/μL) All the reactions were run in triplicate. Changes in gene expression levels are presented as x-fold of relative quantities (RQ) standardized for *actin* (the reference gene) in relation to the control, non-treated plants.

2.7. Statistical Analysis

All experiments were repeated three times in separate runs. The data are shown as mean values ± standard deviation. The Statistica 7 (StatSoft, USA) software was used to estimate statistical significance of the data by means of Student's t test. For expression change analysis one-sample t-test was performed on logFC values. The *p*-values are specified separately for each data set (* $p < 0.05$).

3. Results

3.1. DNA Methylation Patterns of β-1,3-Glucanase and Chitinase Genes

Due to the lack of flax genes' annotation in a database, in the first step we had to extract whole gene sequences for two isoforms of *β-1,3-glucanase* and *chitinase* genes on the basis of known gene fragments to find whole DNA sequences. The obtained sequences were analyzed in silico in order to identify the potential methylation sites. We found 7 CCGG sites in the *β-1,3-glucanase 1* gene (three in the promoter, three in exons, and one in the intron), 12 CCGG sites in the *β-1,3-glucanase 2* gene (eleven in exons and one in the intron), and 8 CCGG sites in exons in the *chitinase* gene. The analysis of methylation profiles of PR genes included all potential CCGG sites. Here we present only data for those sites where differences in a methylation pattern were observed between treatments. The differentiating CCGG sites for the β-1,3-glucanase 1 gene (3 sites) and chitinase gene (3 sites) are shown as x-fold of different methylation of CCGG (CCGG, CCmGG, and CmCmGG) in Figure 1 and as the DNA methylation pattern in Figure S3. We found that the methylation in all analyzed CCGG sequences for the *β-1,3-glucanase 2* gene did not change.

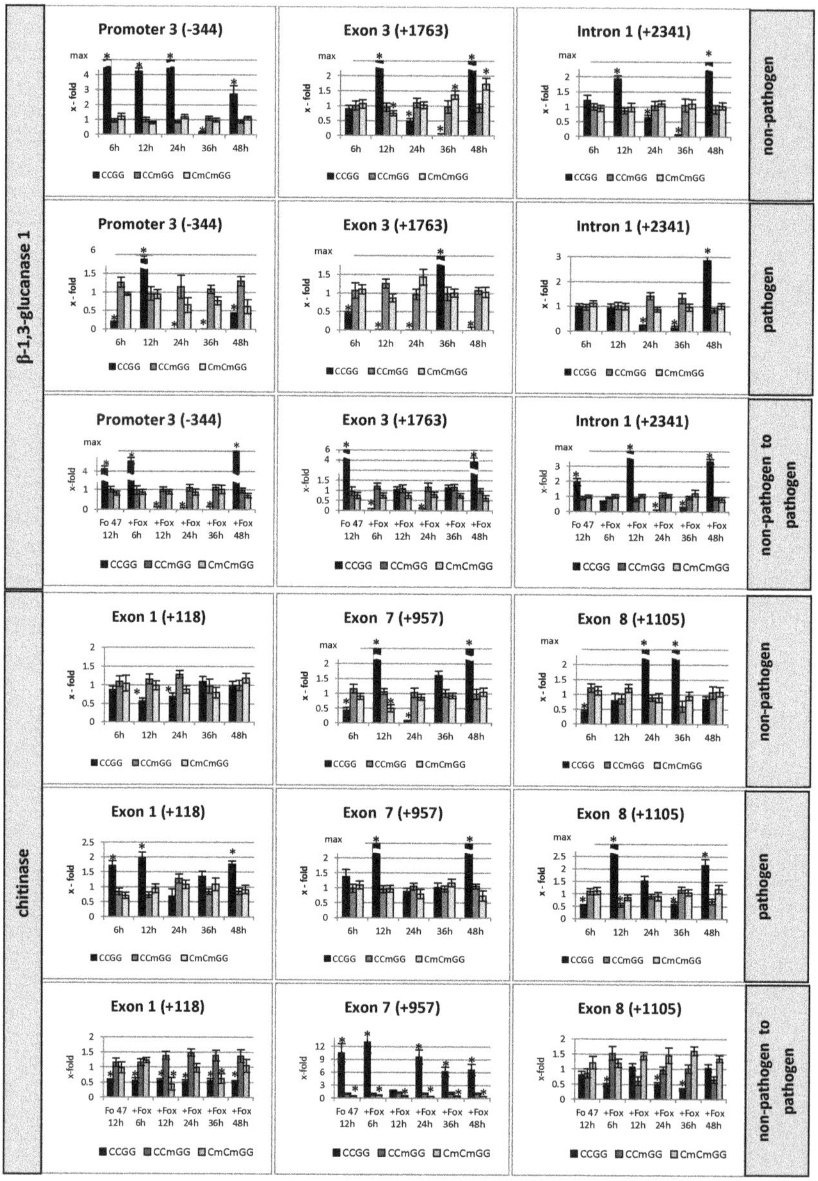

Figure 1. Changes in DNA methylation pattern of *β-glucanase 1*and *chitinase* in flax seedlings treated with non-pathogenic or pathogenic strains of *Fusarium oxysporum* and after sensitizing effects of the

non-pathogenic strain treated with pathogenic Foln3 at 6, 12, 24, 36, and 48 h after inoculation presented as x-fold change in relation to control. The analysis of the third position in the exon, the first in the intron, and the third in the promoter of *β-1,3-glucanase 1* and the first, seventh, and eighth positions in the exon of *chitinase* was performed by the digestion of genomic DNA by the restriction enzymes *HpaII* and *MspI* and then the real-time PCR reaction. The data represent the mean from three biological repetitions. Asterisks mark statistically significant differences ($p < 0.05$) between the treated samples and their own controls.

The DNA methylation pattern of the *β-1,3-glucanase 1* gene in flax incubated with the non-pathogenic strain of *Fusarium oxysporum* was characterized by changes in the level of methylation of internal and/or external cytosine in three CCGG sequences (Figure 1). Analyzing the third CCGG place in the promoter of the *β-1,3-glucanase 1* gene, it was observed that the level of non-methylated cytosines (CCGG) significantly increased over 2.7-fold at the 6th, 12th, 24th, and 48th hour of incubation and decreased to 12% at the 36th hour. The levels of internal methylated cytosine (CCmGG) and internal and external methylated cytosines (CmCmGG) did not change significantly in flax plants incubated with the non-pathogenic strain of *F. oxysporum*. Other changes were shown for the third place in the exon, where a significant increase in non-methylated cytosines (CCGG) was observed at the 12th and 48th hour after incubation, a reduced level to 50% at the 24th hour, and a lack of non-methylated cytosines at the 36th hour. It was followed by a decrease at the 12th hour in the CmCmGG and 1.4-fold and 1.7-fold increases at the 36th and 48th hour, respectively. It was interesting that for CCGG islands located in the intron, the level of CCmGG and CmCmGG remained unchanged, but the level of non-methylated cytosines increased over 1.9-fold at the 12th and 48th hour of incubation and decreased to 64% at the 24th and 7% at the 36th hour.

The analysis of eight CCGG places in the *chitinase* gene revealed that only three of them were characterized by changes in the DNA methylation pattern in flax incubated with the non-pathogenic strain of *Fusarium oxysporum* (Figure 1). In the first CCGG site in the exon, the level of non-methylated cytosines decreased to 56% and 68% at the 12th and 24th hour of incubation with the non-pathogenic strain. An increase in non-methylated cytosines in the CCGG site was noted in the seventh exon (10-fold increase at the 12th hour and more than 6-fold increase at the 48th hour of incubation). It was followed by a decrease in the CCGG to 43% and 8% at the 6th and 24th hour, respectively. In addition, in the eighth site in the exon, the largest increase of the non-methylated cytosines in the CCGG was observed at the 24th and 36th hour after incubation together with the reduced level of the CCGG to 42% at the 6th hour.

In flax plants incubated with the pathogenic strain of *F. oxysporum* the profile status of methylation of cytosines in CCGG sites of the *β-1,3-glucanase 1* gene differed from that determined for flax incubated with the non-pathogenic strain, but the same differentiating CCGG sites remained. We observed a reduced level of non-methylated cytosines at the 6th, 24th, 36^{th,} and 48th hour (a reduction to 20% and to 42% at the 6th and 48th hour and a total reduction at the 24th and 36th hour) and its significantly higher level at the 12th hour in the third CCGG place in the promoter. In the third place in the exon a reduced level of non-methylated cytosines in CCGG was observed at the 6th, 12th, 24^{th,} and 48th hour of incubation (a reduction to 43% at the 6th and a total reduction at the 12th, 24th, and 48th hour) and a significant increase at the 36th hour. In the intron, the decrease in non-methylated cytosines in the CCGG site to 21% and 14% at the 24th and 36th hour after incubation, respectively, was subsequently replaced by an increase (2.9-fold) in non-methylated cytosines at the 48th hour.

The first, seventh, and eighth CCGG sites in the exons in the *chitinase* gene are differentiating sites in flax inoculated with pathogen. A thorough analysis of these CCGG sites in the *chitinase* gene revealed about a two-fold increase in non-methylated cytosines at the 6th, 12th, and 48th hour of incubation in the first CCGG site as well as the largest increase in non-methylated cytosines at the 12th and 48th hour in the seventh site. In the eighth CCGG site, we noted the largest increase in the non-methylated cytosines at the 12th hour of incubation and also a significant but smaller increase at the 48th hour of incubation. Additionally, the lower level of the non-methylated cytosines (a reduction to about 50%) at the 6th and the 36th hour after incubation was observed. The increase in the non-methylated cytosines at the 12th hour was correlated with the decrease in the internal methylated cytosine at this time.

The methylation profile of the *β-1,3-glucanase 1* gene in flax seedlings incubated with the pathogenic strain after previous sensitization by Fo47 was determined based on an analysis of the third place in the promoter, the first place in the intron, and the third place in the exon (Figure 1). The third CCGG site in the promoter was characterized by a total reduced level of non-methylated cytosines at the 12th, 24th, and 36th hour and a significant increase in non-methylated cytosines at the 6th and 48th hour.

Moreover, the methylation profile of the *chitinase* gene was determined by analyzing the methylation status of eight CCGG sites in exons, of which three sites were altered (the first, seventh, and eighth). The level of non-methylated cytosines remained unchanged in relation to the sensitized plants in all time points investigated. However, we observed a reduced level (about 50%) of internal and external methylated cytosines (CmCmGG) at the 12th and 36th hour. The seventh CCGG site in the exon showed a considerable increase in the non-methylated cytosines at all the time-points but the 12th, where the increase was smaller. Constant, reduced levels (about 50%) of internal and external methylated cytosines (CmCmGG) were observed. The eighth site in the exon was characterized by a reduced level of non-methylated cytosines at the 6th, 24th, and 36th hour.

3.2. Level of Total DNA Methylation

After determining the changes in DNA methylation pattern of *β-glucanase 1* and *chitinase* in flax seedlings incubated with non-pathogenic and pathogenic strains of *F. oxysporum* and in flax infected by *Foln3* after previous treatment with the non-pathogenic strain we examined the total level of DNA methylation, and the results are shown in Figure 2.

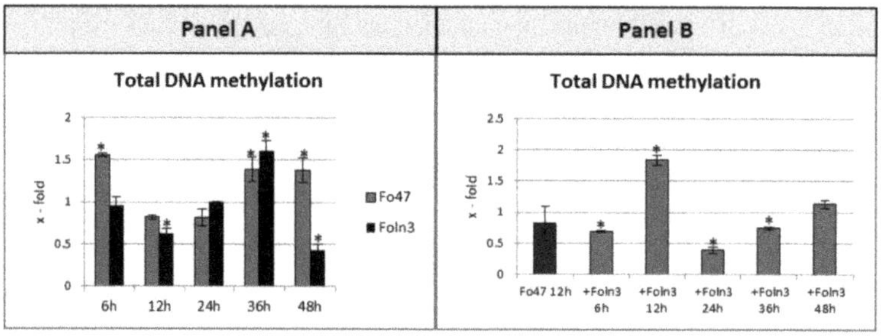

Figure 2. Total DNA methylation in flax seedlings treated with non-pathogenic (Fo47) or pathogenic strains of *Fusarium oxysporum* (Foln3) (panel **A**) and after sensitizing effects of the non-pathogenic strain treated with pathogenic Foln3 (+ Foln3) (panel **B**) at 6, 12, 24, 36, and 48 h after inoculation. The data represent the mean from three biological repetitions. Asterisks mark statistically significant differences ($p < 0.05$) between the treated samples and their own controls.

The level of total DNA methylation increased by about 1.5-fold at the 6th, 36th, and 48th hour of flax seedlings' incubation with the non-pathogen, while at the other analyzed time points only a slight decrease (approximately 20%) was observed when compared to control plants. In the case of the flax plants incubated with the pathogenic strain, a 1.6-fold increase in total DNA methylation at the 36th hour and a reduction to 62% and 43% at the 12th and 48th hour of incubation with the pathogen, respectively, were noted. However, in the flax sensitized with Fo47 the total level of DNA methylation increased 1.8-fold at the 12th hour of incubation, and was reduced to 70%, 40%, and 74% at the 6th, 24th, and 36th hour, respectively.

3.3. Expression Levels of PR Genes

The expression levels of PR genes, two isoforms of *β-1,3-glucanase* and one of *chitinase,* were determined in flax seedlings treated with non-pathogenic and pathogenic strains of *Fusarium oxysporum* as well as in flax seedlings sensitized by the non-pathogenic strain Fo47 and then incubated with the pathogenic strain Foln3. In order to present the precise kinetics of changes, the analyses were performed after 6, 12, 24, 36, and 48 h of flax incubation with appropriate *Fusarium* strains and the obtained results are presented in Figure 3.

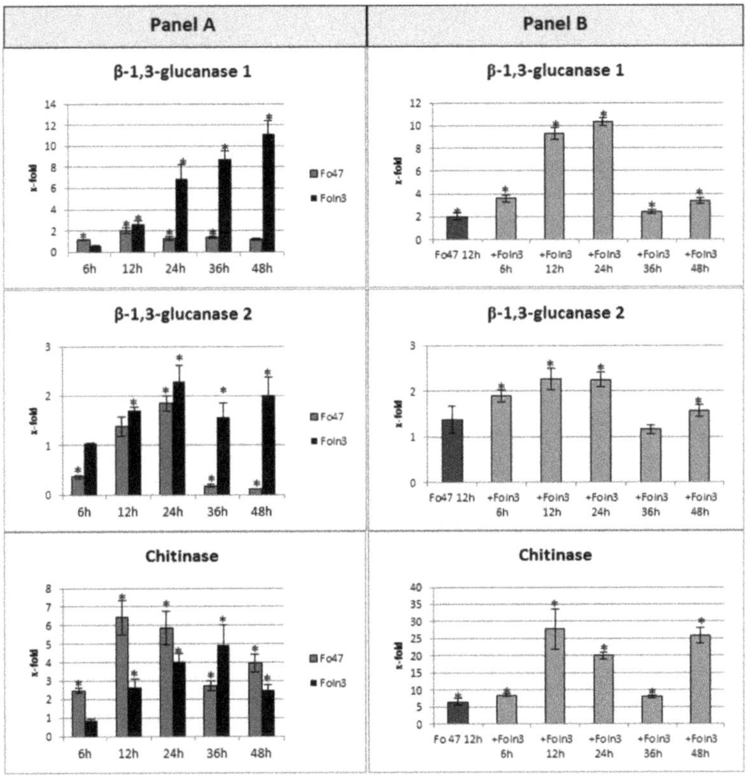

Figure 3. Expression level of *β-glucanase 1*, *β-glucanase 2*, and *chitinase* in flax seedlings treated with non-pathogenic (Fo47) or pathogenic strains of *Fusarium oxysporum* (Foln3) (panel **A**) and after sensitizing effects of the non-pathogenic strain treated with pathogenic Foln3 (+ Foln3) (panel **B**) at 6, 12, 24, 36, and 48 h after inoculation. The data were obtained from real-time RT-PCR analysis. *Actin* was used as a reference gene and the transcript levels were normalized to the control plant. The data represent the mean ± standard deviations from three biological repetitions. Asterisks mark statistically significant differences ($p < 0.05$) between the treated samples and their own controls.

In flax seedlings incubated with the non-pathogenic strain of *Fusarium oxysporum*, the analysis of the expression level of *β-1,3-glucanase 1* gene revealed a two-fold increase at the 12th hour of incubation. An increase in the transcript level was also noted at subsequent time points (the 24th and 36th hour), but it was lower compared to the earlier time (the 12th hour). In the case of flax incubated with the pathogenic strain, the expression level of the *β-1,3-glucanase 1* gene increased continuously with the incubation time (from a 2.6-fold increase at the 12th hour to 11-fold at the 48th hour). However, in the sensitized flax, the transcript level of the *β-1,3-glucanase 1* gene increased during incubation from 3.6-fold at the 6th hour to 10.3-fold at the 24th hour and then decreased to the initial level (after 12 h priming) and then finally a 3.4-fold increase at the 48th hour was observed.

In flax plants incubated with the non-pathogenic strain, the second isoform of the *β-1,3-glucanase* gene was characterized by a 1.85-fold increase in expression level after 24 h of incubation and a decrease to 65%, 20%, and 12% at the 6th, 36th, and 48th hour, respectively. Furthermore, the analysis of this gene in flax seedlings incubated with the pathogenic strain revealed the smallest changes in the expression level as compared to other PR genes. However, in comparison with control plants the mRNA level of this gene increased 1.7-fold, 2.3-fold, 1.6-fold, and 2-fold at the 12th, 24th, 36th, and 48th hour of incubation with pathogenic *F. oxysporum*, respectively. In contrast, the incubation of flax seedlings with the pathogen after previous sensitization by the non-pathogenic strain had a slight effect on the mRNA level of the *β-1,3-glucanase 2* gene. A significant increase in the expression of this gene (about 2-fold) occurred at the 6th, 12th, and 24th hour of incubation with the pathogenic strain and 1.6-fold increase at 48th hour was observed.

For the *chitinase* gene, an increase in the mRNA levels at each time point after incubation with the non-pathogenic strain was observed, with the largest increase at the 12th and 24th hour of incubation (6.4-fold and 5.8-fold increase, respectively). In flax seedlings incubated with the pathogenic strain, the transcript level of *chitinase* increased from 2.6-fold at the 12th hour to 4.9-fold at the 36th hour and then decreased, but still remained higher compared to the control (2.5-fold increase). However, flax treatment with pathogenic Foln3, after earlier sensitization with non-pathogenic Fo47, initially caused a significant 8.5-fold increase in *chitinase* gene expression at the 6th hour, 27.7-fold increase at the 12th hour, and a 20-fold increase at the 24th hour. Then, it decreased to 8-fold of the control at the 36th hour, and finally a large, 25.8-fold increase in the expression of *chitinase* at the 48th hour was observed.

3.4. Expression Levels of Genes Involved in DNA Methylation

Due to the changes in methylation profiles of PR genes and the need to explain such changes, the next stage of the study was to analyze the expression of genes involved in the DNA modifications: Methylation (*chromomethylase 1, CMT1* and *chromomethylase 3, CMT3*) and demethylation (*repressor of silencing 1, ROS1* and *demeter, DME*). Additionally, *argonaute 1* (*AGO1*), *argonaute 4* (*AGO4*), and *RNA dependent RNA polymerase 2* (*RDR2*) genes involved in mechanisms of DNA modification were assessed. Changes in expression levels of genes analyzed in flax seedlings incubated with non-pathogenic and pathogenic strains of *F. oxysporum* and in flax incubated with Foln3 after having a sensitizing effect by Fo47 are shown in Figure 4.

In flax incubated with the non-pathogenic *F. oxysporum*, expression levels of *CMT1* and *CMT3* genes were decreased: A reduction to 40% at the 6th hour of incubation. Additionally, a lower expression level of the *CMT1* gene was maintained until the 36th hour. In contrast, the expression level of the *CMT3* gene was increased at the 12th hour of incubation (up to a level equal to the control), but it again decreased by 79%, 67%, and 55% at the 24th, 36th, and 48th hour, respectively. The *DME* gene exhibited a slight (1.25- to 1.5-fold) increase in the transcript level over time of the flax incubation with the non-pathogenic strain of *Fusarium*, and finally, at the 48th hour, the expression level of this gene increased more than two-fold. An analysis of a second gene involved in DNA demethylation (*ROS1*) revealed a decrease in the expression level of this gene to 63% at the 36th hour and a 2.3-fold increase at the 48th hour compared to the control. Furthermore, the incubation of flax with the non-pathogenic strain of *Fusarium* affected the expression of genes involved in other mechanisms that modify the DNA.

Expression of *AGO4* and *RDR2* genes increased (1.8-fold and 1.7-fold, respectively) at the 48th hour of incubation, while at earlier time points it did not change at all. The *AGO1* gene was characterized by greater variability, wherein the level of expression was increased 2.9-fold, 2.1-fold, and 4.7-fold at the 24th, 36th, and 48th hour, respectively.

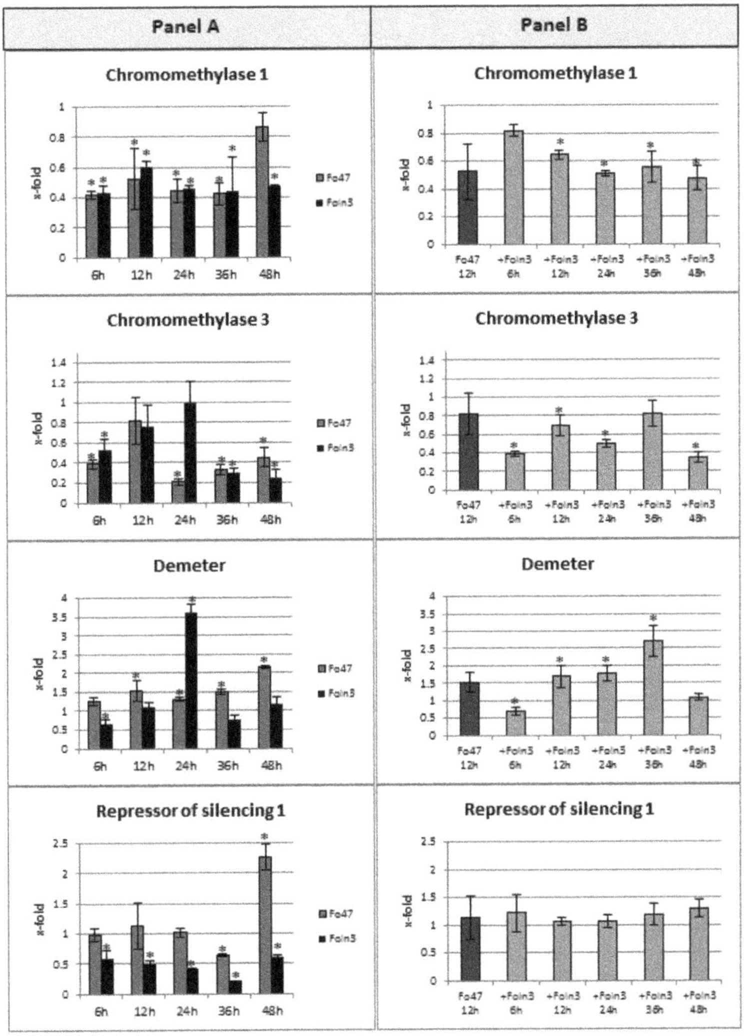

Figure 4. *Cont.*

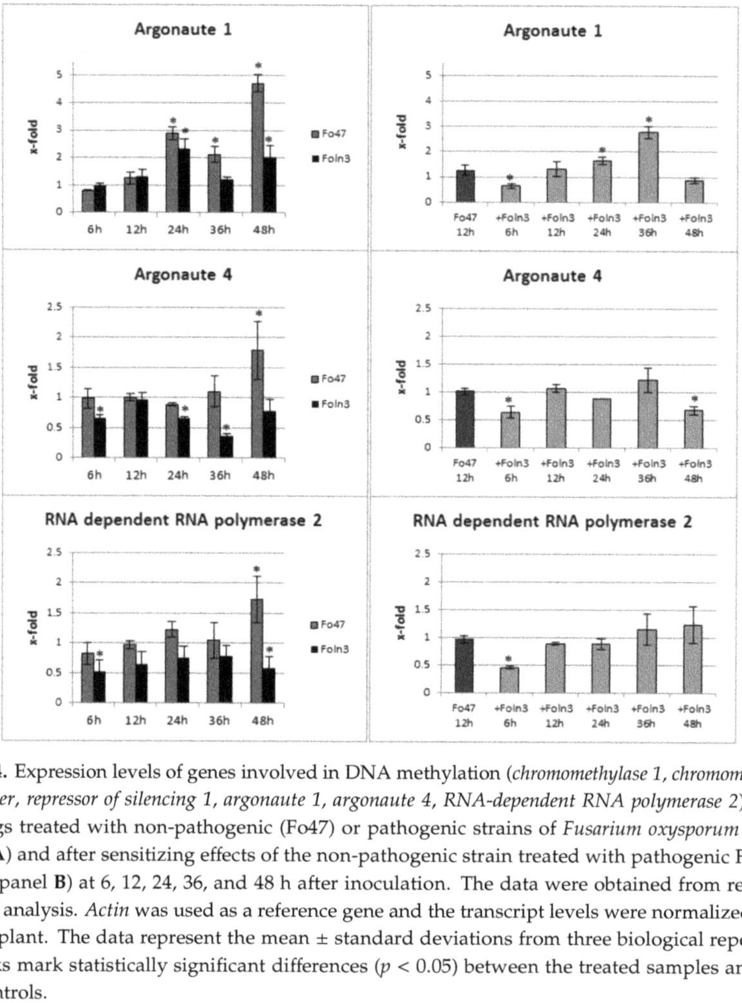

Figure 4. Expression levels of genes involved in DNA methylation (*chromomethylase 1, chromomethylase 3, demeter, repressor of silencing 1, argonaute 1, argonaute 4, RNA-dependent RNA polymerase 2*) in flax seedlings treated with non-pathogenic (Fo47) or pathogenic strains of *Fusarium oxysporum* (Foln3) (panel **A**) and after sensitizing effects of the non-pathogenic strain treated with pathogenic Foln3 (+ Foln3) (panel **B**) at 6, 12, 24, 36, and 48 h after inoculation. The data were obtained from real-time RT-PCR analysis. *Actin* was used as a reference gene and the transcript levels were normalized to the control plant. The data represent the mean ± standard deviations from three biological repetitions. Asterisks mark statistically significant differences ($p < 0.05$) between the treated samples and their own controls.

After incubation of flax plants with the pathogenic *F. oxysporum*, we observed a decrease in the expression of genes involved in the process of DNA methylation. The transcript level of the *CMT1* gene was reduced to approximately 50% for all analyzed times, while the transcript level of the *CMT3* gene was reduced to 50% at the 6th hour, then it returned to the control level at the 12th hour and 24th hour, and then decreased to 30% at the 36th and 48th hour. The DNA demethylation genes responded differently in flax incubated with the pathogen. The expression level of the *ROS1* gene was about half of the control at the majority of time points, with the lowest value (a reduction to 20%) at the 36th hour. The level of expression of the *DME* gene significantly increased by 3.6-fold at the 24th hour, but was also slightly reduced to 65% at the 6th hour. As in the case of the treatment of flax seedlings with the non-pathogen, flax incubation with the pathogen induced similar changes in the expression of *AGO4* and *RDR2* genes and various changes of the *AGO1* gene. The transcript levels of *AGO4* genes were reduced to 50–75% during incubation with pathogen. Only 12 and 48 h after incubation did the mRNA level of the *AGO4* gene remain unchanged. The level of mRNA of the *RDR2* gene was reduced to 51% and 58% at the 6th and 48th hour, respectively. Similarly to the non-pathogenic strain of *Fusarium*, the pathogenic strain caused a 2.3-fold increase in expression of the *AGO1* gene at the 12th hour and a two-fold increase after 48 h of incubation.

In flax incubated with pathogenic Foln3 after prior sensitization by non-pathogenic Fo47 the expression levels of *CMT1* and *CMT3* genes differed significantly from each other, which was not observed in flax after incubation either with pathogenic or with non-pathogenic *F. oxysporum*. The expression of the *CMT1* gene was reduced below 80% in all analyzed time points. The second gene, *CMT3*, showed a decrease in the level of mRNA (to about 40% at the 6th, to 70% at the 12th, to 50% at the 24th, and to 30% at the 48th hour). Among genes participating in DNA demethylation, only the transcript level of the *DME* gene changed, whereas the mRNA level of *ROS1* remained unchanged. The expression level of *DME* declined to 70% at the 6th hour; then showed a 1.7-fold, 1.8-fold and 2.7-fold increase at the 12th, 24th and 36th hour. The mRNA level of the *RDR2* gene was reduced to 40% at the 6th hour, but starting from the 12th hour a return to the state before incubation was observed. A decrease in the expression was also noted for the *AGO4* gene (to 60% and 70% at the 6th and 48th hour, respectively). The *AGO1* gene showed another pattern of expression, wherein initially at the 6th hour of incubation there was a 40% decrease in expression level, then a 1.6-fold and a 2.8-fold increase at the 24th and 36th hour, and a drop back to 90% compared to the control.

3.5. Phenotypic Changes in Flax after Sensitization through Non-Pathogenic Strain of F. Oxysporum

The additional step in evaluation of the sensitizing action of the *Fusarium oxysporum* non-pathogenic strain on flax was based on the phenotypic analysis of plants from in vitro cultures, which were first incubated for two days with the non-pathogenic strain of *Fusarium oxysporum* (for comparison, plants were also incubated without fungi (control plants)) and then through the subsequent days with the pathogenic strain (as previously, appropriate controls were prepared). The first pictures of plants were taken after four and six days (Figure S4).

It is visible at 4 days after inoculation that the prior plant incubation with the non-pathogenic strain reinforced the resistance to pathogen infection compared to plants which had not been pre-incubated with the non-pathogenic strain. These differences increased at later time points. It seems that the selected two-day period of sensitization was optimal for the plants. Plants that have been incubated for the first two days and a further four and six days on medium with the pathogen looked the worst.

The phenotypic analysis of flax after the sensitizing action caused by the non-pathogenic strain of *F. oxysporum* and then after incubation with pathogenic *F. oxysporum* confirmed our supposition that non-pathogenic strains could enhance the flax plants' resistance to the pathogen infection.

4. Discussion

The study of changes in the level of genome methylation mostly accompanied plant development and plant response to biotic and abiotic stressors. In the case of environmental stresses, they can affect methylation in two ways, causing hyper- or hypomethylation. For example, differences in plant methylomes have been found during maize leaf development [24] and precede transcriptional activation of genes leading to cell division and meristem growth in potatoes [25]. Also, endosperm and embryo development [26], vernalization, and fruit ripening are affected by DNA methylation [27]. For example, rice blight pathogen resistance is related to genomic hypermethylation and hypomethylation of resistance-related genes [28]. Also, Infection of tobacco with tobacco mosaic virus (TMV)revealed an increase in genomic methylation and hypomethylation of resistance-related leucine-rich-repeat (LRR)-containing loci [29]. The data thus suggest that DNA methylation affects the plant genome not only at the global level but also at very specific sites, such as individual genes [30]. The identification of differentially methylated CCGG sites within pathogenesis-related (PR) genes: *β-1,3-glucanase* and *chitinase*, the time course of their modification, and the correlation with their expression profile in plants remain largely unexplored. Here we analyzed the methylation state of all CCGG sites of three flax genes: Two *β-1,3-glucanase* (*1* and *2*) and *chitinase* genes. It was found that three CCGG sites of the *β-1,3-glucanase 1* gene (located in promoter, exon, and intron regions) and *chitinase* were changed upon infection and these were further analyzed. The tetramer methylation and expression of *β-1,3-glucanase 2* change slightly upon infection, and served for comparison. The *β-1,3-glucanase 1* and *chitinase* appeared to be strongly activated upon pathogen infection. The analysis revealed that the level of unmethylated sites CCGG ranged from 0 to 19%, CCmGG methylation occurred in 75% on average, and the CmCmGG methylation level was around 15% of all investigated CCGG sites. In contrast to the developing maize leaf where *HpaII/MspI* restriction analysis of the CCGG site revealed that almost half (48.1%) of the loci were unmethylated, full CG methylation was represented by 36.7% and CHG hemimethylation by 15.2% of all investigated CCGG sites [31].

As mentioned, two types of CCGG sites in *β-1,3-glucanases* have been detected: Those whose methylation status is unchanged and those with a highly modified cytosine state. The profile of the latter was changing upon plant growth and progression of the infection. It is in agreement with several observations that plants as sessile organisms have developed prompt response mechanisms to react to rapid environmental changes. It is interesting that the CCGG site methylation level induced by both pathogenic and non-pathogenic *Fusarium* strains is time dependent. After artificial inoculation with both *Fusarium* strains, the methylation pattern of the pathogen-responsive *β-1,3-glucanase* and *chitinase* genes undergoes rapid (6–24 h after inoculation) and dynamic changes in tetramer methylation. It was found that both strains induced a similar time-dependent profile of tetramer methylation but they differ significantly at the beginning and at the end of the profiles. Far earlier pulse demethylation of the *β-1,3-glucanase 1* gene (promoter) occurs and it is maintained longer upon non-pathogenic strain treatment than in the case of the pathogenic one. This suggests that the same CCGG site within the gene might be differentially methylated. Similar to this was the observation that in two cases, two different CCGG sites within the same gene were found to be differentially methylated. In the first one, a conserved C-terminal domain CTC-interacting domain-encoding gene (human ATAXIN2 orthologue) was hypermethylated in two exons, and in the second case a gene encoding the protein kinase PK/UbiC was differentially methylated at two consecutive CCGG sites in an exon and an intron [31].

The *chitinase* gene's tetramers located at the beginning and end of the coding sequences (exons 1, 7, 8) are also pulse demethylated. Notably, in both genes the exon peak demethylation is reached earlier upon pathogen infection. These findings agree with observations that in maize the majority of the differentially methylated CCGG sites that map to a gene lie within an exon and they are not distributed equally throughout the gene body. The highest number of differentially methylated sites was in the first 10% and the last 20% of the gene body. In addition, differential methylation of the promoter and 5′ part of the gene anticorrelated with the gene expression while differential methylation of the central and 3′ part of the gene body or sequences downstream of the gene was unrelated to the gene expression. [31].

In *Arabidopsis thaliana*, promoter-specific methylation occurs in less than 5% of genes, most of which are under tissue-specific control. A surprising result of genome-wide methylation profiling revealed that about one-third of all genes contain CG-specific genic or body methylation patterns within their transcribed regions which are highly expressed [32]. However, it was indicated that in *A. thaliana*, genes with the highest and the lowest transcription level were the least methylated, while moderately transcribed genes were the most frequently methylated [33]. The correlation between expression levels and DNA methylation levels in the gene body region was also observed in *Populus trichocarpa* where the hypermethylation of genes caused the reduced level of their expression while the hypomethylation of genes led to the increased level of their expression [34]. Due to the lack of unambiguous literature data, there are still many questions about the gene body methylation function and its correlation with gene expression.

The levels of *β-1,3-glucanase 1* and *chitinase* gene expression are also significantly altered, while in the case of *β-1,3-glucanase 2* only slight changes were noted. This is consistent with the hypothesis that pathogen infection changes the methylation level of plant genomic DNA and leads to alterations in gene transcription. The peak of demethylation might be a signal for the alteration in gene expression induced by the non-pathogenic strain. In the case of pathogen infection the expression peak lags behind the gene demethylation: Maximal transcript accumulation was detected 24–36 h after the onset of peak methylation.

The result might suggest that flax treatment with the non-pathogenic strain prepares plants for reaction after contact with pathogen. Thus, following non-pathogenic pretreatment, plants were infected with the pathogenic strain, and gene expression was analyzed. Indeed, upon infection, the highest gene expression was reached far earlier in the case of plants pretreated with the non-pathogenic strain. The interesting fact is that the level of gene induction by the pathogen is 2-fold lower in the case of pretreated plants, but even so, they are efficiently protected against infection. While plants infected with the pathogen started to die (at 48 h), the pretreated plants grew normally.

We measured the expression of genes involved in this process and found that both *AGO4* and *RDR2* are activated upon non-pathogenic *Fusarium* treatment while they are suppressed by the pathogenic strain. The gene induction and thus potent target methylation appeared later (at 48 h) than *β-1,3-glucanase* and *chitinase* gene demethylation and their expression increased. We deduced that non-pathogenic strain infection induced at first demethylation and gene up-regulation and at the end, after reaching a certain level of transcript accumulation, activated the re-methylation pathway and thus gene silencing. Temporary demethylation/remethylation serves as memorized experience and helps the plant to overcome pathogenic infection by a mechanism that is similar to immunization.

It was detected that the plant-specific methyltransferase expression profile showed significant suppression upon infection with both *Fusarium* strains. However, the pattern of DNA methylation is the result of co-operative or competing interactions of the methyltransferases and the silencing pathways which involves repressor of silencing (ROS1) and demeter (DME), which encode two closely related DNA glycosylase domain proteins [21]. In *Arabidopsis thaliana*, ROS1 and DME are required for release of transcriptional silencing of a hypermethylated transgene and activate the maternal expression of two genes silenced by methylation [35]. The expression profile of these genes in infected flax differs depending on *Fusarium* strain used and growth time. However, the ratio between the level of methyltransferase and repressor of silencing pathway gene expression promote CCGG sites' demethylation.

In summary, two contrasting effects on the levels of methylation in flax were detected upon infection of both *Fusarium* strains: Genome-wide hypermethylation and hypomethylation of two resistance-related genes, which resulted in an increase of their expression. Plants pretreated with the non-pathogenic strain memorized the hypomethylation pattern and then reacted more efficiently upon pathogenesis (Figure 5). However, the changes in methylation profile and thus the regulation of PR-related gene expression were not the only mechanisms of plant response to stressors. For example, it is known that the argonaute family proteins are core constituents of RNA interference pathways in eukaryotes and mediate gene expression. Guided by sRNAs, AGO proteins bind target sequences through base pairing and following recruitment of cofactors mediate post-transcriptional or transcriptional gene silencing. AGO proteins are also involved in epigenetic modifications of chromatin. In *Arabidopsis thaliana*, AGO4 guided by 24-nt small interfering RNAs (siRNA) recruits DNA methyltransferase for de novo DNA methylation at target loci [36]. Likewise, AGO1 with siRNAs facilitates H3K9 methylation by recruiting H3K9 methyltransferase [37].

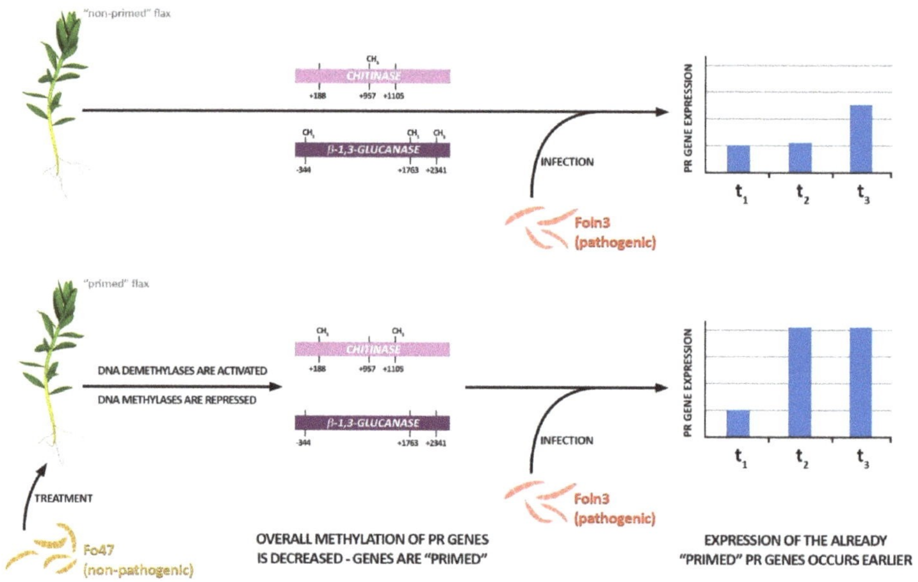

Figure 5. Hypothetical model of priming induced by non-pathogenic *Fusarium* strain. Flax treatment with non-pathogenic *Fusarium* strain (Fo47) results in changes in DNA methylation pattern of *β-1,3-glucanase* and *chitinase* genes which leads to their earlier expression during infection by pathogenic *Fusarium* strain (Foln3).

Very recently, it was reported that direct binding of AGO1, guided by a 21-nt siRNA, to the chromatin of active genes promotes their transcription. Various stimuli, including plant hormones and stresses, specifically trigger AGO1 guided by siRNAs to bind stimulus-responsive genes [38]. Here, we found strong up-regulation of *AGO1* expression in *Fusarium*-treated plants. It thus suggests that in addition to the transcriptional regulation of gene expression, microRNAs and siRNAs, epigenetics occurs as an important mechanism involved in transcriptional regulation of the plant stress response. These phenomena may contribute to the adaptation of plants to the environment and stress situations.

Supplementary Materials: The following are available online at http://www.mdpi.com/2076-2607/7/12/589/s1, Figure S1: A simplified scheme of the treatment of flax plants with the pathogenic and non-pathogenic strain of *F. oxysporum* and the treatment of flax plants with pathogenic strain after sensitizing with a non-pathogenic strain; Figure S2: Schematic diagram of the position of all CCGG sites and the relative position of the primers in *β-1,3-glucanase* 1, *β-1,3-glucanase* 2 and *chitinase* genes that were used to perform qPCR to estimate methylation levels; Figure S3: Changes in DNA methylation pattern of *β-glucanase 1* and *chitinase* genes in flax seedlings treated with non-pathogenic or pathogenic strains of *Fusarium oxysporum* and after sensitizing effects of the non-pathogenic strain treated with pathogenic Foln3 at 6, 12, 24, 36 and 48 h after inoculation. The analysis of the third position in the exon, the first in the intron and the third in the promoter of *β-1,3-glucanase 1*, and the first, seventh and eight positions in the exon of *chitinase* was performed by the digestion of genomic DNA by the restriction enzymes *HpaII* and *MspI* and then the real-time PCR reaction. C, control; N-P, non-pathogen; P, pathogen; CCGG, sequence where cytosines are not methylated; CCmGG, sequence where internal cytosine is methylated; CmCmGG, sequence where two cytosines are methylated. CCGG positions in DNA are counted from ATG. The data represent the mean from three independent experiments. Asterisks mark statistically significant differences ($p < 0.05$) between the treated samples and their own controls; Figure S4: Phenotypic changes in flax after priming of the non-pathogenic strain of *Fusarium oxysporum* treated with pathogenic Foln3. Flax plants from in vitro culture were grown for 2 days on PDA medium or on PDA medium with a non-pathogenic (Fo47) or pathogenic (Foln3), and then were transferred to control PDA medium or PDA medium with Foln3 or Fo47 for 6 days. Analyzed combinations: control to control, control to Foln3, Fo47 to Foln3, Fo47 to Fo47, Foln3 to Foln3; Table S1: Primer sequences designed for real-time PCR: (A) PR genes (*β-1,3-GLU1, β-1,3-glucanase 1; β-1,3-GLU2, β-1,3-glucanase 2; CHIT, chitinase*), genes involved in chromatin modification (*CMT1, chromomethylase 1; CMT3, chromomethylase 3; ROS1, repressor of silencing 1; DME, demeter; AGO1, argonaute 1; AGO4, argonaute 4; RdRp2, RNA-dependent RNA polymerase 2*) and *actin* gene. (B) DNA methylation. E3-G1 (the third position in the exon of *β-1,3-glucanase 1*); I1-G1 (the first position in the intron of *β-1,3-glucanase 1*); P1-G1 (the third position in the promoter of *β-1,3-glucanase 1*); E1-G2 (the first position in the exon of *β-1,3-glucanase 2*); E11-G2 (the eleventh position in the exon of *β-1,3-glucanase 2*); I1-G2 (the first position in the intron of *β-1,3-glucanase 2*); E1-CH (the first position in the exon of *chitinase*); E7-CH (the second position in the exon of *chitinase*); E8-CH (the eighth position in the exon of *chitinase*).

Author Contributions: W.W. performed most of the experiments and wrote the manuscript. A.B. performed the analysis of the expression levels of genes involved in DNA methylation. K.K. performed statistical analyses and graphics. A.K., M.P., and J.S. participated in study design, coordination, and in writing the manuscript. All of the authors read and approved of the final version of the manuscript.

Funding: This work was supported by a grant from the National Science Centre (NCN) [grant number 2013/11/N/NZ1/02378 and 2012/06/A/NZ1/00006].

Conflicts of Interest: The authors declare no conflicts of interest.

References

1. Heller, K.; Andruszewska, A.; Grabowska, L.; Wielgusz, K. Fibre flax and hemp protection in Poland and in the world. *Prog. Plant. Prot.* **2006**, *46*, 88–98.
2. Olivain, C.; Trouvelot, S.; Binet, M.N.; Cordier, C.; Pugin, A.; Alabouvette, C. Colonization of flax roots and early physiological responses of flax cells inoculated with pathogenic and nonpathogenic strains of *Fusarium oxysporum*. *Appl Environ. Microbiol.* **2003**, *69*, 5453–5462. [CrossRef] [PubMed]
3. Fuchs, J.G.; Moënne-Loccoz, Y.; Défago, G. Nonpathogenic *Fusarium oxysporum* Strain Fo47 Induces Resistance to Fusarium Wilt in Tomato. *Plant Dis.* **1997**, *81*, 492–496. [CrossRef]
4. Olivain, C.; Humbert, C.; Nahalkova, J.; Fatehi, J.; L'Haridon, F.; Alabouvette, C. Colonization of tomato root by pathogenic and nonpathogenic *Fusarium oxysporum* strains inoculated together and separately into the soil. *Appl Environ. Microbiol.* **2006**, *72*, 1523–1531. [CrossRef] [PubMed]
5. Fravel, D.; Olivain, C.; Alabouvette, C. *Fusarium oxysporum* and its biocontrol. *New Phytol.* **2003**, *157*, 493–502. [CrossRef]
6. Balasubramanian, V.; Vashisht, D.; Cletus, J.; Sakthivel, N. Plant β-1,3-glucanases: Their biological functions and transgenic expression against phytopathogenic fungi. *Biotechnol. Lett.* **2012**, *34*, 1983–1990. [CrossRef]
7. Shetty, N.P.; Jensen, J.D.; Knudsen, A.; Finnie, C.; Geshi, N.; Blennow, A.; Collinge, D.B.; Jørgensen, H.J.L. Effects of β-1,3-glucan from *Septoria tritici* on structural defence responses in wheat. *J. Exp. Bot.* **2009**, *60*, 4287–4300. [CrossRef]
8. Brunner, F.; Stintzi, A.; Fritig, B.; Legrand, M. Substrate specificities of tobacco chitinases. *Plant J.* **1998**, *14*, 225–234. [CrossRef]
9. Cheong, Y.H.; Kim, C.Y.; Chun, H.J.; Moon, B.C.; Park, H.C.; Kim, J.K.; Lee, S.-H.; Han, C.-d.; Lee, S.Y.; Cho, M.J. Molecular cloning of a soybean class III *β-1,3-glucanase* gene that is regulated both developmentally and in response to pathogen infection. *Plant Sci.* **2000**, *154*, 71–81. [CrossRef]
10. Li, W.L.; Faris, J.D.; Muthukrishnan, S.; Liu, D.J.; Chen, P.D.; Gill, B.S. Isolation and characterization of novel cDNA clones of acidic *chitinases* and *β-1,3-glucanases* from wheat spikes infected by *Fusarium graminearum*. *Theor. Appl. Genet.* **2001**, *102*, 353–362. [CrossRef]
11. Lorenc-Kukula, K.; Zuk, M.; Kulma, A.; Czemplik, M.; Kostyn, K.; Skala, J.; Starzycki, M.; Szopa, J. Engineering flax with the GT family 1 *Solanum sogarandinum* glycosyltransferase SsGT1 confers increased resistance to *Fusarium* infection. *J. Agric. Food Chem.* **2009**, *57*, 6698–6705. [CrossRef] [PubMed]
12. Wrobel-Kwiatkowska, M.; Lorenc-Kukula, K.; Starzycki, M.; Oszmianski, J.; Kepczynska, E.; Szopa, J. Expression of *beta-1,3-glucanase* in flax causes increased resistance to fungi. *Physiological and Molecular Plant. Pathology* **2004**, *65*, 245–256. [CrossRef]
13. Amian, A.A.; Papenbrock, J.; Jacobsen, H.J.; Hassan, F. Enhancing transgenic pea (*Pisum sativum L.*) resistance against fungal diseases through stacking of two antifungal genes (*chitinase* and *glucanase*). *GM Crops* **2011**, *2*, 104–109. [CrossRef]
14. Duijff, B.; Pouhair, D.; Olivain, C.; Alabouvette, C.; Lemanceau, P. Implication of Systemic Induced Resistance in the Suppression of Fusarium Wilt of Tomato by *Pseudomonas fluorescens* WCS417r and by Nonpathogenic *Fusarium oxysporum* Fo47. *European Journal of Plant. Pathology* **1998**, *104*, 903–910. [CrossRef]

15. Aime, S.; Cordier, C.; Alabouvette, C.; Olivain, C. Comparative analysis of PR gene expression in tomato inoculated with virulent *Fusarium oxysporum* f. sp. lycopersici and the biocontrol strain F. oxysporum Fo47. *Physiol. Mol. Plant. Pathol.* **2008**, *73*, 9–15. [CrossRef]
16. Shah, J.; Zeier, J. Long-Distance Communication and Signal Amplification in Systemic Acquired Resistance. *Front. Plant Sci.* **2013**, *4*, 30. [CrossRef]
17. Steven, H.S.; Xinnian, D. How do plants achieve immunity? Defence without specialized immune cells. *Nat. Rev. Immunol.* **2012**, *12*, 89–100.
18. Goll, M.G.; Bestor, T.H. Eukaryotic cytosine methyltransferases. *Annu. Rev. Biochem.* **2005**, *74*, 481–514. [CrossRef]
19. Chan, S.W.; Zilberman, D.; Xie, Z.; Johansen, L.K.; Carrington, J.C.; Jacobsen, S.E. RNA silencing genes control de novo DNA methylation. *Science* **2004**, *303*, 1336. [CrossRef]
20. Regulski, M.; Lu, Z.; Kendall, J.; Donoghue, M.T.; Reinders, J.; Llaca, V.; Deschamps, S.; Smith, A.; Levy, D.; McCombie, W.R.; et al. The maize methylome influences mRNA splice sites and reveals widespread paramutation-like switches guided by small RNA. *Genome Res.* **2013**, *23*, 1651–1662. [CrossRef]
21. Meyer, P. DNA methylation systems and targets in plants. *FEBS Lett.* **2011**, *585*, 2008–2015. [CrossRef] [PubMed]
22. Peng, H.; Zhang, J. Plant genomic DNA methylation in response to stresses: Potential applications and challenges in plant breeding. *Prog.Nat. Sci.* **2009**, *19*, 1037–1045. [CrossRef]
23. Sahu, P.P.; Pandey, G.; Sharma, N.; Puranik, S.; Muthamilarasan, M.; Prasad, M. Epigenetic mechanisms of plant stress responses and adaptation. *Plant Cell Rep.* **2013**, *32*, 1151–1159. [CrossRef] [PubMed]
24. Tolley, B.J.; Woodfield, H.; Wanchana, S.; Bruskiewich, R.; Hibberd, J.M. Light-regulated and cell-specific methylation of the maize PEPC promoter. *J. Exp. Bot* **2012**, *63*, 1381–1390. [CrossRef]
25. Law, R.D.; Suttle, J.C. Transient decreases in methylation at 5′-cCGG-3′ sequences in potato (*Solanum tuberosum* L.) meristem DNA during progression of tubers through dormancy precede the resumption of sprout growth. *Plant Mol. Biol* **2003**, *51*, 437–447. [CrossRef]
26. Choi, Y.; Gehring, M.; Johnson, L.; Hannon, M.; Harada, J.J.; Goldberg, R.B.; Jacobsen, S.E.; Fischer, R.L. DEMETER, a DNA glycosylase domain protein, is required for endosperm gene imprinting and seed viability in *Arabidopsis*. *Cell* **2002**, *110*, 33–42. [CrossRef]
27. Zhong, S.; Fei, Z.; Chen, Y.-R.; Zheng, Y.; Huang, M.; Vrebalov, J.; McQuinn, R.; Gapper, N.; Liu, B.; Xiang, J.; et al. Single-base resolution methylomes of tomato fruit development reveal epigenome modifications associated with ripening. *Nat. Biotechnol.* **2013**, *31*, 154–159. [CrossRef]
28. Sha, A.H.; Lin, X.H.; Huang, J.B.; Zhang, D.P. Analysis of DNA methylation related to rice adult plant resistance to bacterial blight based on methylation-sensitive AFLP (MSAP) analysis. *Mol. Genet. Genom.* **2005**, *273*, 484–490. [CrossRef]
29. Boyko, A.; Kathiria, P.; Zemp, F.J.; Yao, Y.; Pogribny, I.; Kovalchuk, I. Transgenerational changes in the genome stability and methylation in pathogen-infected plants: (virus-induced plant genome instability). *Nucleic Acids Res.* **2007**, *35*, 1714–1725. [CrossRef]
30. Suzuki, M.M.; Bird, A. DNA methylation landscapes: Provocative insights from epigenomics. *Nat. Rev. Genet.* **2008**, *9*. [CrossRef]
31. Candaele, J.; Demuynck, K.; Mosoti, D.; Beemster, G.T.; Inze, D.; Nelissen, H. Differential methylation during maize leaf growth targets developmentally regulated genes. *Plant Physiol.* **2014**, *164*, 1350–1364. [CrossRef] [PubMed]
32. Zhang, X.; Yazaki, J.; Sundaresan, A.; Cokus, S.; Chan, S.W.; Chen, H.; Henderson, I.R.; Shinn, P.; Pellegrini, M.; Jacobsen, S.E.; et al. Genome-wide high-resolution mapping and functional analysis of DNA methylation in *Arabidopsis*. *Cell* **2006**, *126*, 1189–1201. [CrossRef] [PubMed]
33. Zilberman, D.; Gehring, M.; Tran, R.K.; Ballinger, T.; Henikoff, S. Genome-wide analysis of *Arabidopsis thaliana* DNA methylation uncovers an interdependence between methylation and transcription. *Nat. Genet.* **2007**, *39*, 61–69. [CrossRef] [PubMed]
34. Liang, L.; Chang, Y.; Lu, J.; Wu, X.; Liu, Q.; Zhang, W.; Su, X.; Zhang, B. Global Methylomic and Transcriptomic Analyses Reveal the Broad Participation of DNA Methylation in Daily Gene Expression Regulation of *Populus trichocarpa*. *Front. Plant Sci.* **2019**, *10*, 243. [CrossRef]

35. Morales-Ruiz, T.; Ortega-Galisteo, A.P.; Ponferrada-Marin, M.I.; Martinez-Macias, M.I.; Ariza, R.R.; Roldan-Arjona, T. DEMETER and REPRESSOR OF SILENCING 1 encode 5-methylcytosine DNA glycosylases. *Proc. Nat. Acad. Sci. USA* **2006**, *103*, 6853–6858. [CrossRef]
36. Wendte, J.M.; Pikaard, C.S. The RNAs of RNA-directed DNA methylation. *Biochim. et Biophys. Acta (BBA) – Bioenerg.* **2017**, *1860*, 140–148. [CrossRef] [PubMed]
37. Martienssen, R.; Moazed, D. RNAi and heterochromatin assembly. *Cold Spring Harb. Perspect. Biol.* **2015**, *7*. [CrossRef]
38. Liu, C.; Xin, Y.; Xu, L.; Cai, Z.; Xue, Y.; Liu, Y.; Xie, D.; Liu, Y.; Qi, Y. *Arabidopsis* ARGONAUTE 1 Binds Chromatin to Promote Gene Transcription in Response to Hormones and Stresses. *Dev. Cell* **2018**, *44*, 348–361. [CrossRef] [PubMed]

microorganisms

MDPI

Article

Performance of Winter Wheat Cultivars Grown Organically and Conventionally with Focus on Fusarium Head Blight and *Fusarium* Trichothecene Toxins

Tomasz Góral [1,*], Aleksander Łukanowski [2], Elżbieta Małuszyńska [3], Kinga Stuper-Szablewska [4], Maciej Buśko [4] and Juliusz Perkowski [4]

1 Department of Plant Pathology, Plant Breeding and Acclimatization Institute—National Research Institute, Radzików, 05-870 Błonie, Poland
2 Department of Phytopathology and Molecular Mycology, Faculty of Agriculture and Biotechnology, UTP University of Science and Technology, Al. prof. S. Kaliskiego 7, bldg. I, 85-796 Bydgoszcz, Poland; luk-al@utp.edu.pl
3 Department of Seed Science and Technology, Plant Breeding and Acclimatization Institute—National Research Institute, Radzików, 05-870 Błonie, Poland; e.maluszynska@ihar.edu.pl
4 Department of Chemistry, Poznań University of Life Sciences, ul. Wojska Polskiego 75, 60-625 Poznań, Poland; kinga.stuper@up.poznan.pl (K.S.-S.); maciej.busko@up.poznan.pl (M.B.); juliusz.perkowski@up.poznan.p (J.P.)
* Correspondence: t.goral@ihar.edu.pl; Tel.: +48-22-7334-636

Received: 7 September 2019; Accepted: 10 October 2019; Published: 11 October 2019

Abstract: Growing acreage and changing consumer preferences cause increasing interest in the cereal products originating from organic farming. Lack of results of objective test, however, does not allow drawing conclusions about the effects of cultivation in the organic system and comparison to currently preferred conventional system. Field experiment was conducted in organic and conventional fields. Thirty modern cultivars of winter wheat were sown. They were characterized for disease infection including Fusarium head blight, seed sowing value, the amount of DNA of the six species of *Fusarium* fungi as well as concentration of ergosterol and trichothecenes in grain. The intensity Fusarium head blight was at a similar level in both systems. However, *Fusarium* colonization of kernels expressed as ergosterol level or DNA concentration was higher for the organic system. It did not reflect in an increased accumulation of trichothecenes in grain, which was similar in both systems, but sowing value of organically produced seeds was lower. Significant differences between analyzed cropping systems and experimental variants were found. The selection of the individual cultivars for organic growing in terms of resistance to diseases and contamination of grain with *Fusarium* toxins was possible. Effects of organic growing differ significantly from the conventional and grain obtained such way can be recommended to consumers. There are indications for use of particular cultivars bred for conventional agriculture in the case of organic farming, and the growing organic decreases plant stress resulting from intense fertilization and chemical plant protection.

Keywords: Fusarium head blight; *Fusarium* species; soil minerals; mycotoxins; organic farming; sowing value; winter wheat

---

1. Introduction

A way of growing crops is changing because of the geopolitical situation and consumer preferences. In recent years, high interest in organic farming has been observed in Europe (https://ec.europa.eu/eurostat/statistics-explained/pdfscache/5461.pdf). In 2002, organic farming took up 5.0 million hectares, while in 2017 it was 12.6 million hectares. Austria, Estonia, Sweden, Italy, Czech Republic, and Latvia were the countries with the highest share of organic farmland, while the largest areas of organic farmland were in Spain, Italy, France, and Germany. In Poland in 2016 it was 536.6 thousand ha (3.7% of all agricultural land) [1].

This is due to the awareness that in organic farming practices the use of artificial fertilizers as well as pesticides is not allowed. There is limited list of substances, which can be used as natural fungicides to protect crops against fungal diseases. Lack of fungicide protection can result in higher severity of fungal diseases. Chemical seed treatment is not applied which leads to increased incidence of seed borne diseases [2,3]. Thus, seed transmitted diseases are considered the most harmful in organic farming. Leaf diseases (not seed transmitted) and foot rots are less important. Severity of these diseases correlates with high nitrogen doses and high crop density, so under organic farming conditions they are less damaging [4]. Diseases caused by fungi surviving on crop debris (including Fusarium head blight) can be controlled by cultural practices, so they are less damaging than seed borne ones [3]. However, *Fusarium* fungi causing Fusarium head blight are able to produce toxic secondary metabolites–mycotoxins contaminating grain. The main *Fusarium* species causing Fusarium head blight are *F. culmorum*, *F. graminearum*, and *F. avenaceum* [5,6]. Cereal heads are infected mainly during the flowering period [7]. This is the stage where cereals are the most susceptible to infection with *Fusarium* fungi spores. After infection, the fungus develops in infected flower spreading then to other flowers in the spikelet. Then through rachis, the fungus spreads to another spikelet causing necrosis and bleaching individual spikelets [8,9]. The invaded cereal grain, even visually healthy looking, is contaminated with fungal mycotoxins, which are phyto- and zootoxic. *Fusarium* spp. affecting cereals are known as potent producers of type A trichothecenes (T-2 and HT-2 toxins, diacetoxyscirpenol et al.) and of type B (deoxynivalenol, nivalenol et al.) as well as moniliformin, zearalenone, enniatins, beauvericin, and the other toxins [5,6,10].

Avoiding the presence of *Fusarium* mycotoxins in food is very important, thus organic food is perceived as "food without chemistry" of higher quality than conventional [11,12]. In the literature, you can find analyses on this issue comparing wheat form organic and conventional cropping systems [13–17]. Mäder et al. [18] analyzed *Fusarium* metabolites, deoxynivalenol (DON) and nivalenol (NIV), content in wheat grain produced in a 21-year conventional and organic agrosystems. It was found higher concentration of DON in samples from conventional fields in both years of mycotoxin analysis; however, differences were not significant. NIV concentration was similar in both cropping systems. Magkos et al. [19] in their review summarized results of 12 papers on contamination of organic and conventional cereals with *Fusarium* mycotoxins. Organically grown cereals has been reported to be either more, less, or equally contaminated compared with conventional cereals. Authors concluded that this variability resulted from different cultivars, geographical locations of fields and time of harvest in different studies. It makes data not directly comparable.

In the literature, it can be found a number of analyses of effects of organic cultivation of wheat. However, experimental data that can verify the views presented there are still not very numerous. Considering this, we decided to carry out a field experiment on sowing 30 cultivars of winter wheat in the same location, at the same time on conventional and organic plots. The aim of the experiment was a comprehensive comparison of the results obtained for both cropping systems through the analysis in a series of elements that describe the structure of the yield, fungal diseases, presence of *Fusarium* fungi through analysis of the DNA content, production of mycotoxins in grain. The results were subject of the widest possible statistical analysis with the aim of finding relevant or irrelevant differences in both cultivation systems.

2. Material and Methods

2.1. Field Experiments

Thirty cultivars of winter wheat (*Triticum aestivum* L.) were evaluated (Table 1). The cultivars were listed in the Polish National List of the Research Centre for Cultivar Testing (COBORU) and were added to the list between 1998 ('Mewa') and 2009 ('Belenus'). The cultivars were described in detail in the paper of Góral et al. [20]. They differed in the pedigree, morphological characters, and resistance to Fusarium head blight (FHB). Cultivars were grouped in four classes of FHB resistance: susceptible (S), medium susceptible (MS), medium resistant (MR), and resistant (R).

Table 1. List of winter wheat cultivars used in this study.

No.	Cultivar		No.	Cultivar		No.	Cultivar	
1	Akteur	MS *	11	Jenga	MS	21	Naridana	MS
2	Alcazar	S	12	Kampana	S	22	Nateja	R
3	Anthus	MS	13	Kohelia	MR	23	Ostka Strzelecka	MS
4	Batuta	MS	14	Legenda	MR	24	Ostroga	MR
5	Belenus	MS	15	Ludwig	MS	25	Slade	MS
6	Bogatka	MR	16	Markiza	MS	26	Smuga	S
7	Boomer	MR	17	Meteor	MS	27	Sukces	MR
8	Dorota	MR	18	Mewa	MS	28	Tonacja	MR
9	Figura	MS	19	Mulan	MS	29	Türkis	MS
10	Garantus	MS	20	Muszelka	S	30	Zyta	MR

* Group of resistance to Fusarium head blight [20]; S = susceptible, MS = medium susceptible, MR = medium resistant, R = resistant.

Field experiments were established in 2014 in the experimental fields of state-owned research institute—Plant Breeding and Acclimatization Institute (IHAR-PIB) in Radzików, Central Poland. First experiment was sown in the conventional field (GPS coordinates: 52.212517, 20.634765). Pre-crop was oilseed rape. Artificial fertilizers were applied according to standard agricultural practices in IHAR-PIB in particular. In the autumn 3 dt ha^{-1} of 'Polifoska 6' fertilizer was applied (N—18 kg ha^{-1}, P—45 kg ha^{-1}, K—72 kg ha^{-1}). In the spring, after the start of vegetation ammonium nitrate fertilizer was applied in an amount providing 68 kg N ha^{-1}. Weeds and pests were controlled with herbicides and insecticides. Immediately after sowing weeds were controlled with herbicide 'Maraton 375SC' in a dose of 4 L ha^{-1}. In spring weeds were controlled using the herbicide 'Attribut 70GS' in a dose of 60 mg ha^{-1}. Cereal leaf beetle and aphids were controlled with 'Fastac Active 050ME' in a dose of 250 mL ha^{-1}. No fungicides were applied.

Simultaneously the same wheat cultivars were sown in the experimental organic field of IHAR-PIB (GPS coordinates: 52.216319, 20.638653). Wheat was grown according to organic farming practices with no chemical disease control and application of fertilizers. Pre-crop was pea. Weeds were controlled mechanically. No fertilizers or other components allowed in organic farming were applied. Distance between two experimental fields was about 500 m. Single plot size in both experiments was 5 m^2. In both fields, cultivars were sown in three randomized blocks (replications) distant from each other by 2 m.

Heading and full flowering dates for individual plots were recorded. Plant height was measured after the end of heading stage. Fusarium head blight was scored based on the mean percentage of blighted spikelets per infected head (disease severity) and the percentage of infected heads per plot (disease incidence). Fusarium head blight index (FHBi) was calculated as the combination of disease severity and disease incidence.

$$FHB_i = \left(FHB_{severity} \times FHB_{incidence}\right)/100 \quad (1)$$

Presence of other fungal diseases were also recorded. They were as follows: yellow rust (*Puccinia striiformis*), leaf rust (*P. triticina*), Septoria tritici blotch (*Zymoseptoria tritici*), Stagonospora nodorum blotch (*Parastagonospora nodorum*) and tan spot (*Pyrenophora tritici-repentis*). These diseases were scored according to percentage of leaf area per plot with symptoms of disease—necrosis and/or sporulation.

2.2. *Analysis of Mineral Elements in Soil*

In spring, soil samples were collected from conventional and organic fields. Twenty soil cores were taken from experimental plots in both fields using soil sampler. Soil cores from plots were mixed thoroughly.

The material was mineralized with a CEM Mars 5 Xpress (CEM, Matthews, NC, USA) microwave mineralization system (55 mL vessels) using 8 mL HNO_3 (65%) and 2 mL H_2O_2, according to the program comprising three stages: First stage—power 800 W, time 10 min, temperature 120 °C; second stage-power 1600 W, time 10 min, temperature 160 °C; third stage-power 1600 W, time 10 min, temperature 200 °C [21]. Materials after digestion were filtered through 45 mm filters (Qualitative Filter Papers Whatman, Grade 595: 4–7 μm; GE Healthcare, Buckinghamshire, UK), and filtrate completed with deionized water from Milli-Q Academic System (non-TOC (Total Organic Carbon); Millipores. A.S., Molsheim, France) to a final volume of 50 mL. Concentration of particular trace elements was analyzed by the flame atomic absorption spectrometry (Cd, Cu, Mn, Cr, Co, Si, Ni, and Zn), atomic emission spectrometry (Mg, Ca, Na, K, B) using an AA Duo—AA280FS/AA280Z spectrometer (Agilent Technologies, Mulgrave, Victoria, Australia), equipped with a Varian hollow-cathode lamp (HCL; Varian, Mulgrave, Victoria, Australia). Calibration curves were prepared in four replicates per each trace element concentration. Detection limit for the analyzed metals was, mg kg^{-1}: Ca 0.015, Na 0.10, K 0.09, Mg 0.003, B 0.06, Cu 0.18, Zn 0.06, Cr 0.005, Mn 0.005, Co 0.011, Si 0.12, Ni 0.005, Cd 0.01.

2.3. *Seed Quality*

For the evaluation of germination ability, 3 × 50 seeds from each experimental plot (180 samples) were sown in plastic boxes between two layers of moistened (to 60% WR) filter paper. After sowing, the samples were prechilled at 7 °C for 3 days and placed in Sanyo growth chamber (Sanyo Electric Co., Ltd., Osaka, Japan) at constant temperature 20 °C. After four days, first count (germination energy) was made. The normal seedlings were counted and share in percent was evaluated. According to present International Seed Testing Association Rules [22] after eight days, the final count (germination capacity) and evaluation of normal seedlings, abnormal seedlings (AS), dead seeds (DS), and fresh ungerminated (FUS) seeds were made.

2.4. *Fusarium DNA Quantification with Real-Time PCR*

2.4.1. Isolation of Total DNA from Grain

DNA was extracted according to Doyle and Doyle [23] protocol with modifications of Department of Phytopathology and Molecular Mycology UTP.

Ten grams of grain was homogenized to fine powder and 100 mg of such prepared sample was taken for DNA isolation. Samples were transferred into 2.0 mL tubes and poured with 600 μL of the extraction buffer containing CTAB 5.0%, EDTA 0.5 M, NaCl 5.0 M, Tris-HCl (pH 8.0) 1.0 M, β-mercaptoethanol, PVP (2.0%), and water. DNA was purified by addition of phenol:chloroform: isoamyl alcohol mixture (25:24:1) followed by centrifugation and taking the upper phase (supernatant) to the new tube, where equal volume of chloroform:isoamyl alcohol mixture (24:1) was added, mixed by inverting and centrifuged. Supernatant was taken and DNA was precipitated with cold ethanol. DNA pellet was washed with 70% cold ethanol, left to dry for 25–30 min and poured with TE buffer or sterile water to dissolve. Samples were stored at −20 °C for further analyses.

2.4.2. Preparation of Standard Curve

Material for preparation of standard curve was a series of 10-fold dilutions of DNA isolated from pure culture of researched six *Fusarium* species (*F. avenaceum*, *F. culmorum*, *F. graminearum*, *F. langsethiae*, *F. poae* and *F. sporotrichioides*). Pure fungal cultures were grown on PDA medium (Difco, Becton, Dickinson and Company Sparks, MD, USA) on Petri dish and DNA was isolated from scraped and lyophilized mycelium using the same protocol as for grain.

2.4.3. Preparation of DNA Samples for Real-Time PCR

Concentrations of DNA obtained from kernels were measured with Quantus fluorometer (Promega, Madison, WI, USA). All samples were diluted in sterile deionized water to 10 ng·µL^{-1}. Final concentration of *Fusarium* DNA in a sample was expressed in picograms per 100 ng of total DNA.

2.4.4. Real-Time PCR Reaction Conditions

Amplification was performed with LightCycler 480II (Roche, Basel, Switzerland) using LightCycler 480 SYBR Green I Master (Roche, Basel, Switzerland) in a volume of 10 mL per sample (5.5 µL premix + 4.5 µL DNA) in 45 cycles according to thermal profiles specific to each *Fusarium* species. The primers used for each researched *Fusarium* species were shown in Table 2.

Table 2. Sequences and names of *Fusarium* species specific primers

Fusarium Species	Primer Name	Sequence (5′–3′)	Source
F. avenaceum	JIAf	GCTAATTCTTAACTTACTAGGGGCC	[24]
	JIAr	CTGTAATAGGTTATTTACATGGGCG	
F. culmorum	Fc01F	ATGGTGAACTCGTCGTGGC	[25]
	Fc01R	CCCTTCTTACGCCAATCTCG	
F. graminearum	Fg16F	CTCCGGATATGTTGCGTCAA	[25]
	Fg16R	GGTAGGTATCCGACATGGCAA	
F. langsethiae	FlangF3	CAAAGTTCAGGGCGAAAACT	[26]
	LanspoR1	TACAAGAAGACGTGGCGATAT	
F. poae:	Fp82F	CAAGCAAACAGGCTCTTCACC	[27]
	Fp82R	TGTTCCACCTCAGTGACAGGTT	
F. sporotrichioides	FsporF1	CGCACAACGCAAACTCATC	[26]
	LanspoR1	TACAAGAAGACGTGGCGATAT	

2.5. *Analysis of Trichothecenes*

Grain samples (60) were analyzed for the presence of trichothecenes according to Perkowski et al. [28]. Subsamples (10 g) were extracted with acetonitrile:water (82:18) and purified on a charcoal column (Celite 545/charcoal Draco G/60/activated alumina neutral 4:3:4 (*w*/*w*/*w*).

Type A trichothecenes (HT-2 toxin (HT-2), T-2 toxin (T-2), T-2 tetraol, T-2 triol, diacetoxyscirpenol (DAS), scirpentriol (STO)) were analyzed as TFAA (trifluoroacetic anhydride) derivatives. To the dried sample, 100 µL of trifluoroacetic acid anhydride were added. After 20 min, the reacting substance was evaporated to dryness under nitrogen. The residue was dissolved in 500 µL of isooctane and 1 µL was injected onto a gas chromatograph-mass spectrometer (GC/MS, Hewlett Packard GC 6890, Waldbronn, Germany). Type B trichothecenes (DON, NIV, 3-acetyldeoxynivalenol (3-AcDON), 15-acetyldeoxynivalenol (15-AcDON), fusarenon X (FUS-X)) were analyzed as TMS (trimethylsilylsilyl ethers) derivatives. To the dried extract, the amount of 100 µL of TMSI/TMCS (trimethylsilyl imidazole/trimethylchlorosilane; *v*/*v* 100/1) mixture was added. After 10 min 500 µL of isooctane were added and the reaction was quenched with 1 mL of water. The isooctane layer was used for the analysis and 1 µL of the sample was injected on a GC/MS system.

The analyses were run on a gas chromatograph (Hewlett Packard GC 6890, Waldbronn, Germany) hyphenated to a mass spectrometer (Hewlett Packard 5972 A, Waldbronn, Germany), using an HP-5MS,

0.25 mm × 30 m capillary column. The injection port temperature was 280 °C, the transfer line temperature was 280 °C and the analyses were performed with programmed temperature, separately for type A and type B trichothecenes. The type A trichothecenes were analyzed using the following programmed temperatures: Initial 80 °C held for 1 min, from 80 °C to 280 °C at 10 °C min^{-1}, the final temperature being maintained for 4 min. For the type B trichothecenes initial temperature of 80 °C was held for 1 min, from 80 °C to 200 °C at 15 °C min^{-1} held for 6 min and from 200 °C to 280 °C at 10 °C min^{-1}, with the final temperature being maintained for 3 min. The helium flow rate was held constant at 0.7 mL min^{-1}.

Quantitative analysis was performed in the single ion monitored mode (SIM) using the following ions for the detection of STO: 456 and 555; T-2 tetraol 455 and 568; T-2 triol 455 and 569 and 374; HT-2 455 and 327; T-2 327 and 401. DON: 103 and 512; 3-AcDON: 117 and 482; 15-AcDON: 193 and 482; NIV: 191 and 600. Qualitative analysis was performed in the SCAN mode (100–700 amu). Recovery rates for the analyzed toxins were as follows: STO 82 ± 5.3%; T-2 triol 79 ± 5.1%; T-2 86 ± 3.8%; T-2 tetraol 88 ± 4.0%; HT-2 91 ± 3.3%; DON 84 ± 3.8%; 3AcDON 78 ± 4.8%; 15 AcDON 74 ± 2.2%; and NIV 81 ± 3.8%. The limit of detection was 0.01 μg kg^{-1}.

2.6. Chemical Analysis of Ergosterol

Ergosterol (ERG) in 60 grain samples was determined by HPLC as described by Young [29] with modifications [30,31]. A detailed evaluation of the method was given by Perkowski et al. [31]. Samples containing 100 mg of ground grains were placed into 17-mL culture tubes, suspended in 2 mL of methanol, treated with 0.5 mL of 2 M aqueous sodium hydroxide and tightly sealed. The culture tubes were then placed within 250-mL plastic bottles, tightly sealed and placed inside a microwave oven (Model AVM 401/1WH, Whirlpool, Sweden) operating at 2450 MHz and 900 W maximum output. Samples were irradiated (370 W) for 20 s and after about 5 min for an additional 20 s. After 15 min the contents of culture tubes were neutralized with 1 M aqueous hydrochloric acid, 2 mL MeOH were added and extraction with pentane (3 × 4 mL) was carried out within the culture tubes. The combined pentane extracts were evaporated to dryness in a nitrogen stream. Before analysis samples were dissolved in 4 mL of MeOH, filtered through 13-mm syringe filters with a 0.5 mm pore diameter (Fluoropore Membrane Filters, Millipore, Ireland) and evaporated to dryness in a N_2 stream. The sample extract was dissolved in 1ml of MeOH and 50 μL were analyzed by HPLC. Separation was performed on a 150 × 3.9 mm Nova Pak C-18, 4 mm column and eluted with methanol/acetonitrile (90:10) at a flow rate of 0.6 mL min^{-1}. Ergosterol was detected with a Waters 486 Tunable Absorbance Detector (Milford, MA, USA) set at 282 nm. The presence of ERG was confirmed by a comparison of retention times and by co-injection of every tenth sample with an ergosterol standard.

2.7. Statistics

The statistical analysis was performed using Microsoft® Excel 2016/XLSTAT© Ecology (Version 18.0738413, Addinsoft, Inc., Brooklyn, NY, USA). Differences between variable means for the two experimental variants were compared using parametric Student't t-test (XLSTAT procedure: Two-sample t and z tests). Variables distribution in samples from the two experimental variants were compared using the Kruskal–Wallis one-way analysis of variance (XLSTAT procedure: Comparison of k samples—Kruskal–Wallis, Friedman). The Kruskal–Wallis test was selected because some of the variables did not follow normal distribution.

The relationships between FHBi, seed quality and concentration of ergosterol, mycotoxins and *Fusarium* DNA were investigated by Pearson correlation tests (XLSTAT procedure: *Correlation tests*). Prior to analysis, data that did not follow normal distribution was log10 transformed to normalize residual distributions. Multivariate data analysis method was applied to the data on FHB (FHBi, DS, mycotoxin concentrations, *Fusarium* DNA concentrations) resistance. Principal component analysis (XLSTAT procedure: Principal Component Analysis PCA) was used to show how wheat cultivars within two experimental variants (60 observations) are distributed with respect to the main variation described in the first two components and how variables (FHBi, DS, ERG, sum of type A trichothecenes, sum of type B trichothecenes, *Fusarium* DNA) influence the construction of the two components. PCA results also revealed associations among variables measured by the angle between variable vectors.

Differences between two variants for all variables were analyzed using multidimensional tests (XLSTAT procedure: Multidimensional tests (Mahalanobis)) and multivariate analysis of variance (XLSTAT procedure: *MANOVA*).

Cultivars in the organic field were grouped according to their resistance to infection of heads with *Fusarium* fungi measured by FHB index, dead seeds proportion, ERG, sum of type A trichothecenes, sum of type B trichothecenes, *Fusarium* DNA. K-means clustering procedure of XLSTAT was applied. Results were visualized using Discriminant analysis procedure of XLSTAT. Classes obtained from K-means analysis were applied as a qualitative depended variable in DA analysis.

3. Results

3.1. Concentration of Mineral Elements in Soil

In order to determine soil conditions in both experimental fields analysis of a number of elements that occur in these environments was made (Table S1). For the most of 13 analyzed compounds significant differences between organic and conventional fields were found. Only for Co, concentration difference was not significant. In soil of conventional field, concentration of K, Mg, Cd, Cr, Cu, Ni, and Zn was higher than in soil of organic field. The highest differences were found for Zn (7-fold) and Cd (3-fold). On the other hand, in soil of organic field, concentration of Ca, Na, Si, B, and Mn was higher than in soil of conventional field.

3.2. Phenotypic Data and Fungal Diseases

Wheat cultivars differed in heading and flowering time. In the conventional field, heading time was 29.7 days from 1 May, at a range 24.0 ('Smuga')—35.0 days ('Sukces') (Table S2). Flowering time was on average 31.6 days from 1 May, at a range 26.0 ('Smuga')—35.0 days ('Sukces', 'Boomer'). In the organic field, heading time was 28.0 days from 1 May, at a range 24.0 ('Smuga')—33.0 days ('Sukces'). Flowering time was on average 29.8 days from 1 May, at a range 26.0 ('Smuga', 'Ludwig')—35.0 days ('Sukces'). Heading and flowering time were significantly earlier in organic field than in conventional one (Table 3).

Table 3. Phenotypic characters, grain yield and Fusarium head blight (FHB) infection of 30 wheat cultivars grown in conventional and organic field.

Variant	Heading (Days from 1st May)	Flowering (Days from 1st May)	Plant Height (cm)	Grain Yield Per Plot (kg)	FHBi (%)
		Conventional			
Mean	29.7 b	31.6 b	97.8 a	5.0 a	0.74 a
Std. deviation	2.53	2.39	12.22	0.89	1.00
		Organic			
Mean	28.0 a	29.8 a	99.0 a	5.1 a	0.66 a
Std. deviation	2.39	2.55	10.39	0.76	0.77

Values within the same column followed by the different letters are significantly different at the level of probability < 0.01.

On average, plant height of wheat cultivars did not differ between organic and conventional fields (Table 3). In organic field plant height ranged between 73.7 cm ('Muszelka') and 114.3 cm ('Akteur') (Table S2). In conventional field, this parameter ranged between 76.3 cm ('Alcazar') and 118.7 cm ('Ludwig').

Fusarium head blight severity was low and average values for conventional and organic fields did not differ significantly (Table 3, Figure 1). In conventional field, FHB index range was from 0 to 4.4%.

Cultivars 'Nateja' and 'Legenda' showed no symptoms of FHB and cultivars 'Kampana', 'Muszelka' and 'Slade' were the most infected (FHBi = 4.4%, 3.5%, and 2.1%, respectively) (Table S2). In organic field, FHB index range was from 0 to 3.2%. Cultivars 'Nateja' and 'Mewa' showed no symptoms of FHB and cultivars 'Slade', 'Kampana', 'Turkis' and 'Belenus' were the most infected (FHBi = 3.2%, 2.3%, 1.8%, and 1.8%, respectively). FHB indexes for conventional and organic fields correlated significantly ($r = 0.776$ at $p < 0.001$).

Figure 1. FHB symptoms on heads of wheat grown in organic field. Clockwise from top left: 'Figura', 'Muszelka', 'Kampna', 'Slade'. Wheat at early milk growth stage (BBCH 73).

Heading and flowering time were not correlated with FHB severity. In both fields plant height significantly negatively correlated with FHB indexes ($r = -0.519$ for organic field and $r = -0.589$ for conventional field; at $p < 0.001$).

Symptoms of leaf diseases in both experimental plots were observed starting from half of May, when yellow rust was detected. Seventeen cultivars were fully resistant to yellow rust and showed no symptoms of disease (Table S3). On average yellow rust severity was slightly higher in conventional field; however, difference with organic field was not significant. Leaf rust severity was low. On average, it was 0.7% in organic field and 1.0% in conventional one. (Table S3). Symptoms of Septoria tritici blotch were observed on most (28) cultivars in conventional field and on 17 cultivars in organic field. Symptoms of Stagonospora nodorum blotch were observed on 12 cultivars in conventional field and on 23 cultivars in organic field. Tan spot was observed only in organic field with average severity 2.0%. This disease affected fourteen cultivars.

Grain yield per plot was similar for both field and do not differ statistically significantly (Table 3). In organic field grain yield ranged between 3.1 kg ('Nateja') and 6.7 kg ('Jenga') (Table S1). In conventional field, this parameter ranged between 2.9 kg ('Nateja') and 6.6 kg ('Boomer'). In both field grain yield was significantly negatively correlated with yellow rust severity ($r = -0.517$ for organic field and $r = 0.647$ for conventional field; at $p < 0.001$) and not correlated with FHB indexes.

3.3. *Characteristic of Seed Germination*

Sowing quality of seeds from conventional field was significantly higher than those from organic one were (Table 4). The mean value for the germination energy for the conventional seeds was much higher (87%) than for the organic seeds (63.2%). Similar was found for the germination capacity. In organic seed material lower percent's share of normal seedlings, but higher number of abnormal seedlings, dead seeds and fresh ungerminated seeds was observed.

Table 4. Germination characteristic of 30 wheat cultivars grown in conventional and organic field.

Variant	Germination Energy (%)	Germination Capacity (%)	Abnormal Seedlings (%)	Dead Seeds (%)	Fresh, Ungerminated Seeds (%)
			Conventional		
Mean	87.0 b ***	93.4 b ***	3.6 a **	2.5 a **	0.6 a
Std. deviation	10.86	3.81	2.26	1.79	0.63
			Organic		
Mean	63.2 a ***	89.3 a ***	5.3 b **	4.2 b **	1.2 a
Std. deviation	24.00	4.81	2.10	3.01	1.91

Values within the same column followed by the different letters are significantly different at the level of probability *** $p < 0.001$ or ** $p < 0.01$.

Values of the germination energy ranged from 54.0% ('Mewa') to 98.0% ('Nateja') in conventional samples and from 11.5% ('Belenus') to 93.0% ('Mewa') in organic samples (Table S4). The germination capacity ranged from 83.5% ('Slade') to 98.5% ('Nateja') in conventional material and from 75.0% ('Belenus') to 96.0% ('Batuta') in organic material. The percent shares of abnormal seedlings as well as dead seeds were significantly higher in organic samples (Table 4). These variables ranged from 0.5% ('Markiza') to 8.5% ('Alcazar', 'Garantus') and from 0.5% ('Batuta') to 15.0% ('Belenus') in organic field and 0 ('Figura', 'Belenus') to 7.5% ('Mewa', 'Ostroga', 'Slade') and from 0 ('Batuta', 'Nateja') to 7.5% ('Jenga') in conventional field. Additionally, percentage of fresh, ungerminated seeds was twice higher in organic material than in conventional. However, difference was not significant. It was the highest in organic seed material of cultivars 'Belenus' (7.5%), 'Akteur' (6.0%), and 'Ostroga' (5.0%).

3.4. *Concentration of Ergosterol and Trichothecenes*

Concentration of ergosterol in grain was significantly higher in samples from organic field than from conventional one (Table 5). Level of ERG varied from 0.26 ('Ostka Strzelecka') to 1.85 mg kg^{-1} ('Mulan') in conventional field and from 0.26 ('Boomer') to 3.46 mg kg^{-1} ('Akteur') in organic field (Table S4).

Amount of type B trichothecenes was low and varied from 8.9 to 460.2 µg kg^{-1} in conventional field and from 10.1 to 384.5 µg kg^{-1} in organic field (Table 5). On average, more trichothecenes were present in grain from conventional field; however, difference was statistically insignificant. Regarding specific toxins, only concentration of 3-AcDON was significantly higher in conventional samples. Concentration of NIV was higher in samples from organic field; however, difference was not significant. Distributions for FUS-X and 3-AcDON in organic and conventional samples were significantly different. In conventional samples, these toxins were detected in higher amounts in single samples whereas they were more evenly distributed in organic samples.

Table 5. Concentrations of ergosterol (mg kg^{-1}) and type B trichothecenes (μg kg^{-1}) in grain of 30 wheat cultivars grown in conventional and organic fields.

Variant	ERG	DON	FUS-X	3-AcDON	15-AcDON	NIV	Total TCT B
				Conventional			
Mean	0.74 a **	84.8 a	0.9 a	7.3 b ***	1.5 a	5.6 a	100.0 a
Range	0.26–1.85	5.8–444.4	0–11.6	2.3–30.3	0–14.3	0–19.0	8.9–460.2
Std. deviation	0.39	97.3	2.4	5.9	2.6	5.0	101.4
				Organic			
Mean	1.42 b **	63.7 a	0.9 a	3.1 a ***	1.1 a	7.4 a	76.2 a
Range	0.26–3.46	2.2–348.4	0–2.9	0–6.2	0–3.3	0–29.5	10.1–384.5
Std. deviation	0.87	86.2	1.0	1.5	1.2	7.3	93.6

Values within the same column followed by the different letters are significantly different at the level of probability *** $p < 0.001$ or ** $p < 0.01$. ERG—ergosterol; DON—deoxynivalenol; FUS-X—fusarenon X; 3-AcDON—3-acetyldeoxynivalenol; 15-AcDON—15-acetyldeoxynivalenol; NIV—nivalenol; Total TCT B— total type B trichothecenes concentrations.

The highest concentrations of type B trichothecenes were found in grain of cultivars 'Anthus', 'Ostroga', and 'Garantus' in conventional field (460.2 μg kg^{-1}, 321.1 μg kg^{-1}, 308.9 μg kg^{-1}, respectively) and in grain of 'Alcazar', 'Kampana', 'Muszelka' and 'Anthus' (384.5 μg kg^{-1}, 278. μg kg^{-1}, 257.6 μg kg^{-1}, 244.6 μg kg^{-1}, respectively) in organic field (Table S5).

Amount of type A trichothecenes was very low and similar in conventional and organic samples (5.5 and 5.1 μg kg^{-1}, respectively) (Table S6). It varied from 0.7 ('Boomer') to 30.5 μg kg^{-1} ('Garantus') in conventional samples and from 1.0 ('Figura') to 14.5 μg kg^{-1} ('Zyta') in organic samples. Average concentrations of type A trichothecenes were similar in both groups, and they not differ significantly. Only average concentration of DAS was significantly higher in conventional samples.

3.5. *Fusarium* Species

Presence of biomass of six *Fusarium* species was detected in wheat grain. *Fusarium langsethiae* was detected only in six samples in trace amounts. On average, the highest amount of DNA was found as follows for *F. poae*, *F. graminearum*, *F. sporotrichioides*, *F. culmorum*, and *F. avenaceum* (Table 6). It was true in organic field. In conventional field concentration of *F. culmorum* DNA was higher than *F. graminearum* and *F. sporotrichioides* DNA.

Table 6. Concentration of DNA (pg 100 ng^{-1} of wheat DNA) of five *Fusarium* species in grain of 30 wheat cultivars grown in conventional and organic fields.

Variant	*F. a.* DNA	*F. c.* DNA	*F. g.* DNA	*F. p.* DNA	*F. sp.* DNA	*Fusarium* DNA
			Conventional			
Mean	10.8 a ***	23.3 a	22.7 a ***	34.7 a ***	15,1 a *	106.6 a ***
Range	0–106.7	0–346.2	0.7–79.5	9.4–132.4	0–113.3	15.4–405.2
Std. deviation	19.61	63.14	21.58	24.82	29.10	90.91
			Organic			
Mean	30.2 b ***	41.1 a	67.0 b ***	98.2 b ***	50.5 b *	285.7 b ***
Range	0.2–184.6	0–415.5	0.5–280.0	14.1–222.0	0–350.0	15.3–1205.8
Std. deviation	44.88	85.73	62.92	56.44	94.3	244.83

Values within the same column followed by the different letters are significantly different at the level of probability *** $p < 0.001$ or * $p < 0.05$. *F. a.*—*F. avenaceum*, *F. c.*—*F. culmorum*, *F. g.*—*F. graminearum*, *F. p.*—*F. poae*, *F. sp.*—*F. sporotrichioides*, *Fusarium* DNA—total DNA of five species.

Total *Fusarium* DNA concentration in organic samples was more than twice higher than in conventional samples. It ranged from 15.4 ('Batuta') to 405.2 pg 100 ng^{-1} ('Figura') in conventional samples and from 15.3 ('Batuta') to 1205.8 pg 100 ng^{-1} ('Slade') in organic samples. The difference was statistically significant. Similarly, significantly higher (about three times) were concentrations of *F. poae*, *F. graminearum*, *F. avenaceum*, and *F. sporotrichioides* in organic samples. Amount of *F. culmorum* was about twice higher in organic samples; however, difference was not statistically significant. The most *Fusarium* colonized was grain of 'Figura', 'Kampana', and 'Alcazar' cultivars in conventional field and 'Muszelka', 'Kampana', 'Turkis', 'Ostroga', 'Meteor', 'Bogatka', 'Alcazar', and 'Slade' in organic field (Table S7).

3.6. Correlation Between Experimental Components

Result's correlations of the evaluation of conventional wheat material showed that the proportion of the dead seed was poorly correlated with the germination energy, but highly negatively correlated with germination capacity (Table 7). In the case of organic seed, the dependence was the same, except that the negative values of the correlation coefficients were higher (Table 8). The proportion of abnormal seedlings was significantly negatively correlated with the energy and germination capacity, as well as the number of dead seeds in conventional material. In contrast to the results of organic material, where the relevant dependencies for these traits were not found. However, organic material has demonstrated a highly significant negative relationship between the proportion of fresh ungerminated seeds and the energy and germination capacity and a positive correlation between fresh ungerminated and dead seed.

In conventional and ecological material, the same negative relationship between FHBi and germination was found, and the positive relationship between FHBi and the share of dead seeds occurred. These two parameters in conventional samples correlated also significantly with concentration of type B trichothecenes in grain.

We observed significant effect of colonization of kernels with *Fusarium* species on seed quality (Tables 7 and 8). The proportion of dead seed in conventional material was highly correlated with the quantity of the DNA of *F. graminearum*, while in organic material with the amount of DNA of *F. poae*, *F. sporotrichioides*, and total *Fusarium* DNA.

There was lack of correlation between Fusarium head blight index and amount of ERG and type A and B trichothecenes in grain in both variants (Tables 7 and 8). However, in conventional samples positive tendency FHBi versus type B trichothecenes was observed and the same was found for FHBi versus type A trichothecenes in organic samples. Fusarium head blight index correlated significantly with concentration of DNA of three *Fusarium* species—*F. avenaceum*, *F. graminearum*, and *F. poae* in both variants. In organic variant FHBi correlated significantly also with *F. sporotrichioides* DNA concentration. No correlation was found with *F. culmorum* DNA. Ergosterol content in grain did not correlated with type A or B trichothecenes as well as with DNA concentration of *Fusarium* species.

In samples from conventional field amount of type B trichothecenes correlated highly significantly with *F. graminearum* DNA but not with *F. culmorum* DNA. Contradictory, in organic samples *F. graminearum* did not correlate with type B trichothecenes and for *F. culmorum* there was found positive tendency however not statistically significant. Regarding specific toxins in conventional samples *F. graminearum* correlated significantly with DON amount ($r = 0.531$) and for *F. culmorum* some positive tendency was observed for FUS-X and 15-AcDON. In organic samples, only correlation of *F. culmorum* with 3-AcDON ($r = 0.421$) was found. There was no significant correlation between amount of type A trichothecenes and potentially producing species *F. sporotrichioides* and *F. poae*.

Amounts of DNA of three *Fusarium* species (*F. avenaceum*, *F. graminearum* and *F. poae*) in grain form both variants correlated statistically significantly (Tables 7 and 8). *Fusarium sporotrichioides* DNA concentration correlated with *F. avenaceum* and *F. poae DNA*. *Fusarium culmorum* DNA concentration did not correlated with the other species.

Table 7. Coefficients of correlations between seed quality parameters (germination energy—GE, germination capacity—GC, dead seeds, abnormal seedling—AS, fresh ungerminated seeds—FUS), *Fusarium* head blight index (FHBi), ergosterol (ERG) and type A and B trichothecenes concentrations (Total TCT A, Total TCT B) as well as amount of DNA of five *Fusarium* species in grain samples from conventional field.

Variables	GE	GC	Dead Seeds	AS	FUS	FHBi	ERG	Total TCT B	Total TCT A	*F. a.* DNA	*F. c.* DNA	*F. g.* DNA	*F. p.* DNA	*F. sp.* DNA
Final count	0.584 ***													
Dead seeds	−0.419 *	−0.803 ***												
AS	−0.553 **	−0.777 ***	0.436 *											
FUS	−0.112	−0.320	0.048	0.012										
FHBi	−0.125	−0.476 **	0.545 **	0.219	0.058									
ERG	−0.022	0.222	−0.111	−0.190	−0.162	−0.064								
Total TCT B	−0.019	−0.431 *	0.369 *	0.312	0.160	0.290	−0.163							
Total TCT A	−0.216	−0.252	0.185	0.373 *	0.044	0.096	0.011	0.573 ***						
F. a DNA	0.258	−0.221	0.412 *	0.095	−0.170	0.430 *	0.016	–						
F. c DNA	−0.052	−0.215	0.168	0.002	0.289	0.011	−0.325	0.174	–	−0.013				
F. g DNA	−0.204	−0.502 **	0.661 ***	0.288	0.010	0.586 ***	−0.104	0.501 **	–	0.583 ***	0.347			
F. p DNA	0.259	−0.169	0.393 *	−0.020	0.010	0.388 *	−0.285	0.202	0.108	0.531 **	0.053	0.497 **		
F. sp DNA	0.181	0.016	0.197	0.125	−0.564 ***	0.193	−0.270	–	0.162	0.451 *	−0.241	0.215	0.447 *	
Total DNA	0.124	−0.297	0.565	0.040	−0.107	0.481 **	−0.267	0.201	0.063	0.657 ***	0.343	0.712 ***	0.762 ***	0.515 **

F.a.—*F. avenaceum*, *F. c.*—*F. culmorum*, *F. g.*—*F. graminearum*, *F. p.*—*F. poae*, *F. sp.*—*F. sporotrichioides*; coefficients significant at $p \leq 0.001$ –***; 0.01 – **; 0.05 *.

Table 8. Coefficients of correlations between seed quality parameters (germination energy—GE, germination capacity—GC, dead seeds, abnormal seedling—AS, fresh ungerminated seeds—FUS) and *Fusarium* head blight index (FHBi), ergosterol (ERG) and type A and B trichothecenes concentrations (Total TCT A, Total TCT B) as well as amount of DNA of five *Fusarium* species in grain samples from organic field.

Variables	GE	GC	Dead Seeds	AS	FUS	FHBi	ERG	Total TCT B	Total TCT B	*F. a.* DNA	*F. c.* DNA	*F. g.* DNA	*F. p.* DNA	*F. sp.* DNA
Final count	0.639 ***													
Dead seeds	−0.511 **	−0.874 ***												
AS	−0.059	−0.327	0.203											
FUS	−0.538 **	−0.579 ***	0.470 **	−0.346										
FHBi	−0.274	−0.533 **	0.456 *	0.360	0.002									
ERG	−0.083	−0.056	0.080	0.257	−0.152	0.036								
Total TCT B	−0.104	−0.219	0.207	0.309	−0.158	0.173	0.182							
Total TCT A	0.097	0.027	0.158	0.056	−0.225	0.315	−0.072	0.162						
F. a DNA	−0.023	−0.479 **	0.557 **	0.357	0.117	0.511 **	0.192	–	–					
F. c DNA	−0.052	0.068	0.006	−0.202	−0.111	0.077	0.165	0.235	–	−0.210				
F. g DNA	−0.127	−0.379 *	0.459 *	0.370 *	−0.003	0.461 **	−0.031	−0.006	–	0.506 **	−0.039			
F. p DNA	−0.382 *	−0.577 ***	0.653 ***	0.220	0.282	0.508 **	−0.055	0.208	0.163	0.550 **	0.006	0.546 **		
F. sp DNA	−0.338	−0.590 ***	0.593 ***	0.302	0.264	0.584 ***	0.098	–	0.106	0.607 ***	−0.085	0.323	0.530 **	
Total DNA	−0.201	−0.475 **	0.557 ***	0.323	0.089	0.636 ***	0.004	0.216	0.243	0.640 ***	0.214	0.760 ***	0.838 ***	0.623 ***

F. a.—*F. avenaceum*, *F. c.*—*F. culmorum*, *F. g.*—*F. graminearum*, *F. p.*—*F. poae*, *F. sp.*—*F. sporotrichioides*; coefficients significant at $P \leq 0.001$ -***; 0.01—**; 0.05 *.

3.7. Multivariate Principal Component Analysis

Multivariate principal component analysis showed significant difference between two studied populations (wheat cultivars in two environments) in terms of FHB infection (Figure1). However, this difference was caused by only some cultivars, which showed higher *Fusarium* infection (measured with different parameters) in organic or conventional field. Cultivars from organic field had higher FHB index, proportion of dead seeds and *Fusarium* DNA content. In conventional field, the most infected cultivars had higher toxin content in the grain but moderate FHB index, dead seeds proportion and *Fusarium* biomass amount in kernels. The exception was cultivar 'Kampana' (C_12) (Figure 2).

There were also carried out other tests—Multidimensional Wilks' Lambda test and Fisher distances test. They pointed to the significance of the separation between the analyzed growing systems at the significance level of $p < 0.0001$.

There was also compared which source of variation had higher effect on the obtained results (i.e., FHBi, DS, Total TCT B, Total TCT A, and *Fusarium* DNA concentration) using multivariate analysis of variance (MANOVA). Both sources statistically significantly affected the results; however, experimental variant (conventional vs organic field) had much higher significance ($p < 0.0001$). It means that Fusarium head blight infection and its effect on grain quality, toxins concentration and *Fusarium* biomass in kernels depended mainly on wheat growing environment. Resistance of cultivars to FHB was less important ($p < 0.025$).

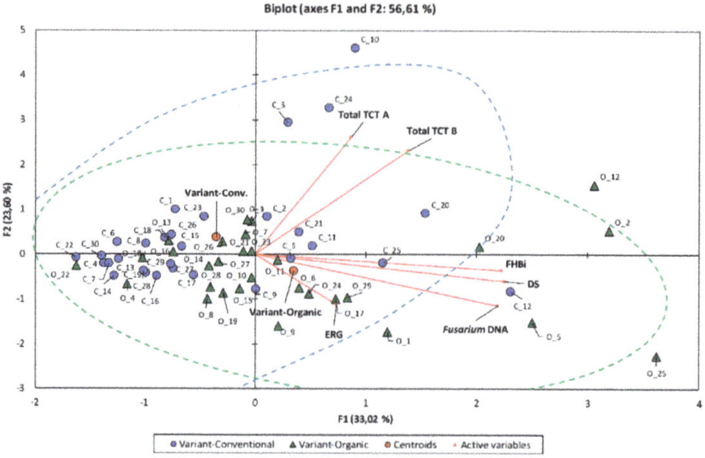

Figure 2. Biplot of the principal component analysis for 30 winter wheat cultivars grown in conventional (C) and organic field (O). Two first components explained 55.90% of variability of Fusarium head blight index (FHBi), dead seeds proportion (DS), ergosterol (ERG) and type A (Total TCT A) and type B (Total TCT B) trichothecenes content, and concentration of DNA (*Fusarium* DNA) of five *Fusarium* species in grain. Samples from conventional field marked with circles and from organic field with triangles.

Cultivars grown in organic field were compared for their overall performance under such conditions with respect to resistance to *Fusarium* infection. Multivariable analysis (K-means, discriminant analysis) made it possible to divide cultivars into three groups depending on their resistance to head infection, number of dead seeds, accumulation of ergosterol and *Fusarium* toxins in the grain as well as contamination of grain with *Fusarium* fungi (Figure 3, Table 8).

The most infected five cultivars were in the second group (Figure 3, Table 9). They could be described by the highest FHB index, high number of dead seeds, high accumulation of *Fusarium* toxins and the highest concentration of *Fusarium* biomass in kernels. Only amount of ERG was medium in grains of the cultivars of the group 2. The other 25 cultivars were in two close groups 1 and 3. They mainly differed in amount of ERG in grain, which was the highest in the group 1 while the lowest in the group 3.

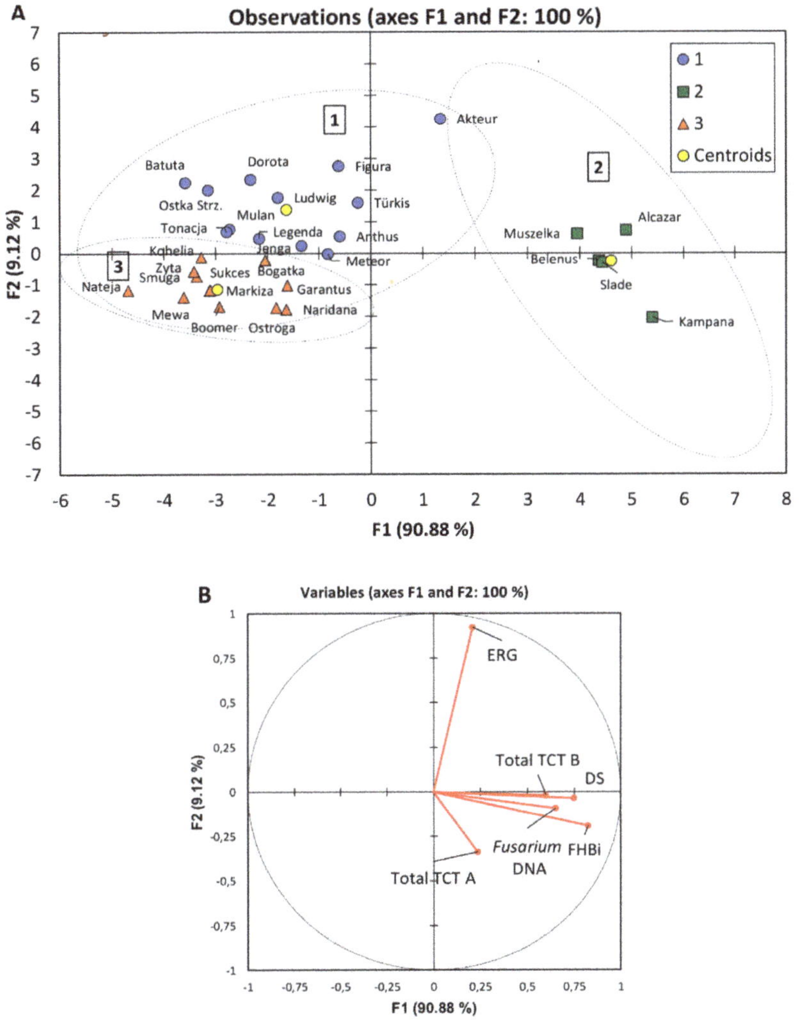

Figure 3. Discriminant analysis of 30 cultivars grown in organic field for Fusarium head blight index (FHBi), dead seeds proportion (DS), ergosterol (ERG), type A (Total TCT A) and type B (Total TCT B) trichothecenes content, and concentration of DNA of five *Fusarium* species in grain (*Fusarium* DNA): (**A**) Observations on the factor axes with marked groups 1–3; (**B**) Correlation circle.

The lowest overall infection showed cultivars from the group 3: 'Nateja', 'Mewa', and 'Markiza'. Regarding only accumulation of trichothecenes, it was the lowest in grain of 'Nateja' and 'Ostroga' from the group 3. It was low also in grain of cultivars 'Figura', 'Dorota', 'Mulan', 'Batuta', 'Tonacja' from the group 1, and surprisingly in grain of cultivar 'Belenus' from group of the most infected cultivars. Among the low-toxin accumulating cultivars, 'Ostroga', 'Batuta', 'Mulan', and 'Figura' were in the group of cultivars showing the highest grain yield per plot in organic field (Table S1). The lowest infected cultivar 'Nateja' had low grain yield caused by high yellow rust infection.

Table 9. Average values for groups shown in Figure 3A.

Group	Number of Cultivars	FHBi (%)	DS (%)	ERG (mg kg^{-1})	Total TCT B (μg kg^{-1})	Total TCT A (μg kg^{-1})	*Fusarium* DNA (pg 100 ng^{-1})
1	13	0.4	3.8	2.10	63	4	206
2	5	2.0	8.7	1.48	191	7	610
3	12	0.4	2.8	0.67	43	6	237

4. Discussion

Until recently, the issue of organic farming was considered marginal. However, the constantly increasing acreage crops grown in this system and fast increasing percentage of consumers interested in obtaining the organic food encouraged a detailed address of this issue, which, so far, was recognized only partially. Thus, it was decided to provide field experiment for a representative sample of winter wheat cultivars for both systems of crops growing-conventional and organic under the same environmental conditions (location, time, and weather). Of course, not all growing conditions were the same. This was mainly related to soil conditions. This was reflected in the studies presented by the analysis of the mineral elements. In most cases, we found significant differences in the concentrations occurring in the soil. A higher concentration in organic soil was found for Ca, Na, Si, B, and Mn. Lower for K, Mg, Cd, Co, Cr, Cu, Ni, and Zn. This is certainly due to organic cultivation resulting from the lack of use of mineral fertilizers. This leads to the relative impoverishment of the soil. However, we found higher concentration of Ca, Na, B and Si in soil from organic field. In review paper by Romero et al. [32] the authors found that silicon shows the beneficial effects on growth, development, and health of crops. It activates the defense mechanisms of plants and increases tolerance to fungal diseases [33]. Concentrations of all analyzed metals (despite Mn) was lower in organic soil. Differences were significant for Cr, Cu, Ni, and Zn. Depletion of the soil in organic cultivation system has an impact on the amount of soil microorganisms [34,35]. They are responsible for the biochemical processes, and thus, consequently, the resistance mechanism of plants. Studies on the reduction of the content of alkaline metals showed it leads to its acidification, which promotes the growth of micro-organisms for which the acidic (pH 4–5) is beneficial, including microscopic fungi and among them plant pathogenic species [36,37]. With this widespread phenomenon, we have to deal with in our research. These confirms our results indicating a higher concentration of K, Mg, Zn, or Cd detected on fertilized, conventional plots. Much higher concentration of Zn and Cd was especially interesting.

The presented experiments concerned 30 cultivars of winter wheat, which were examined comprehensively for several years under conventional conditions to determine their susceptibility to Fusarium head blight and were divided into 4 groups as shown in Table 1 [20,38]. They constitute a complete cross-section of widely grown wheat cultivars in Poland, which gives the basis for determining them as model cultivars. In conditions of experimental year, Fusarium head blight severity was low and average values for conventional and organic field did not differ significantly. Heading and flowering time were significantly earlier in organic field than in conventional. On average, plant height of wheat cultivars did not differ between two analyzed cropping systems. Effect of plant height of FHB severity was cultivar-depended and similar in both systems. Taller cultivars were less FHB infected.

Sowing quality measured as germination energy and germination capacity of conventional material was significantly higher than for the seeds from organic cultivation system. The percentage's share of abnormal seedlings as well as share of dead seeds was significantly higher in organic seed material. It confirms results obtained for sowing value of conventional and organic oats [39]. Additionally, percentage's share of fresh ungerminated seeds was twice higher in organic seed material than in conventional one. However, difference was not statistically significant.

It is evident that there were significant differences in the seed quality obtained from both cultivation systems. For all parameters, the differences were important to the detriment of organic farming. Particularly large was the difference in germination energy, which for material from the ecological field was almost 25% lower than for the conventional. The above differences may also result from higher colonization of kernels by mycobiota obtained from the ecological field. This was indicated by twice-higher ergosterol content (the total fungal quantity meter) and a three-fold higher *Fusarium* biomass content in organic grain. The content of ergosterol in grain depends on the type of grain (hulled, hull-less), cereal species and on the level of contamination of the grain with microscopic fungi, both pathogenic strains and native mycobiota [31,40,41]. Its content is affected also by the method of cultivation resulting from the use of fertilizers and treatments related to the use of plant protection chemicals [30].

Plant protection causes a disturbance of the natural homeostasis of the microbiota of kernel's surface, resulting in development of more expansive microbes that can dominate the environment. Thus, unfortunately, probiotic microorganisms, which are a natural barrier to pathogens, are completely removed [42,43]. The logical consequence is that, because of conventional agro-technical management, such crop is more susceptible to colonization by mycobiota. On the other hand, this increases competitiveness with pathogenic fungi producing specific fungal metabolites. Consequently, this has to do with the detection of a twice-higher concentration of ergosterol in organic material. Among the detected types of microscopic fungi, pathogenic ones represent a small percentage. This is related to the presence of large amounts of nonpathogenic mycobiota, which is a competition for pathogens [44,45].

Presence of DNA of six *Fusarium* species was detected in wheat grain. *F. langsethiae* was detected in six samples only in trace amounts. The highest amount of DNA was found as follows for *F. poae*, *F. graminearum*, *F. sporotrichioides*, *F. culmorum* and *F. avenaceum*. It was true in organic field. In conventional field, concentration of *F. culmorum* DNA was higher than *F. graminearum* and *F. sporotrichioides* DNA. The composition of *Fusarium* species was similar to that observed in last years in Europe [5,46–48].

Total *Fusarium* DNA concentration in organic samples was more than twice higher than in conventional samples, what can be explained by the cultivation system. In the case of organic cultivation, an environmental niche with a stabilized microorganism population is formed, which is enriched with probiotic organisms. Living in a symbiosis of microorganisms contribute to the improvement of soil condition, and thus naturally strengthen the resistance mechanisms of plants through i.a. mycorrhiza. In the conventional case, there are stress related to fertilization or the use of pesticides. Some fungi are eliminated, others often having a strong pathogenic remain. An analysis of the DNA content in the grain also gives a lot of interesting information. In the grain of organic farming, almost three times more DNA was found and stronger links between the contents of single species were identified. This was not true for *F. culmorum*, which could be related to the presence of *F. graminearum* in grain species being a competitor in the biosynthesis of type B trichothecenes.

Another legitimate conclusion here is that the testing of the DNA content of the grain is a much more accurate test method than the determination of the fungal concentration by an ERG analysis [49–51]. At the same time, it is emphasized that the amount of ERG gives a full image of the level of contamination with microscopic fungi. This is confirmed by the correlation factors for the ERG. They are, in all cases insignificant what confirms the above argument. Pathogenic fungi in the grain produce various metabolites and among them mycotoxins. This also occurs in the case of fungi of the genus *Fusarium*, which synthesize trichothecene toxins. This was also the case in the analyzed samples.

Based on the above results and conclusions is imposed another one. In grain of organic farming theoretically, the concentration of trichothecenes should be significantly higher. However, it was found that type A and B trichothecenes concentrations in both cases were similar and the differences were not significant. Vanova et al. [52] found higher concentration of DON in grain of wheat grown in three conventional systems. It was significantly higher in two systems where no chemical protection against FHB was applied. Similar tendency was found in barley and oats from organic and traditional farming [53–56]. In their review, Brodal et al. [56] concluded that contamination with *Fusarium* toxins of organically produced cereal grains was similar and sometimes lower than conventionally produced ones.

The established correlation coefficients for both groups were significant for the conventional system only. This is probably due to the fact, that for this system, more fungal biomass of *F. graminearum* was found in the grain. That resulted in a higher correlation with the sum of type B trichothecenes which *F. graminearum* is an important producer.

The concentration of detected toxins was relatively small. Concentration of DON and T-2/HT-2 toxins was below the European limit and recommendation (Commission Regulation No. 1126/2007 of 28 September 2007; Commission Recommendation No. 2013/165/EU of 27 March 2013). Comparing the two cultivation systems, however, it is evident that in grain the average concentration of type B trichothecenes was lower in the case of organic trials. Differences were not statistically significant, but at the same time, concentration of *Fusarium* DNA was almost 3 times higher in organic grain. Although it can be also found similar data in other papers [52,57–60] it is a positive result. This proves once again that in the organic system determines the community of co-existing microorganisms is established. The most pathogenic and toxigenic are not predominant and environmental stress is not as harmful as the stress associated with significant doses of artificial fertilizers and pesticides as well as simplified rotations [61,62].

When considering the concentration of specific toxins, only concentration of 3-AcDON was significantly higher in conventional samples. Concentration of NIV was higher in samples from organic field; however, difference was not significant. Distributions for FUS-X and 3-AcDON in organic and conventional samples were significantly different. In conventional samples, these toxins were detected in higher amounts in single samples whereas they were more evenly distributed in organic samples. Amount of type A trichothecenes was very low and similar in conventional and organic samples, and they not differ significantly. Only average concentration of DAS was significantly higher in conventional samples.

Comparison of the sum of trichothecene toxins of groups A and B indicate environmental effects. Important correlations were obtained for the conventional system, by the fact that a strong pathogenic species *F. graminearum* stood out, being the most important producer of such toxins as DON, its derivatives and to a lesser extent (depending on the chemotype) NIV. For organic farming, the established coexistence of species was confirmed and no dominance of *F. graminearum* was found. When analyzing occurring mycobiota using ERG as a measure, no significant correlation was found for both environments with the other characteristics. In the case of *Fusarium* DNA testing, such correlations were found with the stronger link found for organic farming both between species (except *F. culmorum*) and other studied traits (mainly FHBi). The data presented is a significant contribution to understanding the philosophy of cultivation system and its effects. A similar method of reasoning and application may be found in paper of Lazzaro et al. [63].

By summarizing this aspect of the research, it is possible to identify clearly the relationship between the analyzed factors in the case of organic cultivation as stronger (Table 7). The above statements was confirmed by a comprehensive statistical analysis. It included a number of tests comparing analyzed populations based on factors such as FHBi, DS, ERG, Total TCT A and B, and *Fusarium* DNA concentration. The designated p-value for the multidimensional Wilk's test had the value < 0.0001. p-value was similar for Fisher distances. It gives clear grounds for supporting the above conclusions indicating the different mechanism of reaction of plants on environmental stress of both cultivation systems. The MANOVA test was conducted to further validate these conclusions. It clearly showed that its effects depended on the type of cropping system in a very important way ($p < 0.0001$), and to a much lesser extent on the cultivars used in the experiment ($p < 0.025$). Similar observations can be found in the work of e.g., Newton et al. [64].

The issue of wheat cultivars applied in cultivation is often raised [3,4,65,66]. The most important question is whether the same cultivar can be used in both systems. During the study we wanted also to deepen this issue using 30 different cultivars with varying resistance to FHB. The possibility of successfully applying the same cultivar in both systems is becoming increasingly important, also for breeding reasons. Biplot of the principal component analysis shown in Figure 2 indicates the effects of cultivar on the results of the experiment.

It can be concluded that the results indicate a diversified behavior of cultivars, which was characterized by varying distances between cropping effects in two systems. Determined by multidimensional scaling (MDS) method the average distances between pairs in the conventional and organic systems were for resistant cultivars (R) 0.576; medium resistant (MR) 2.335; medium susceptible (MS) 2.819 and susceptible (S) 3.547. This result is unambiguous and indicates that it is possible to use the cultivars used in conventional crops for organic farming [64,67].

The final stage of the study was comparison of the overall performance of cultivars grown under organic field conditions with respect to resistance to *Fusarium* infection. Using multivariable analysis (K-means, discriminant analysis), it was possible to divide cultivars into 3 groups depending on traits tested as indicated in Table 8 and Figure 3. The division on the three groups finds its justification both in the values shown in the table and separation because of their FHB susceptibility. For five cultivars ('Alcazar' (S), 'Muszelka' (S), 'Kampana' (S), 'Belenus' (MS), 'Slade' (MS)), significantly higher values (excluding ERG) have been obtained for all experimental traits. Discriminant analysis confirmed the condition of these cultivars, which already in other experiments showed low resistance after artificial inoculation of heads when they had high head infection and DON accumulation [20,38].

All the results presented indicate the usefulness of the above studies for the recommendation of individual cultivars to a particular growing method. Such studies requires evidently multiyear or multi location experiments to be fully reliable. Results show differences in effects of the conventional and organic system. The interesting preliminary results obtained, in the meaning of authors, will contribute to a better understanding of the processes of growth and development and effect of cereal farming in certain environmental conditions. They also allow for an objective look at organic farming and perhaps contribute to its rapid growth, as the idea of sustainable cultivation and avoidance of plant stress should gain new supporters.

Supplementary Materials: The following are available online at http://www.mdpi.com/2076-2607/7/10/439/s1.

Author Contributions: Conceptualization, J.P. and T.G.; methodology, T.G., A.Ł., E.M., K.S.-S., M.B. and J.P; formal analysis, T.G.; investigation, T.G., A.Ł., E.M., K.S.-S. and M.B.; writing—original draft preparation, T.G. and J.P.; writing—review and editing, T.G., A.Ł., E.M. and J.P.; supervision, J.P.; project administration, J.P.

Funding: This research was financed by National Science Centre (NCN), Poland. Project OPUS4 No. 2012/07/B/NZ9/02385.

Conflicts of Interest: The authors declare no conflict of interest.

References

1. *Anonym Raport o Stanie Rolnictwa Ekologicznego w Polsce w Latach 2015–2016 [The Report on Organic Farming in Poland in 2015–2016]*; Agricultural and Food Quality Inspection (GIJHARS): Warszawa, Poland, 2017.
2. Borgen, A. Strategies for regulation of seed borne diseases in organic farming. *Seed Test. Int. ISTA News Bull.* **2004**, *127*, 19–21.
3. Wolfe, M.S.; Baresel, J.P.; Desclaux, D.; Goldringer, I.; Hoad, S.; Kovacs, G.; Löschenberger, F.; Miedaner, T.; Østergård, H.; Lammerts Van Bueren, E.T. Developments in breeding cereals for organic agriculture. *Euphytica* **2008**, *163*, 323–346. [CrossRef]
4. Letourneau, D.; van Bruggen, A.H.C. Crop protection in organic agriculture. In *Organic Agriculture: A Global Perspective*; Kristiansen, P., Acram, T., Reganold, J., Eds.; CABI Publishing: Wallingford, UK, 2006; pp. 93–121. ISBN 0 643 09090 8.
5. Bottalico, A.; Perrone, G. Toxigenic Fusarium species and mycotoxins associated with head blight in small-grain cereals in Europe. *Eur. J. Plant Pathol.* **2002**, *108*, 611–624. [CrossRef]
6. Bottalico, A. Fusarium diseases of cereals: Species complex and related mycotoxin profiles, in Europe. *J. Plant Pathol.* **1998**, *80*, 85–103.
7. Wagacha, J.M.; Muthomi, J.W. Fusarium culmorum: Infection process, mechanisms of mycotoxin production and their role in pathogenesis in wheat. *Crop. Prot.* **2007**, *26*, 877–885. [CrossRef]
8. Parry, D.W.; Jenkinson, P.; McLeod, L. Fusarium ear blight (scab) in small grain cereals—A review. *Plant Pathol.* **1995**, *44*, 207–238. [CrossRef]

9. Dweba, C.C.; Figlan, S.; Shimelis, H.A.; Motaung, T.E.; Sydenham, S.; Mwadzingeni, L.; Tsilo, T.J. Fusarium head blight of wheat: Pathogenesis and control strategies. *Crop. Prot.* **2017**, *91*, 114–122. [CrossRef]
10. Bennett, J.W.; Klich, M. Mycotoxins. *Clin. Microbiol. Rev.* **2003**, *16*, 497–516. [CrossRef]
11. Juan, C.; Ritieni, A.; Mañes, J. Occurrence of *Fusarium* mycotoxins in Italian cereal and cereal products from organic farming. *Food Chem.* **2013**, *141*, 1747–1755. [CrossRef] [PubMed]
12. Vrček, I.V.; Čepo, D.V.; Rašić, D.; Peraica, M.; Žuntar, I.; Bojić, M.; Mendaš, G.; Medić-Šarić, M. A comparison of the nutritional value and food safety of organically and conventionally produced wheat flours. *Food Chem.* **2014**, *143*, 522–529. [CrossRef] [PubMed]
13. Malmauret, L.; Parent-Massin, D.; Hardy, J.-L.; Vergey, P. Contaminants in organic and conventional foodstuffs s in France. *Food Addit. Contam.* **2002**, *19*, 524–532. [CrossRef] [PubMed]
14. Marx, H.; Gedek, B.; Kollarczik, B. Vergleichende Untersuchungen zum mykotoxikologischen Status von ökologisch und konventionell angebautem Getreide. *Z. Lebensm. Unters. Forsch.* **1995**, *201*, 83–86. [CrossRef] [PubMed]
15. McKenzie, A.J.; Whittingham, M.J. Birds select conventional over organic wheat when given free choice. *J. Sci. Food Agric.* **2010**, *90*, 1861–1869. [CrossRef] [PubMed]
16. Champeil, A.; Fourbet, J.F.; Doré, T.; Rossignol, L. Influence of cropping system on Fusarium head blight and mycotoxin levels in winter wheat. *Crop. Prot.* **2004**, *23*, 531–537. [CrossRef]
17. Harcz, P.; De Temmerman, L.; De Voghel, S.; Waegeneers, N.; Wilmart, O.; Vromman, V.; Schmit, J.F.; Moons, E.; Van Peteghem, C.; De Saeger, S.; et al. Contaminants in organically and conventionally produced winter wheat (*Triticum aestivum*) in Belgium. *Food Addit. Contam.* **2007**, *24*, 713–720. [CrossRef] [PubMed]
18. Mäder, P.; Hahn, D.; Dubois, D.; Gunst, L.; Alföldi, T.; Bergmann, H.; Oehme, M.; Amadò, R.; Schneider, H.; Graf, U.; et al. Wheat quality in organic and conventional farming: Results of a 21 year field experiment. *J. Sci. Food Agric.* **2007**, *87*, 1826–1835. [CrossRef]
19. Magkos, F.; Arvaniti, F.; Zampelas, A. Organic Food: Buying More Safety or Just Peace of Mind? A Critical Review of the Literature. *Crit. Rev. Food Sci. Nutr.* **2006**, *46*, 23–56. [CrossRef] [PubMed]
20. Góral, T.; Stuper-Szablewska, K.; Buśko, M.; Boczkowska, M.; Walentyn-Góral, D.; Wiśniewska, H.; Perkowski, J. Relationships between genetic diversity and *Fusarium* toxin profiles of winter wheat cultivars. *Plant Pathol. J.* **2015**, *31*, 226–244. [CrossRef] [PubMed]
21. Ostrowska-Kolodziejczak, A.; Stuper-Szablewska, K.; Kulik, T.; Busko, M.; Rissmann, I.; Wiwart, M.; Perkowski, J. Concentration of fungal metabolites, phenolic acids and metals in mixtures of cereals grown in organic and conventional farms. *J. Anim. Feed Sci.* **2016**, *25*, 74–81. [CrossRef]
22. ISTA. *International Rules for Seed Testing 2019*; International Seed Testing Association—ISTA: Bassersdorf, Switzerland, 2019; ISSN 2310-3655.
23. Doyle, J.J.; Doyle, J.L. Isolation of plant DNA from fresh tissue. *Focus (Madison)* **1990**, *12*, 13–15.
24. Turner, A.S.; Lees, A.K.; Rezanoor, H.N.; Nicholson, P. Refinement of PCR-detection of *Fusarium avenaceum* and evidence from DNA marker studies for phenetic relatedness to *Fusarium tricinctum*. *Plant Pathol.* **1998**, *47*, 278–288. [CrossRef]
25. Nicholson, P.; Simpson, D.R.; Weston, G.; Rezanoor, H.N.; Lees, A.K.; Parry, D.W.; Joyce, D.; Centre, J.I.; Lane, C. Detection and quantification of *Fusarium culmorum* and *Fusarium graminearum* in cereals using PCR assays. *Physiol. Mol. Plant Pathol.* **1998**, *53*, 17–37. [CrossRef]
26. Wilson, A.; Simpson, D.; Chandler, E.; Jennings, P.; Nicholson, P. Development of PCR assays for the detection and differentiation of *Fusarium sporotrichoides* and *Fusarium langsethiae*. *FEMS Microbiol. Lett.* **2004**, *233*, 69–76. [CrossRef] [PubMed]
27. Parry, D.W.; Nicholson, P. Development of a PCR assay to detect *Fusarium poae* in wheat. *Plant Pathol.* **1996**, *45*, 383–391. [CrossRef]
28. Perkowski, J.; Kiecana, I.; Kaczmarek, Z. Natural occurrence and distribution of *Fusarium* toxins in contaminated barley cultivars. *Eur. J. Plant Pathol.* **2003**, *109*, 331–339. [CrossRef]
29. Young, J.C. Microwave-assisted extraction of the fungal metabolite ergosterol and total fatty acids. *J. Agric. Food Chem.* **1995**, *43*, 2904–2910. [CrossRef]
30. Perkowski, J.; Wiwart, M.; Busko, M.; Laskowska, M.; Berthiller, F.; Kandler, W.; Krska, R. *Fusarium* toxins and total fungal biomass indicators in naturally contaminated wheat samples from north-eastern Poland in 2003. *Food Addit. Contam.* **2007**, *24*, 1292–1298. [CrossRef] [PubMed]

31. Perkowski, J.; Buśko, M.; Stuper, K.; Kostecki, M.; Matysiak, A.; Szwajkowska-Michałek, L. Concentration of ergosterol in small-grained naturally contaminated and inoculated cereals. *Biologia* **2008**, *63*, 542–547. [CrossRef]
32. Romero, A.; Munévar, F.; Cayón, G. Silicon and plant diseases. A review. *Agron. Colomb.* **2011**, *29*, 473–480.
33. Dorneles, K.R.; Dallagnol, L.J.; Pazdiora, P.C.; Rodrigues, F.A.; Deuner, S. Silicon potentiates biochemical defense responses of wheat against tan spot. *Physiol. Mol. Plant Pathol.* **2017**, *97*, 69–78. [CrossRef]
34. Wasilkowski, D.; Mrozik, A.; Piotrowska-Seget, Z.; Krzyzak, J.; Pogrzeba, M.; Plaza, G. Changes in enzyme activities and microbial community structure in heavy metal-contaminated soil under in situ aided phytostabilization. *Clean Soil Air Water* **2014**, *42*, 1618–1625. [CrossRef]
35. Ahmad, I.; Hayat, S.; Ahmad, A.; Inam, A.; Samiullah, I. Effect of heavy metal on survival of certain groups of indigenous soil microbial population. *J. Appl. Sci. Environ. Manag.* **2005**, *9*, 115–121.
36. Martyniuk, S.; Martyniuk, M. Occurrence of *Azotobacter* spp. in some Polish soils. *Polish J. Environ. Stud.* **2003**, *12*, 371–374.
37. Malik, A.A.; Puissant, J.; Buckeridge, K.M.; Goodall, T.; Jehmlich, N.; Chowdhury, S.; Gweon, H.S.; Peyton, J.M.; Mason, K.E.; van Agtmaal, M.; et al. Land use driven change in soil pH affects microbial carbon cycling processes. *Nat. Commun.* **2018**, *9*, 1–10. [CrossRef] [PubMed]
38. Góral, T.; Walentyn-Góral, D. Variation for resistance to Fusarium head blight in winter and spring wheat cultivars studied in 2009–2016. Short communication. *Biul. IHAR* **2018**, *284*, 3–12.
39. Małuszyńska, E.; Mańkowski, D.R. Seed sowing value and response to drought stress of organic and conventional oat (*Avena sativa* L.) seeds during 5 years of storage. *Acta Sci. Pol. Agric.* **2016**, *15*, 27–36.
40. Wiwart, M.; Perkowski, J.; Budzyński, M.; Suchowilska, E.; Buśko, M.; Matysiak, A. Concentrations of ergosterol and trichothecenes in the grains of three *Triticum* species. *Czech J. Food Sci.* **2011**, *29*, 430–440. [CrossRef]
41. Buśko, M.; Stuper, K.; Jeleń, H.; Góral, T.; Chmielewski, J.; Tyrakowska, B.; Perkowski, J. Comparison of volatiles profile and contents of trichothecenes group B, ergosterol, and ATP of bread wheat, durum wheat, and triticale grain naturally contaminated by mycobiota. *Front. Plant Sci.* **2016**, *7*, 1243. [CrossRef]
42. Karlsson, I.; Friberg, H.; Kolseth, A.K.; Steinberg, C.; Persson, P. Organic farming increases richness of fungal taxa in the wheat phyllosphere. *Mol. Ecol.* **2017**, *26*, 3424–3436. [CrossRef]
43. Lori, M.; Symnaczik, S.; Mäder, P.; De Deyn, G.; Gattinger, A. Organic farming enhances soil microbial abundance and activity—A meta-analysis and meta-Regression. *PLoS ONE* **2017**, *12*, 1–25. [CrossRef]
44. Sapkota, R.; Knorr, K.; Jørgensen, L.N.; O'Hanlon, K.A.; Nicolaisen, M. Host genotype is an important determinant of the cereal phyllosphere mycobiome. *New Phytol.* **2015**, *207*, 1134–1144. [CrossRef] [PubMed]
45. Hartman, K.; van der Heijden, M.G.A.; Wittwer, R.A.; Banerjee, S.; Walser, J.C.; Schlaeppi, K. Cropping practices manipulate abundance patterns of root and soil microbiome members paving the way to smart farming. *Microbiome* **2018**, *6*, 1–14.
46. Lukanowski, A.; Sadowski, C. *Fusarium langsethiae* on kernels of winter wheat in Poland—Occurrence and mycotoxigenic abilities. *Cereal Res. Commun.* **2008**, *36*, 453–457. [CrossRef]
47. Stenglein, S.A. *Fusarium poae*: A pathogen that needs more attention. *J. Plant Pathol.* **2009**, *91*, 25–36.
48. Xu, X.M.; Parry, D.W.; Nicholson, P.; Thomsett, M.A.; Simpson, D.; Edwards, S.G.; Cooke, B.M.; Doohan, F.M.; Brennan, J.M.; Moretti, A.; et al. Predominance and association of pathogenic fungi causing Fusarium ear blightin wheat in four European countries. *Eur. J. Plant Pathol.* **2005**, *112*, 143–154. [CrossRef]
49. Reischer, G.H.; Lemmens, M.; Farnleitner, A.; Adler, A.; Mach, R.L. Quantification of *Fusarium graminearum* in infected wheat by species specific real-time PCR applying a TaqMan probe. *J. Microbiol. Methods* **2004**, *59*, 141–146. [CrossRef]
50. Nicholson, P.; Turner, A.S.; Edwards, S.G.; Bateman, G.L.; Morgan, L.W.; Parry, D.W.; Marshall, J.; Nuttall, M. Development of stem-base pathogens on different cultivars of winter wheat determined by quantitative PCR. *Eur. J. Plant Pathol.* **2002**, *108*, 163–177. [CrossRef]
51. Justesen, A.F.; Hansen, H.J.; Pinnschmidt, H.O. Quantification of *Pyrenophora graminea* in barley seed using real-time PCR. *Eur. J. Plant Pathol.* **2008**, *122*, 253–263. [CrossRef]
52. Váňová, M.; Klem, K.; Míša, P.; Matušinsky, P.; Hajšlová, J.; Lancová, K. The content of *Fusarium* mycotoxins, grain yield and quality of winter wheat cultivars under organic and conventional cropping systems. *Plant Soil Environ.* **2008**, *54*, 395–402. [CrossRef]

53. Ibáñez-Vea, M.; González-Peñas, E.; Lizarraga, E.; López De Cerain, A. Co-occurrence of aflatoxins, ochratoxin A and zearalenone in barley from a northern region of Spain. *Food Chem.* **2012**, *132*, 35–42. [CrossRef]
54. Edwards, S.G. *Fusarium* mycotoxin content of UK organic and conventional barley. *Food Addit. Contam. Part A Chem. Anal. Control. Expo. Risk Assess.* **2009**, *26*, 1185–1190. [CrossRef]
55. Edwards, S.G. *Fusarium* mycotoxin content of UK organic and conventional oats. *Food Addit. Contam. Part A Chem. Anal. Control. Expo. Risk Assess.* **2009**, *26*, 1063–1069. [CrossRef] [PubMed]
56. Remža, J.; Lacko-Bartošová, M.; Kosík, T. *Fusarium* mycotoxin content of Slovakian organic and conventional cereals. *J. Cent. Eur. Agric.* **2016**, *17*, 164–175. [CrossRef]
57. Brodal, G.; Hofgaard, I.S.; Eriksen, G.S.; Bernhoft, A.; Sundheim, L. Mycotoxins in organically versus conventionally produced cereal grains and some other crops in temperate regions. *World Mycotoxin J.* **2016**, *9*, 755–770. [CrossRef]
58. Hoogenboom, L.A.P.; Bokhorst, J.G.; Northolt, M.D.; van de Vijver, L.P.L.; Broex, N.J.G.; Mevius, D.J.; Meijs, J.A.C.; Van der Roest, J. Contaminants and microorganisms in Dutch organic food products: A comparison with conventional products. *Food Addit. Contam. Part A Chem. Anal. Control. Expo. Risk Assess.* **2008**, *25*, 1195–1207. [CrossRef] [PubMed]
59. Bernhoft, A.; Clasen, P.-E.; Kristoffersen, A.B.; Torp, M. Less *Fusarium* infestation and mycotoxin contamination in organic than in conventional cereals. *Food Addit. Contam. Part A Chem. Anal. Control. Expo. Risk Assess.* **2010**, *27*, 842–852. [CrossRef]
60. Meister, U. *Fusarium* toxins in cereals of integrated and organic cultivation from the Federal State of Brandenburg (Germany) harvested in the years 2000–2007. *Mycotoxin Res.* **2009**, *25*, 133–139. [CrossRef] [PubMed]
61. Schmid, F.; Moser, G.; Müller, H.; Berg, G. Functional and structural microbial diversity in organic and conventional viticulture: organic farming benefits natural biocontrol agents. *Appl. Environ. Microbiol.* **2011**, *77*, 2188–2191. [CrossRef] [PubMed]
62. Bernhoft, A.; Torp, M.; Clasen, P.-E.-E.; Løes, A.-K.; Kristoffersen, A.B. Influence of agronomic and climatic factors on *Fusarium* infestation and mycotoxin contamination of cereals in Norway. *Food Addit. Contam. Part A Chem. Anal. Control. Expo. Risk Assess.* **2012**, *29*, 1129–1140. [CrossRef] [PubMed]
63. Lazzaro, I.; Moretti, A.; Giorni, P.; Brera, C.; Battilani, P. Organic vs conventional farming: differences in infection by mycotoxin-producing fungi on maize and wheat in Northern and Central Italy. *Crop. Prot.* **2015**, *72*, 22–30. [CrossRef]
64. Newton, A.C.; Guy, D.C.; Preedy, K. Wheat cultivar yield response to some organic and conventional farming conditions and the yield potential of mixtures. *J. Agric. Sci.* **2017**, *155*, 1045–1060. [CrossRef]
65. Mason, H.E.; Spaner, D. Competitive ability of wheat in conventional and organic management systems: a review of the literature. *Can. J. Plant Sci.* **2006**, *86*, 333–343. [CrossRef]
66. Osman, A.M.; Almekinders, C.J.M.; Struik, P.C.; Lammerts van Bueren, E.T. Adapting spring wheat breeding to the needs of the organic sector. *NJAS Wageningen J. Life Sci.* **2016**, *76*, 55–63. [CrossRef]
67. Scholten, O.E.; Steenhuis-Broers, G.; Timmermsns, B.; Osman, A. Screening for resistance to Fusarium head blight in organic wheat production. In *Proceedings of the COST SUSVAR Fusarium Workshop: Fusarium Diseases in Cereals—Potential Impact from Sustainable Cropping Systems, Velence, Hungary, 1–2 June 2007*; Vogelgsang, S., Jalli, M., Kovacs, G., Vida, G., Eds.; Risø National Laboratory: Roskilde, Denmark, 2007; pp. 20–23.

microorganisms

MDPI

Article

Three-Dimensional Study of *F. graminearum* Colonisation of Stored Wheat: Post-Harvest Growth Patterns, Dry Matter Losses and Mycotoxin Contamination

Xavier Portell [1], Carol Verheecke-Vaessen [2], Rosa Torrelles-Ràfales [2], Angel Medina [2], Wilfred Otten [1], Naresh Magan [2] and Esther García-Cela [2,3,*]

1 School of Water, Energy and Environment, Cranfield University, Cranfield, Bedfordshire MK43 0AL, UK; xavier.portell@cranfield.ac.uk (X.P.); wilfred.otten@cranfield.ac.uk (W.O.)
2 Applied Mycology Group, School of Water, Energy and Environment, Cranfield University, Cranfield, Bedfordshire MK43 0AL, UK; c.verheecke@cranfield.ac.uk (C.V.-V.); rosaatorrelles@gmail.com (R.T.-R.); a.medinavaya@cranfield.ac.uk (A.M.); n.magan@cranfield.ac.uk (N.M.)
3 School of Life and Medical Sciences, University of Hertfordshire, Hatfield, Hertfordshire AL10 9AB, UK
* Correspondence: e.garcia-cela@herts.ac.uk; Tel.: +44-1707284596

Received: 6 July 2020; Accepted: 30 July 2020; Published: 1 August 2020

Abstract: *Fusarium* causes significant post-harvest quality losses and mycotoxin contamination in stored wheat but the colonisation dynamics of the grain and how this may be affected by the initial inoculum position in the grain mass is poorly understood. This study examined the 3D growth kinetics and mycotoxin production (deoxynivalenol and zearalenone) by *F. graminearum* during hyphal colonisation from different initial inoculum positions in wheat microcosms (top-centre, bottom-centre, and bottom-side) maintained at two water activities (a_w; 0.95 and 0.97). Clear jars were used to visually follow the colonisation dynamics. Fungal respiration and associated dry matter loss (DML) and ergosterol were also quantified. Colonisation dynamics was shown to be affected by the inoculation position. At the end of the colonisation process, fungal respiration and DML were driven by the inoculation position, and the latter also by the prevailing a_w. Fungal biomass (ergosterol) was mainly affected by the a_w. The initial inoculum position did not affect the relative mycotoxin production. There was a positive correlation between respiration and ergosterol, and between mycotoxin production and colonisation indicators. We suggest that spatially explicit predictive models can be used to better understand the colonisation patterns and mycotoxin contamination of stored cereal commodities and to aid more effective post-harvest management.

Keywords: Cereals; silo; fungi; modelling; 3D colonisation; respiration; ergosterol; zearalenone; deoxynivalenol; trichothecenes.

1. Introduction

Wheat is susceptible to *Fusarium* head blight (FHB) world-wide which causes both yield and quality losses, in addition to contamination of harvested grain with mycotoxins, particularly zearalenone (ZEN) and type B trichothecenes such as deoxynivalenol (DON). There are strict regulations in place for maximum contamination levels of these two toxins to reduce accumulation in the human and animal food chains. Both mycotoxins can be produced pre- and post-harvest. Although *Fusarium graminearum* requires higher moisture content levels compared to other more xerotolerant or xerophilic spoilage fungi (e.g., *Penicillium* and *Aspergillus* species, respectively), delayed or inefficient drying and unsafe storage practices can allow fungal growth and potential mycotoxin accumulation.

Recent studies have shown that under conducive moisture contents and over a wide range of temperatures, *F. graminearum* colonisation of stored wheat grain can cause significant dry matter losses (DMLs) and result in DON and ZEN contamination exceeding the EU maximum limits for unprocessed cereals for human food consumption (DON: 1.750 and ZEN: 100 μgKg^{-1}) [1] and animal feed (DON: 8.000 and ZEN: 2.000 μgKg^{-1}) [2]. Overall, <1% DML can result in significant impacts on quality losses and toxin contamination exceeding the EU limits [3–5].

Stored cereals in silos represent an ecosystem in which there are significant interactions between the abiotic and biotic factors [6]. Indeed, Limay-Rios et al. [7] showed that the main sources for contamination of stored wheat in grain silos are at the top of the silo, where the inlet for grain filling is placed, and the outlet, where the grain is removed from the silo. These areas and the silo walls represent areas where higher moisture content can occur due to moisture migration, leading to increased contamination with mycotoxins.

However, there is very little information on how fungal growth is initiated and the way in which mycelial colonisation occurs, especially in three dimensions. Most previous studies have been conducted by measuring colonisation in two directions or dimensions only [8,9]. However, in reality, filamentous fungi such as *F. graminearum* grow in stored grain in three dimensions extending their mycelium in all directions to obtain nutrients from the rich wheat substrate. Du et al. [10] and Vidal-Diez de Ulzurrun et al. [11] recently reported novel lattice-based models for a three dimensional study of fungal growth, but there are no publications that have examined *Fusarium* colonisation in stored cereal grain.

Measurement of ergosterol, fungal biomass and colony diameter have been positively correlated on agar media previously suggesting that the three parameters may be useful for primary modelling [12–14]. Colony diameter is extensively used to assess fungal growth. However, in aged colonies (>30 days) ergosterol did not increase exponentially, suggesting that despite the colony expansion in terms of area, there was a decrease in the rate of ergosterol accumulation [12]. Similarly, fungal diameter measurements may not be a perfect indicator of fungal biomass, especially in aging colonies where biomass may increase despite limitation in actual colony extension [13]. Other studies have used ergosterol content as a measure of the overall mould contamination levels of commodities and related this to safe or unsafe storage conditions [15,16]. In addition, the use of ergosterol was found to be an effective biochemical marker for early indications of fungal contamination as well as of predicted DON contamination levels in stored cereals [17,18].

However, few studies, if any, have examined the colonisation of cereals by a mycotoxigenic species in three dimensions to try to understand the growth kinetics and the relationship between growth and toxin production. There is also a lack of knowledge of how temporal fungal colonisation occurs in a 3D matrix such as stored grains via the intergranular spaces, and how this may be affected by the initial inoculum position or the prevailing environmental factors.

The objectives of this study were to examine the kinetics of 3D spatial colonisation of stored wheat grain by *F. graminearum* and the accumulation of DON and ZEN when inoculated in different positions and under different water availabilities in laboratory microcosm storage systems. This allowed information to be obtained on (a) spatial patterns of colonisation depending on the different initial inoculum positions, (b) the changes in respiration and associated DMLs, (c) quantification of fungal biomass, using ergosterol, (d) the temporal DON and ZEN mycotoxin accumulation during colonisation and (e) the correlation between these different biological and physical parameters.

2. Materials and Methods

2.1. Fungal Isolate and Spores Preparation

F. graminearum strain Fg 08/111 isolated from wheat in the UK was cultivated on V8 vegetable juice plates (V8®, 175 mL; $CaCO_3$, 3 g; $ZnSO_4 \cdot 7H_2O$, 0.01g; $CuSO_4 \cdot 5H_2O$, 0.005g; agar, 20 $g \cdot L^{-1}$) and incubated in the dark at 25 °C for ten days to obtain heavily sporulating cultures [19]. Following incubation, 3 mL of sterile Tween 80® (0.05%) was poured on the Petri plates previously incubated and then gently scratched with a sterile Drigalsky's spatula. The liquid was transferred to a second Petri plate containing fungi and the procedure repeated with the remaining 4 Petri plates. The resulting liquid was recovered and collected in a sterile falcon tube. After homogenizing, the number of spores per mL was determined using a Thoma counting chamber (Celeromics, Grenoble, France) and the final concentration adjusted to 1×10^4 spores·mL^{-1} in sterile water + Tween 80 (0.005%).

2.2. Grain Preparation, Inoculation and Incubation

Dry wheat grain (0.71 water activity (a_w), 13.5 % moisture content (m.c.)) was irradiated with 12–15 kGy (Synergy Health, Swindon, UK) to reduce microbial contaminants while maintaining germinative capacity of the grain, and stored at 4 °C until use. The grain a_w levels were modified to 0.95 and 0.97 by adding known amounts of sterile water and reference to an adsorption curve for this batch of wheat grain, which was detailed in Garcia-Cela et al. [20], and verified using the Aqualab 4TE a_w meter.

Square jars of 136 mL capacity (Pattersons Glass Ltd., Grimsby, UK) (see supplementary Figure S1) with a 9 mm diameter hole in the jar lid, which was closed with a cotton wool plug to allow gaseous exchange while avoiding contamination, were autoclaved at 121 °C for 20 min. Single grains inoculated with a drop of 5 µL of the 10^4 spores mL^{-1} suspension (approx. 50 spores per grain) were placed in distinct positions to provide initial inoculum of *F. graminearum* for initiating growth in the different positions within the jars: (a) top-centre, (b) bottom-side and (c) bottom-centre. For the bottom inoculations, an 8 µL drop of organic clear nail polish was used to fix the grain on the jar bottom. Fixed grains were inoculated with the stated spore solution drop and 1 min later, the remaining grains were added to the jar. For the top-centre position, the spore solution drop was placed over a single grain in the centre of the top layer of grains. In all cases, 15 jars per treatment combination were used, including 3 samples without *F. graminearum* inoculated per a_w tested.

Inoculated jars were incubated at 25 °C for 10 days. To maintain atmospheric equilibrium relative humidity, jars were placed in food storage containers (10 L volume) containing beakers of glycerol–water solution at the treatment a_w level (6 × 100 mL per container). Glycerol–water solutions were renewed once (day 5) during the experimental period.

2.3. Colonisation Pattern Assessment

Fungal colonisation of the wheat grain was followed by recording the superficial mycelial extension visible through the six sides of the square clear jars (supplementary Figure S2). Clear labels with a mesh of 0.5 × 0.5 cm were aligned and pasted on the bottom and sides of the jars. To avoid external contamination of the grains, jar lids were opened under sterile conditions and a plastic sheet with the same mesh was aligned on the top. Visual fungal colonisation was followed every 24 h until the grain was fully colonised.

The colonisation data were used to compute volumetric colonisation (cm^3) by *F. graminearum* from the three inoculation treatment positions. In two dimensional studies, fungal growth is known to expand in circular colonies following a radial pattern [21]. In the present study, we expanded the approach assuming that mycelial colonisation followed an ellipsoidal shape before being constrained by the cubic shape of the jar. The detailed procedure followed to compute the volumetric colonisation is presented in Appendix A.

2.4. Respiration Determination and Dry Matter Loss Estimation

Every 24 h, samples were sealed for 1 h at 25 °C with injection lids that contained polytetrafluoroethylene (PTFE)/silicone septa (20 mm diameter × 3.175 mm thickness) to allow for CO_2 accumulation in the headspace. Headspace volumes of jars containing grain at 0.95 or 0.97 a_w were, respectively, 90.67 and 88.33 mL. In both growth conditions, 5 mL of headspace air were withdrawn with a syringe and directly inserted into the Gas Chromatograph (GC) for CO_2 analysis. An Agilent 6890N Network GC (Agilent Technologies, Cheshire, UK) with a Thermal Conductivity Detector and helium as a carrier gas was used. A Chromosorb 103 packed column was usedand the data were analysed using Agilent Chemstation Software (Agilent Technologies, Cheshire, UK).

GC was calibrated with a standard gas bottle of 10.18% CO_2 and 2% O_2 in nitrogen (British Oxygen Company, Guilford, Surrey, UK). *F. graminearum* CO_2 production was obtained by subtracting the sample respiration of a blank culture that had not been inoculated with the fungal species. The percentages of CO_2 production were used to calculate the Respiration (R) in mg CO_2 (kg·h^{-1}), total cumulative production CO_2 and total DML as described in Mylona and Magan [3].

2.5. Mycotoxins Extraction and Analysis

Before mycotoxin extraction, samples were dried at 60 °C for 24 h and then milled in a laboratory blender (Waring Commercial, Christian, UK). Mycotoxins accumulation of three replicate samples per combination of inoculation position and a_w on alternate days (days 2, 4, 6, 8 and 10) were extracted by adding 500 µL of acetonitrile:water:formic acid (79:20.9:0.1, *v:v:v*) to 100 (± 10) mg of milled wheat and agitated for 90 min at 300 rpm at 25 °C on a rotary shaker (miniShaker VWR, Leighton Buzzard, UK). Then, samples were centrifuged for 10 min at 22,600 g (Centrifuge 5417S Eppendorf, Stevenage, UK), and 1 or 3 µL of the supernatant was injected into an Exion LC series HPLC linked to a 6500+ qTRAP-MS system in Electrospray Ionisation (ESI) mode (Sciex Technologies, Warrington, UK). An ACE 3-C18 column (2.1 × 100 mm, 3 µm particle size; Hichrom) with guard cartridge (4 × 3 mm, Gemini, Agilent) was conditioned at 40 °C. The elution gradient used water:acetic acid (*v:v*, 99:1, solvent A) and methanol:acetic acid (*v:v*, 99:1) (solvent B), both supplemented with 5 mM ammonium acetate. The 15 min-long gradient included: 0 min, 90% A; 0–2.0 min, 90–60% A; 2.0–10.0 min, 0% A and 10.0–11.50 min, 0% A; 11.50–12.0 min, 90% A; 12.00–15.0 min, 90% A at a 0.3 mL·min^{-1} flow rate. 10 msec of dwell time per daughter ion (2 per metabolites) was used in an unscheduled Multiple Reaction Monitoring (MRM) in both positive and negative mode using the parameters listed in Table 1.

Table 1. Overview of the metabolite studies in the mycotoxin analysis. The first Q3 for each was used for quantification. DP: Declustering potential, CE: Collision Energy, CXP: Cell Exit potential.

Analyte	Retention Time (min)	Q1 (m/z)	DP (V)	Q3 (m/z)	CE (V)	CXP (V)
3-AcetylDeoxynivalenol	3.86	397.3	−70	−59.2/−307.1	−38/−20	−8/−7
15-AcetylDeoxynivalenol	3.84	339.1	91	137.2/321.2	17/13	8/18
Deoxynivalenol	2.60	355.1	−70	−59.2/−265.2	−40/−22	−13/−10
Zearalenone	7.33	317.1	−110	−175/−121.1	−34/−42	−13/−8

The source conditions used included: Curtain gas 40%, Collision Gas Medium, IonSpray voltage 4500 V/ 5500 V, Temperature 400 °C, both ion sources gas at 60 psi, with a 10 V Entrance Potential for all compounds. Data acquisition was conducted with Analyst® Data Acquisition version 1.6.3, and quantification through MultiQuant™ version 3.0.3.

The analyte recovery, the limit of detection (LOD) and the limit of quantification (LOQ) were calculated using recommendations stated elsewhere [22]. The calculated recoveries were used as correction factors to calculate the absolute concentration of the analytes. The LODs were 0.26 and 2.29 ng·g^{-1} and the LOQs 0.85 and 7.63 ng·g^{-1} for ZEN and DON, respectively.

2.6. Ergosterol Analysis

The replicates and treatments were oven-dried at 60 °C for 48 h, milled in a laboratory blender (Waring Commercial, Christian, UK), homogenised, and the ergosterol content analysed using an adaptation of the rapid ultrasonic method developed previously [23]. This involved adding 2 mL of deionised water to a 5 g of sample. After 15 min, 10 mL methanol:ethanol (*v:v*, 4:1) was added and samples were stored at 4 °C for 2 h. Then, 20 mL hexane:propan-2-ol (*v:v*, 98:2) was added to each sample and they were ultrasonicated at 150 W in an ice bath with water for 3 min 20 s. After 30 s,

2 mL of the supernatant was centrifuged at 7000× *g* for 10 min, from which 1.5 mL was used for HPLC analysis.

The HPLC system was a Shimadzu Class-VP, which consisted of SCL-10Avp controller, SPD-10Avp detector and LC-10ADvp pump (Shimadzu Corporation, Japan). An amount of 100 μL was injected into a Lichrosorb Phenomenex column (250 × 4.6 mm, 10 μm) at 35 °C. Run time for samples was 15 min with ergosterol being detected at about 9.20 min. The flow rate of the mobile phase (hexane:propan-2-ol; *v:v*, 98:2) was 1.4 mL·min^{-1}.

Calibration standards of 5, 25, 50, 100 and 150 μg·g^{-1} ergosterol in hexane:propan-2-ol (*v:v*, 98:2) were used as an external calibration. A general recovery test was conducted from spiked samples using 4 replicates. The procedure was the same as explained above, with two changes: (a) 1 mL of internal standard was added after the deionised water and samples were allowed to stand for 15 min, (b) 19 mL of hexane:propan-2-ol (*v:v*, 98:2) was added instead of 20 mL. Average recovery rates for 50, 100, 500 and 1000 μg·g^{-1} ergosterol from wheat grain matrix were 89.18 ± 14.44%, 104.35 ± 12.37%, 106.37 ± 9.81% and 94.41 ± 5.97%, respectively.

2.7. Data Analysis

Data were analysed with JMP Pro (JMP®, version 14. SAS Institute, Inc., Cary, NC, USA). All data were assessed for normality and homoscedasticity. Growth rate data, DML, respiration and ergosterol, were normally distributed, homoscedastic and independent. Therefore, data sets were analysed using two-way Analysis of Variance (ANOVA) for the determination of the significance between a_w treatments and inoculation position. Then, Student's t- or Tukey's HSD test were used to identify significant differences between groups. Mycotoxins accumulation data were not normally distributed, therefore a Kruskal–Wallis test for the determination of the significance between a_w treatments and inoculation position was undertaken. Then, non-parametric comparisons for each pair was conducted using the Wilcoxon test. The statistical relationship between variables was assessed by Pearson or Spearman correlations. A signification level of 5% was assumed for all statistical analysis.

The "fit_growthmodel" function of the "growthrates" R package [24] was used to fit the colonisation data to a two-phase linear model of the form:

$$C_t = (t > \lambda) * R_C * (t - \lambda), \quad (1)$$

where C_t is the colonisation (cm^2 or cm^3) at time t (days), λ a lag phase with 0 colonisation (days) and R_c the colonisation rate (cm^2·day^{-1} or cm^3·day^{-1}).

3. Results

3.1. Fungal Colonisation

F. graminearum showed a rapid colonisation rate at both tested a_w levels with all of the wheat grain visually completely colonised after 5–6 days. A comparative example of the recorded superficial growth showed distinct colonisation patterns of *F. graminearum* depending on the initial inoculum position (Figure 1). Superficial colonisation data were well described using a two-phase linear model (Figure 2) providing estimates of the lag time prior to colonisation and the superficial colonisation rate (cm^2·day^{-1}) shown in Table 2. The lag phase length prior to superficial colonisation was shown to be significantly affected by the inoculation position but not by a_w, with mean lag phase times higher for the bottom-side position (2.97 ± 0.22 h) than the bottom-centre (2.92 ± 0.31 h) and top-centre (2.94 ± 0.32 h) positions. No significant differences (α = 0.05) in the colonisation rate were detected among treatments. As a consequence of this increased lag time, the bottom-side inoculations were able to completely colonise the grain surface only by day 6 of the experiment. For the other inoculum positions (top-centre and bottom-centre) this occurred after 5 days (Figure 3a,b).

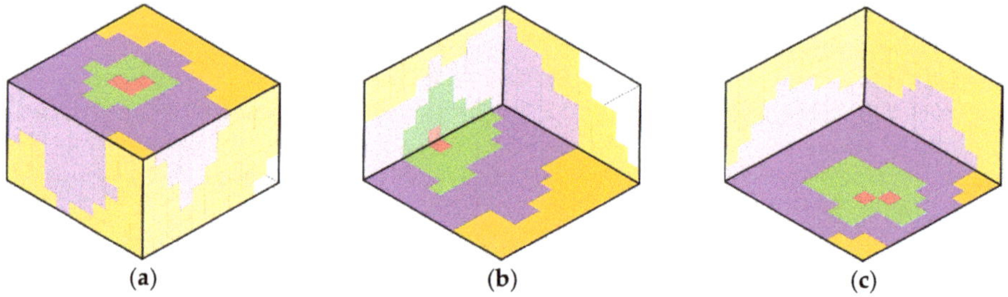

Figure 1. Example of the visual observation of the fungal colonisation dynamics at 0.95 a_w starting from different inoculation positions: (**a**) top-centre, (**b**) bottom-side and (**c**) bottom-centre. In the figure, colours depict colonisation after day 2 (●), day 3 (●), day 4 (●), day 5 (●) and day 6 (□).

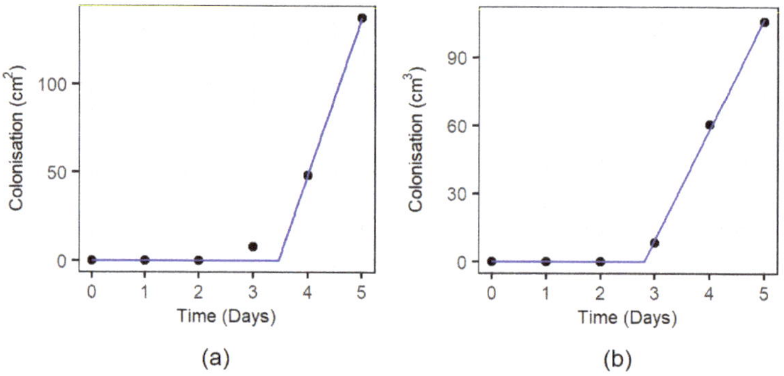

Figure 2. Example of the two-phase linear model fit (blue line) for the colonisation data (dots) recorded for the sample grown on 0.95 a_w, bottom-centre inoculation, second replication. (**a**) Model fit to surface colonisation data, and (**b**) model fit to volume colonisation data.

Table 2. Superficial and volumetric colonisation rates. Lag times and colonisation rates were obtained by adjusting the data to a two-phase linear model.

Water Activity (a_w)	Inoculation	Replicate	Superficial Colonisation		Volumetric Colonisation	
			Lag Time (Days)	Rate ($cm^2 \cdot Day^{-1}$)	Lag Time (Days)	Rate ($cm^3 \cdot Day^{-1}$)
0.95	Top-centre	1	2.71	59.50	2.66	48.93
0.95	Top-centre	2	3.46	45.30	2.75	48.76
0.95	Top-centre	3	2.90	64.25	3.25	60.67
0.95	Bottom-side	1	3.35	51.25	3.18	39.08
0.95	Bottom-side	2	3.04	47.50	3.40	59.93
0.95	Bottom-side	3	2.79	43.67	3.33	59.80
0.95	Bottom-centre	1	2.69	63.63	2.54	46.22
0.95	Bottom-centre	2	2.88	63.50	2.80	48.73
0.95	Bottom-centre	3	3.51	92.50	2.65	47.78
0.97	Top-centre	1	2.82	63.75	2.94	94.35
0.97	Top-centre	2	2.61	60.87	2.83	50.91
0.97	Top-centre	3	2.83	64.50	2.93	78.37
0.97	Bottom-side	1	2.85	43.93	3.27	54.07
0.97	Bottom-side	2	2.75	44.15	3.01	51.18
0.97	Bottom-side	3	3.04	47.88	3.46	64.12
0.97	Bottom-centre	1	2.65	65.50	2.87	75.23
0.97	Bottom-centre	2	2.92	66.38	2.84	81.10
0.97	Bottom-centre	3	2.79	65.62	2.89	76.33

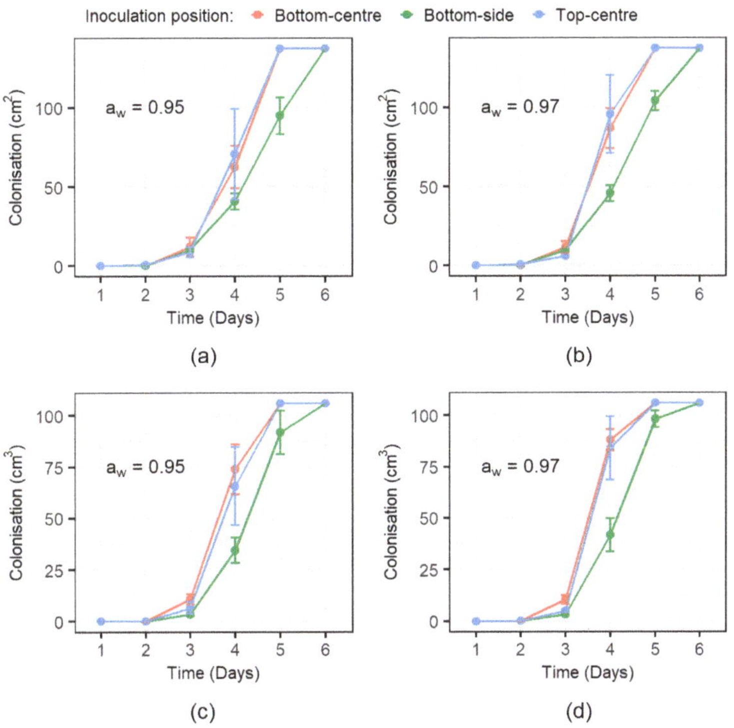

Figure 3. Temporal colonisation of the wheat grain by *F. graminearum* in both a_w treatments from the three different initial inoculum positions. (**a**) Superficial colonisation for 0.95 a_w treatment, (**b**) superficial colonisation for 0.97 a_w treatment, (**c**) volumetric colonisation for 0.95 a_w treatment, and (**d**) volumetric colonisation for 0.97 a_w treatment. Bars show standard deviation of the three replicates.

Volumetric fungal colonisation (cm^3) was obtained from the visible colonisation assuming that the fungal growth of the grain was consistent with a partial ellipsoidal expansion of the mycelia constrained by the cubic volume of the jar (see Appendix A). Volumetric colonisation (Figure 3b,c) was also well adjusted to a simple two-phase linear model, obtaining colonisation lag times and volumetric colonisation rate ($cm^3 \cdot day^{-1}$) estimates shown in Table 2. The length of the lag phase was shown to be significantly affected by the inoculation position (p-value = 0.003), with the bottom-side inoculation having an extended lag period (3.28 ± 0.16 h) distinguishable from that of the top-centre (2.89 ± 0.21 h) and bottom-centre (2.76 ± 0.14 h) inoculum positions. The volumetric colonisation rate were significantly affected by a_w, and was more rapid at 0.97 a_w (69.52 ± 15.24 cm^3/day) than at 0.95 a_w (51.10 ± 7.42 $cm^3 \cdot day^{-1}$).

3.2. Indirect Indicators of Fungal Growth

3.2.1. Fungal Respiration Dynamics

Cumulative fungal respiration for the 10 day experimental period is shown in Figure 4a,b. In all inoculation positions and a_w treatments, respiration increased over time as the *F. graminearum* mycelia colonised the wheat grain. The accumulated respiration at the end of the colonisation process (days 6 for bottom-side, and day 5 for top-centre and bottom centre inoculations) was significantly affected by the initial inoculation position (p-value = 0.022). Fungal colonisation from the bottom-side proceeded more slowly but produced more CO_2 (4.96 ± 0.84 $g \cdot CO_2 \cdot kg^{-1}$) than the top-centre (3.83 ± 0.79 $g \cdot CO_2 \cdot kg^{-1}$)

and bottom-centre (3.23 ± 0.52 g·CO_2·kg^{-1}). Bottom-side respiration was significantly higher (α = 0.05) than the bottom-centre respiration. However, these two treatments were indistinguishable from the top-centre cumulative respiration.

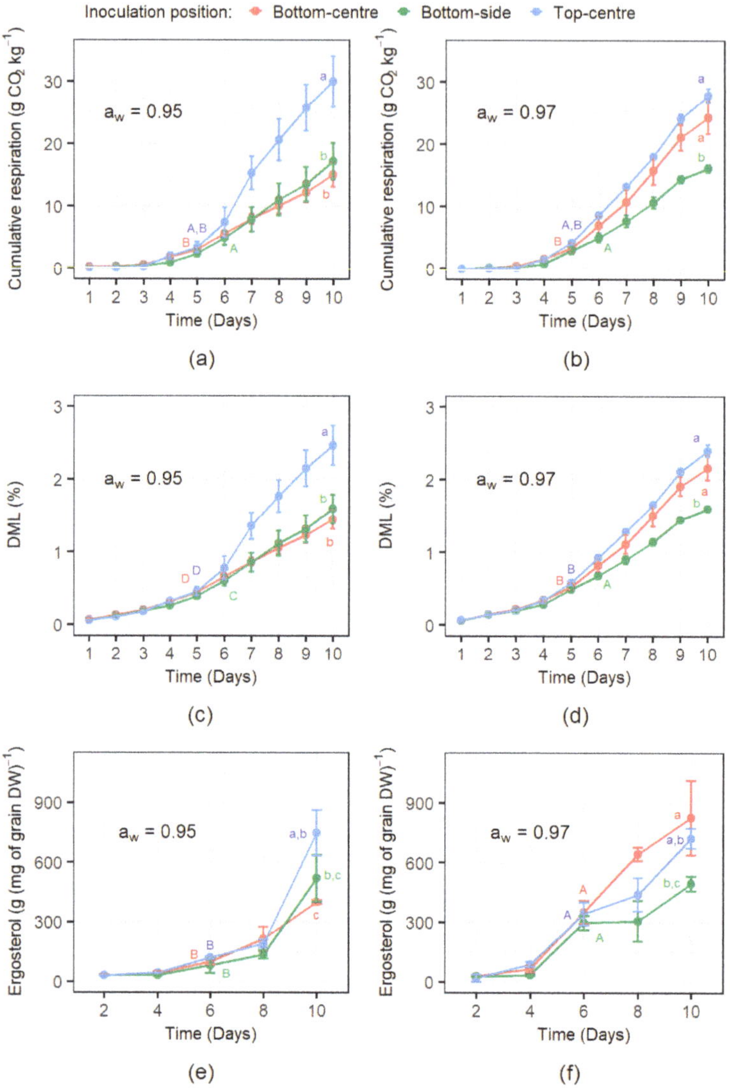

Figure 4. Temporal evolution of the indirect indicators of fungal growth from three different inoculum positions. Cumulative respiration of *F. graminearum* at (**a**) 0.95 a_w and (**b**) 0.97 a_w. The dry matter loss (DML; %) of wheat grain inoculated with *F. graminearum* at (**c**) 0.95 a_w and (**d**) 0.97 a_w. Ergosterol content of *F. graminearum* inoculated grain at (**e**) 0.95 a_w, and (**f**) 0.97 a_w. Bars show standard deviation of the three replicates. Upper case and lower case letters show statistically significant differences among treatments at the end of the colonisation and at the end of the experiment, respectively.

The ANOVA analysis at the end of the experiment (day 10), suggested a significant effect of the inoculation treatment position (p-value < 0.001) and an interaction amongst the inoculation positions and a_w factors (p-value = 0.003). The 0.95 a_w top-centre, the 0.97 a_w top-centre, and 0.97 a_w bottom-centre treatments were similar. The other treatments: 0.95 a_w bottom-side, 0.97 a_w bottom-side, and 0.95 a_w bottom-centre, were similar but significantly lower than the previous group (Figure 4a,b).

3.2.2. Dry Matter Loss Dynamics

The DML followed a similar temporal increase at both a_w levels depending on the inoculum position (Figure 4c,d) and reflected the respiration rates (see Figure 4a,b). However, there was a more linear increase earlier than that for the respiration rate. At the end of the colonisation period (days 6 for bottom-side, and day 5 for top-centre and bottom-centre inoculations) the DML was significantly affected by a_w levels (p-value = 0.001) and the inoculation position (p-value = 0.002). The DMLs were higher at 0.97 a_w than in the slightly drier 0.95 a_w treatment (0.60 ± 0.07 vs. 0.51 ± 0.09% respectively). As observed for respiration, grain colonisation initiated from the bottom-side position proceeded more slowly and generated more DML (0.64 ± 0.06%) by the time the grain was completely colonised than the top-centre (0.53 ± 0.08%) and bottom-centre (0.49 ± 0.06%) treatments.

The ANOVA analysis at the end of the experiment (day 10) suggested a significant effect of both a_w (p-value = 0.018) and inoculation position (p-value < 0.0001) treatments, and for the interaction between these two factors (p-value = 0.003). The 0.95 a_w top-centre, 0.97 a_w top-centre and 0.97 a_w bottom-centre treatments were very similar. The remaining treatments again grouped together and had significantly lower DMLs than the other group of treatments.

3.2.3. Ergosterol Production Dynamics

During the colonisation process, ergosterol content of all of the treatment combinations appeared to be mainly driven by the storage a_w (Figure 4e,f). By day 6, close to the end of the colonisation of the wheat grain, a_w had a significant effect (p-value < 0.001) on ergosterol content, although was interestingly not influenced by the inoculation position (p-value = 0.07). Overall, there was a significantly higher ergosterol content at 0.97 a_w (331.56 ± 50.76 g mg^{-1}) than at 0.95 a_w (100.97 ± 27 g·mg^{-1}). At the end of the experimental period, ergosterol content was statistically affected by the a_w (p-value = 0.027), inoculation position (p-value = 0.005) and the interaction between these two factors (p-value = 0.004). The 0.97 a_w bottom-centre produced a higher ergosterol content than the bottom-side inoculation and 0.95 a_w bottom-centreinoculation. Top-centre inoculations were very similar statistically to the 0.97 a_w bottom-centre treatment and the bottom-side inoculations. Similarly, bottom-side inoculations could not be distinguished from the 0.95 a_w bottom-centre treatment.

3.3. *Mycotoxin Production Dynamics*

Mycotoxin production was significantly impacted by the a_w treatment (p-value < 0.0001 and p-value = 0.0331 for DON and ZEN, respectively) with higher production occurring at 0.97 a_w. DON production could be quantified from day 4 for the wetter 0.97 a_w (Figure 5b) condition and from day 6 in the drier 0.95 a_w treatment (Figure 5a). For ZEN production, this was predominantly found from day 6 onwards in both a_w treatments (Figure 5). ZEN increased markedly by the end of the experimental period (10 days). Unlike for colonisation of the wheat grain, the inoculum position did not significantly impact either DON or ZEN production.

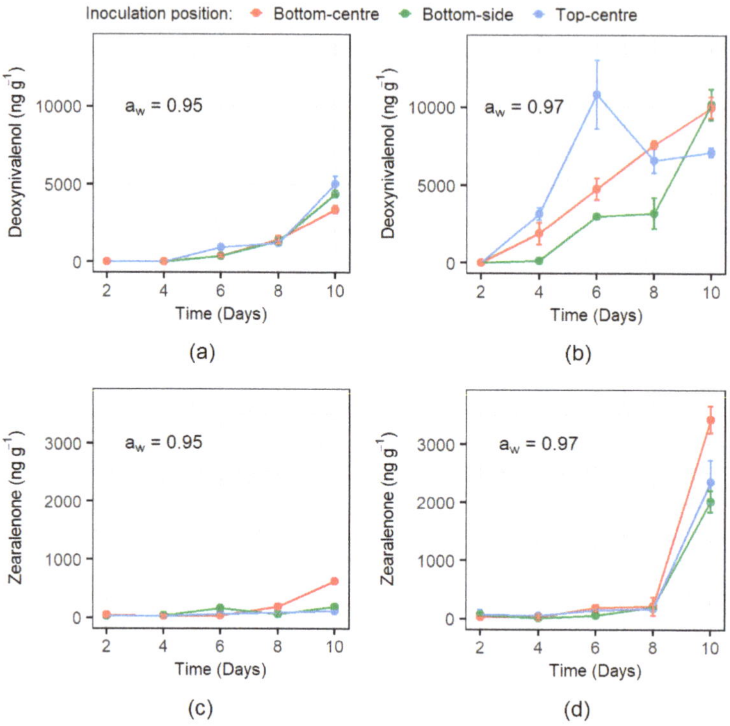

Figure 5. Temporal deoxynivalenol (DON) and zearalenone (ZEN) mycotoxin production from each inoculation position at the two water activity treatments examined. (**a**) DON at 0.95 a_w; (**b**) DON at 0.97 a_w; (**c**) ZEN at 0.95 a_w; and (**d**) ZEN at 0.97 a_w. Bars show the standard deviation of the mean.

3.4. Correlation between Experimental Measures

3.4.1. Correlation between Fungal Respiration and Ergosterol Content

There was a significant correlation between ergosterol content and cumulative respiration data for *F. graminearum* inoculation in wheat grain. Spearman rank order correlation suggested a strong positive correlation between these two variables (Table 3). When considering individual a_w treatments the correlation was stronger during the colonisation period but this decreased when both a_w levels are considered together. The overall correlation coefficient among water activities for the length of the experiment was 88.58%.

Table 3. Spearman rank order correlation coefficient between cumulative fungal respiration and ergosterol content. *p*-values of all coefficients were < 0.001.

Water Activity Levels	0.95 a_w	0.97 a_w	0.95 + 0.97 a_w
Days 2, 4 and 6	0.9193	0.9329	0.7796
Days 2, 4, 6, 8 and 10	0.8785	0.9404	0.8858

3.4.2. Correlation between Mycotoxin Contamination Levels and Indicators of Colonisation

Correlations between the growth indicators used in this study and the mycotoxin production is shown in Table 4. For DON, all of the indicators used (i.e., volume, ergosterol and cumulative DML) showed a positive correlation. The correlations ranged from 0.8149 to 0.9528 and were dependent on the a_w treatment, with higher correlations at 0.95 a_w than 0.97 a_w. The cumulative DML and the ergosterol content were found to be the indicators most correlated to DON contamination level.

Table 4. Correlation (Spearman's) factors of the mycotoxin content with volumetric colonisation, ergosterol, and Cumulative dry matter loss. DON: Deoxynivalenol, ZEN: Zearalenone, DML: dry matter loss. Correlation to the volume was performed using data from the days 2, 4 and 6 of growth, time when the samples were fully colonised.

Mycotoxins	Fungal Growth Indicator	0.95 a_w		0.97 a_w	
		Correlation	*p*-Value	Correlation	*p*-Value
DON (ng/g)	Volume (cm^3)	0.8149	<0.0001	0.8528	<0.0001
	Ergosterol ($mg{\cdot}g^{-1}$)	0.9528	<0.0001	0.8972	<0.0001
	Cumulative DML	0.9220	<0.0001	0.8660	<0.0001
ZEN (ng/g)	Volume (cm^3)	0.4150	0.0313	0.7490	<0.0001
	Ergosterol ($mg{\cdot}g^{-1}$)	0.6686	<0.0001	0.7856	<0.0001
	Cumulative DML	0.5971	<0.0001	0.8193	<0.0001
DON (ng/g)	ZEN ($ng{\cdot}g^{-1}$)	0.6417	<0.0001	0.7604	<0.0001

For ZEN, all indicators except for the volume of growth obtained at 0.95 a_w correlated well with the mycotoxin content. The average correlation ranged from 0.4150 to 0.8193. Higher correlation factors were observed at 0.97 a_w when compared to 0.95 a_w. The best correlated indicators for ZEN were the cumulative DML at 0.97 a_w and both cumulative DML and Ergosterol at 0.95 a_w.

Overall, DON and ZEN production showed a good correlation at 0.95 a_w (0.6417) and 0.97 a_w (0.7604).

4. Discussion

This study found differences in the colonisation patterns from different initial inoculum positions for a mycotoxigenic filamentous fungus such as *F. graminearum*. The inoculation position affected the length of the lag phase but not the actual rates of mycelial colonisation. This suggests that the geometry of the mycelia expanding through the rich nutritional matrix had an effect on the fungal expansion via the intergranular spaces. To our knowledge, this is the first attempt to follow and model the volumetric colonisation of such stored commodities. An assumption was made that colonisation followed an ellipsoidal front. Volumetric data obtained was adjusted assuming a two phase linear model. Being a non-linear transformation, one would expect the non-constrained volumetric colonisation to be better fitted by a non-linear relationship. Nonetheless, in our conditions, this could be explained by the fact that the colonisation was constrained by the finite cubic shape of the storage jars shortly after initial growth. Alternative techniques for examining the volume of colonised grains could include visualisation techniques that allow exploration of internal fungal growth, such as resin impregnation and thin slicing [25], or non-invasive Computed Tomography (CT). Visualisation of the fungal colonisation using X-ray CT is currently challenging due to the similar absorption properties of fungi and organic matter, although it may be possible for fungi accumulating minerals and colonisation in wood [26].

Unlike the superficial colonisation rate, volumetric colonisation rates were significantly different between a_w levels. Significant differences in the volumetric colonisation pattern between a_w treatments were found in this study that followed similar (pseudo) 2D fungal growth patterns where mycelial colonisation by *F. graminearum* of layers of wheat grain was more rapid in wetter grain [8,9]. However, the differences in colonisation patterns in the former studies only examined superficial colonisation rates and did not take account of a volumetric 3D grain matrix with multiple grain layers as opposed to a single layer of grain. The exploratory nature of this filamentous fungus is thus able to more effectively utilise the nutrients from the wheat grain in different directions.

Analysis of the cumulated respiration at the end of the colonisation period suggested that (i) fungi required more energy to colonise the grain when the colonisation was slower, and (ii) that the a_w

levels tested do not significantly affect this process. Differences in the grain colonisation dynamics from different initial inoculum positions may be related to the ability of the hyphae to grow into the intergranular spaces to colonise other wheat grains and produce the necessary extracellular enzymes to exploit the rich nutritional substrate effectively. This may have resulted in the differences observed in respiration dynamics with faster colonisation (bottom-centre and top-centre) than other inoculum positions. The low respiration levels found in the bottom-centre colonisation at 0.95 a_w may thus have been due to lower oxygen levels in the intergranular spaces and the more stressful water availability treatment. Overall, the present results show that fungal colonisation rates can be accurately estimated by the amount of CO_2 produced in different stored cereals [3,5,27,28].

In the present study, fungal respiration rates and DML obtained by CO_2 measurements supported the colonisation data sets. Fungal respiration rates have been correlated with DML, as fungal growth produces CO_2 due to the oxidation of carbohydrates and production of water vapour and heat during aerobic respiration [29]. DML has been previously used as a quality indicator of stored grains [29,30]. Therefore, the differences among contamination points reported in the present study suggests that contamination source might have an impact on quality losses in a storage silo.

In our study, ergosterol as a fungal biomass measurement was shown to be affected by both a_w level and the inoculation position. Considering that the analysis performed at day 6 did not allow the comparison of the ergosterol amounts at the exact end of the colonisation, the statistical analyses showed the effect of inoculum position on the fungal biomass produced based on ergosterol. This supports the hypothesis that the fungal biomass composition changes depending on the rates of colonisation. A higher ergosterol production in the wetter conditions (0.97 a_w) was probably due to the increased utilisation of the carbohydrates in the wheat grain by the *F. graminearum*. Overall, the present results are similar to the ergosterol levels found in wheat cultivars infected by *F. graminearum* by Stuper-Szablewska et al. [31], although they used a different ergosterol analysis method. Despite the absence of a widely accepted analysis method for ergosterol as a biomass indicator in cereals, it has been directly correlated with fungal colonisation of cereals and DON contamination [18,32,33]. However, as it is a destructive method it cannot be performed in real time. Thus, few attempts have been made in food science to develop a high-recovery method for this fungal biomass indicator in food commodities [34–36].

This study showed that respiration rates and ergosterol content have a highly significant positive correlation. Previously, DML has been successfully correlated with *Fusarium* mycotoxin levels [3,5] and can be calculated from measured respiration rates. Therefore, the correlation between respiration rate and ergosterol found suggests that further research should be conducted to examine the relationship between ergosterol content, respiration rates and mycotoxin contamination levels.

Finally, mycotoxin production (both DON and ZEN) was found to be unaffected by the inoculation position but was higher in the wetter growth condition. The higher DON and ZEN production at 0.97 a_w when compared to 0.95 a_w was due to the effect of the relative water stress the mycotoxigenic species was exposed to. *F. graminearum* is more sensitive to drier conditions [9]. The present study showed earlier production of DON at day 4 at 0.97 a_w compared to day 6 at 0.95 a_w. Our results also showed ZEN production starting at day 6 independently of the a_w tested and an increase in production by day 10. Our study also showed that production of these mycotoxins occurred within 6 days, with the inoculum position having no effect on relative DON and ZEN production patterns. Previously, strains of *F. graminearum* from Argentina were shown to produce higher amounts of DON at 0.97 a_w (43 $ng{\cdot}g^{-1}$) compared to none at 0.95 a_w in wheat gains after 7 days incubation [8]. They also found that at 0.95 a_w production of DON only occurred after 14 days. Ezekiel et al. [37] monitored the production of ZEN in wheat at 0.95 a_w every 6 days and showed that production only occurred after day 12 followed by a steady increase in ZEN production.

The lack of effect of the inoculation position in mycotoxin content may be related to the fact that the whole sample was homogenized before the mycotoxin extraction. One previous report showed larger toxin clusters and stronger spatial autocorrelation in the outer grain layers in a silo, in which the higher humidity and more favourable oxygen availability resulted in better fungal development, while the presence of toxins in deeper locations in the stored grain was related to the influence of gravity [38]. This highlights that more information is needed about the differences between outer and inner layers within a large grain mass in terms of fungal growth and toxin contamination.

This study showed a strong correlation between cumulative DML and DON production found at both a_w levels (0.9220 at 0.95 a_w; 0.8669 at 0.97 a_w), similar to that of Mylona et al. [4] who found a 0.9572 spearman correlation between DML and DON production on wheat at three different a_w levels (0.89, 0.94, 0.97 a_w) at 15 to 30 °C. The present study also showed a significant correlation between ZEN production and DML (0.5971 at 0.95 a_w; 0.8193 at 0.97 a_w). Similar correlations were observed by Garcia-Cela et al. [5] and Mylona et al. [4]. Both correlations for DON and ZEN production provide effective information to develop post-harvest management tools to be integrated for improved Decision Support Systems (DSS).

5. Conclusions

In this study, the behaviour of *F. graminearum* in stored wheat in terms of grain colonisation and mycotoxin production (DON and ZEN) was evaluated in a 3D volume for a period of ten days. Primary inoculum position affected the initial growth significantly and therefore the colonised grain. Modern silos are currently monitored at different spatial levels, consequently, spatial modelling could be used to predict the level of risk. Respiration and DML indicators seem to be as reliable as ergosterol measurements (to indicate fungal colonisation) but with the advantage that they can be monitored in real-time. Thus, to perform efficient silo management, different approaches must be tailored to each of the spatial areas covered by the sensors in which the alert was detected.

The results of this study revealed that understanding the fungal growth pattern and the diffusion of multiple mycotoxins is essential for the development of accurate predictive models that can support effective post-harvest management of grain. This is critical as grain is traded on a wet weight basis and very slight changes in the moisture can lead to an increase in the activity of mycotoxigenic spoilage moulds and mycotoxin contamination.

Supplementary Materials: The following are available online at http://www.mdpi.com/2076-2607/8/8/1170/s1, Figure S1: Clear square jars used in this experiment, Figure S2: Labelling system used to follow 3D colonisation of wheat grains by *F. graminearum*.

Author Contributions: Conceptualization, X.P. and E.G.-C.; Formal analysis, X.P., C.V.-V. and E.G.-C.; Funding acquisition, E.G.-C.; A.M. and N.M.; Investigation, R.T.-R.; Methodology, X.P., C.V.-V., A.M., W.O. and E.G.-C.; Supervision, X.P., A.M., W.O. and E.G.-C.; Writing–original draft, X.P., C.V.-V., R.T.-R. and E.G.-C.; Writing–review & editing, X.P., C.V.-V., R.T.-R, A.M., W.O., N.M. and E.G.-C. All authors have read and agreed to the published version of the manuscript.

Funding: This research was supported by European Union's Horizon 2020 research and innovation programme under grant agreement No. 678012 (MyToolBox), and BBSRC-SFI research grant (BB/P001432/1) between the Applied Mycology Group at Cranfield University and the School of Biology and Environmental Science, University College Dublin, Ireland.

Acknowledgments: We thank Simon Edwards, Harper Adams University, for providing the strain used in this study. The authors would like to thank Dominique Vaessen for helping in the preparation of the jars for the experiment.

Conflicts of Interest: The authors declare no conflict of interest. The funders had no role in the design of the study; in the collection, analyses, or interpretation of data; in the writing of the manuscript, or in the decision to publish the results.

Appendix A. Computation of the Grain Colonised Volume

Three-dimensional fungal colonisation was assumed to follow an ellipsoidal shape of radii a, b and c (Figure A1). After the initial stages of colonisation, the ellipsoidal expansion was constrained by the finite volume of the jar and a number of assumptions were required. The procedure followed for the centre and side positions is described below.

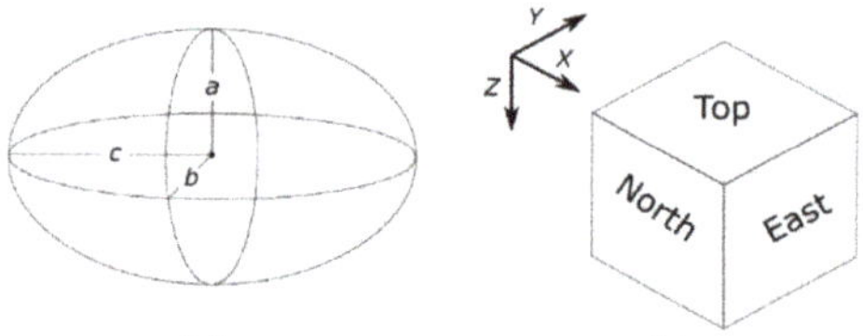

(a) (b)

Figure A1. Radii and spatial configuration assumed in the study. (**a**) Ellipsoid showing radii of the fungal colonisation, and (**b**) coordinate system and faces of the medium.

Appendix A.1. Top-Centre and Bottom-Centre Positions

Volume computation could be generalized in two main cases:

1. Mycelial expansion just observable on the top (bottom) side. Volumetric colonisation at the time t (V_t) was computed assuming a half ellipsoidal shape (Equation (A1))

$$V_t = \frac{2}{3}\pi \cdot a \cdot b \cdot c, \tag{A1}$$

where a, b and c are elliptic radii. Radii b and c where computed as the maximum minus the minimum coordinates showing colonisation along X and Y axis divided by two, and a was assumed to be the $0.75 \cdot \frac{b+c}{2}$. This assumes an upwards (downwards) radial reduction due to the fact that the mycelial expansion proceeds through a more tortuous path that the surface as observed experimentally [39].

2. Mycelial expansion was observable on the vertical sides of the grain volume. Colonised volume was approximated as cubic shape and a half ellipsoid (Equation (A2)) as follows:

$$V_t = L_{\hat{X}} \cdot L_{\hat{Y}} \cdot L_{\hat{Z}} + \frac{2}{3}\pi \cdot \frac{L_{\hat{X}}}{2} \cdot \frac{L_{\hat{Y}}}{2} \cdot a, \tag{A2}$$

where $L_{\hat{X}}$, $L_{\hat{Y}}$ and $L_{\hat{Z}}$ are mean distances obtained along axes X, Y and Z, respectively, showing colonisation. Radii a (in cm) was computed as $3.5 - L_{\hat{Z}}$ if the face opposite to the inoculation point (bottom and top, respectively, for top-centre and bottom-centre inoculations, respectively) showed any sign of colonisation, or as the minimum value between $0.75 \cdot \frac{L_{\hat{X}} - L_{\hat{Y}}}{2}$ and $3.5 - L_{\hat{Z}}$ if not.

Appendix A.2. Bottom-Side Position

Fungi was inoculated at the bottom-side position of the jar. The vertical face of the jar where the fungi was inoculated was denoted as North face, their adjacent faces as East and West, and the opposed vertical face denoted South. Colonisation patterns fell into three main cases:

1. First mycelial expansion not observable on either the East or West faces. The colonisation was assumed to follow a quarter of ellipsoid (Equation (A3)):

$$V_t = \frac{1}{3}\pi \cdot a \cdot b \cdot c, \tag{A3}$$

where a, b and c are elliptic radii. Radius a was obtained on the North face as the mean value of the colonisation shown along the Z-axis. Radius c was obtained on the North face as the maximum minus the minimum coordinates showing colonisation along the X-axis divided by two. Radius b was obtained on the bottom face as the mean value of the colonisation shown along the Y-axis, assuming a minimum value of 1 square.

2. Mycelial expansion observable on either the East or West faces. The growth was assumed to follow a cylindrical shape (V_c) and an ellipsoid shape (V_e) as expressed in Equation (A4):

$$V_t = \frac{V_e}{8} + \frac{V_c}{4} = \frac{1}{12}\pi \cdot a \cdot b \cdot \left(c - \frac{L_W}{2}\right) + \frac{1}{4}\pi \cdot a \cdot b \cdot \frac{L_W}{2}, \quad \text{(A4)}$$

where L_W is the wide of the jar (5.5 cm).

3. Mycelial expansion observable on the top or South faces. Colonisation front was assumed to follow a cylindrical shape until the end of the colonisation (Equation (A5)) according the expression:

$$V_t = L_W \cdot a \cdot L_T + L_W \cdot L_S \cdot (b - L_T) + \frac{1}{4}\pi \cdot (b - L_T) \cdot (a - L_S) \cdot L_W \quad \text{(A5)}$$

where L_T, and L_N are the colonisation length attained by the fungi on the top and South faces, respectively. These were computed as the mean value of the colonisation shown along the Y-axis of the top face (L_T) and the Z-axis of the South face (L_S).

References

1. European Commisison. Commission Regulation (EC) No 1881/2006 of 19 December 2006 setting levels for certain contaminants in foodstuffs. *Off. J. Eur. Union* **2009**, *2006*, 11–21.
2. European Commisison. Commission Recommendation No 2006/576 of 17 August 2006 on the presence of deoxynivalenol, zearalenone, ochratoxin A, T-2 and HT-2 and fumonisins in products intended for animal feeding. *Off. J. Eur. Union* **2006**, *229*, 7–9.
3. Mylona, K.; Magan, N. *Fusarium langsethiae*: Storage environment influences dry matter losses and T2 and HT-2 toxin contamination of oats. *J. Stored Prod. Res.* **2011**, *47*, 321–327. [CrossRef]
4. Mylona, K.; Sulyok, M.; Magan, N. Relationship between environmental factors, dry matter loss and mycotoxin levels in stored wheat and maize infected with *Fusarium* species. *Food Addit. Contam. Part A Chem. Anal. Control Expo. Risk Assess.* **2012**, *29*, 1118–1128. [CrossRef]
5. Garcia-Cela, E.; Magan, N.; Elsa, K.; Sulyok, M.; Medina, A. *Fusarium graminearum* in Stored Wheat: Use of CO_2 Production to Quantify Dry Matter Losses and Relate This to Relative Risks of Zearalenone Contamination under Interacting Environmental Conditions. *Toxins* **2018**, *10*, 86. [CrossRef] [PubMed]
6. Magan, N.; Aldred, D. Post-harvest control strategies: Minimizing mycotoxins in the food chain. *Int. J. Food Microbiol.* **2007**, *119*, 131–139. [CrossRef] [PubMed]
7. Limay-Rios, V.; Miller, J.D.; Schaafsma, A.W. Occurrence of *Penicillium verrucosum*, ochratoxin A, ochratoxin B and citrinin in on-farm stored winter wheat from the Canadian Great Lakes Region. *PLoS ONE* **2017**, *12*, 1–22. [CrossRef] [PubMed]
8. Ramirez, M.L.; Chulze, S.; Magan, N. Temperature and water activity effects on growth and temporal deoxynivalenol production by two Argentinean strains of *Fusarium graminearum* on irradiated wheat grain. *Int. J. Food Microbiol.* **2006**, *106*, 291–296. [CrossRef] [PubMed]
9. Hope, R.; Aldred, D.; Magan, N. Comparison of environmental profiles for growth and deoxynivalenol production by *Fusarium culmorum* and *F. graminearum* on wheat grain. *Lett. Appl. Microbiol.* **2005**, *40*, 295–300. [CrossRef]
10. Du, H.; Perré, P. A novel lattice-based model for investigating three-dimensional fungal growth on solid media. *Phys. A Stat. Mech. Appl.* **2020**, *541*, 123536. [CrossRef]
11. de Ulzurrun, G.V.-D.; Baetens, J.M.; den Bulcke, J.V.; De Baets, B. Modelling three-dimensional fungal growth in response to environmental stimuli. *J. Theor. Biol.* **2017**, *414*, 35–49. [CrossRef] [PubMed]
12. Marín, S.; Cuevas, D.; Ramos, A.J.; Sanchis, V. Fitting of colony diameter and ergosterol as indicators of food borne mould growth to known growth models in solid medium. *Int. J. Food Microbiol.* **2008**, *121*, 139–149. [CrossRef] [PubMed]
13. Garcia, D.; Ramos, A.J.; Sanchis, V.; Marín, S. Modeling kinetics of aflatoxin production by *Aspergillus flavus* in maize-based medium and maize grain. *Int. J. Food Microbiol.* **2013**, *162*, 182–189. [CrossRef] [PubMed]
14. Marín, S.; Ramos, A.J.; Sanchis, V. Comparison of methods for the assessment of growth of food spoilage moulds in solid substrates. *Int. J. Food Microbiol.* **2005**, *99*, 329–341. [CrossRef]

15. Gourama, H.; Bullerman, L.B. Relationship between aflatoxin production and mold growth as measured by ergosterol and plate count. *LWT-Food Sci. Technol.* **1995**, *28*, 185–189. [CrossRef]
16. Tothill, I.E.; Harris, D.; Magan, N. The relationship between fungal growth and ergosterol content of wheat grain. *Mycol. Res.* **1992**, *96*, 965–970. [CrossRef]
17. Rao, B.S.; Rao, V.S.; Ramakrishna, Y.; Bhat, R.V. Rapid and specific method for screening ergosterol as an index of fungal contamination in cereal grains. *Food Chem.* **1989**, *31*, 51–56. [CrossRef]
18. Lamper, C.; Teren, J.; Bartok, T.; Komoroczy, R.; Mesterházy, Á.; Sagi, F. Predicting DON contamination in *Fusarium*-infected wheat grains via determination of the ergosterol content. *Cereal Res. Commun.* **2000**, *28*, 337–344. [CrossRef]
19. Booth, C. Chapter II: Fungal culture media. In *Methods in Microbiology*; Academic Press: London, UK, 1971; Volume 4, pp. 49–94.
20. Garcia-Cela, E.; Kiaitsi, E.; Medina, A.; Magan, N. Interacting environmental stress factors affects targeted metabolomic profiles in stored wheat and that inoculated with F graminearum. *Toxins (Basel)* **2018**, *10*, 56. [CrossRef]
21. Riquelme, M.; Reynaga-Peña, C.G.; Gierz, G.; Bartnicki-García, S. What determines growth direction in fungal hyphae? *Fungal Genet. Biol.* **1998**, *24*, 101–109. [CrossRef]
22. Malachová, A.; Sulyok, M.; Beltrán, E.; Berthiller, F.; Krska, R. Optimization and validation of a quantitative liquid chromatography–tandem mass spectrometric method covering 295 bacterial and fungal metabolites including all regulated mycotoxins in four model food matrices. *J. Chromatogr. A* **2014**, *1362*, 145–156. [CrossRef] [PubMed]
23. Ruzicka, S.; Norman, M.D.P.; Harris, J.A. Rapid ultrasonic method to determine ergosterol concentration in soil. *Soil Biol. Biochem.* **1995**, *27*, 1215–1217. [CrossRef]
24. Petzoldt, T. Growthrates: Estimate Growth Rates from Experimental Data, R Package Version 0.8.1. 2019. Available online: https://cran.r-project.org/package=growthrates (accessed on 6 July 2020).
25. Juyal, A.; Otten, W.; Falconer, R.; Hapca, S.; Schmidt, H.; Baveye, P.C.; Eickhorst, T. Combination of techniques to quantify the distribution of bacteria in their soil microhabitats at different spatial scales. *Geoderma* **2019**, *334*, 165–174. [CrossRef]
26. Van den Bulcke, J.; Boone, M.; Van Acker, J.; Van Hoorebeke, L. Three-dimensional x-ray imaging and analysis of fungi on and in wood. *Microsc. Microanal.* **2009**, *15*, 395–402. [CrossRef] [PubMed]
27. Maier, D.E.; Channaiah, L.H.; Martínez-Kawas, A.; Lawrence, J.; Chaves, E.; Coradi, P.; Fromme, G. Monitoring carbon dioxide concentration for early detection of spoilage in stored grain. In Proceedings of the 10th International Working Conference on Stored Product Protection, Estoril, Portugal, 27 June–2 July 2010; pp. 505–509, Julius Kühn Institut, Bundesforschungsinstitut für Kulturpflanzen. [CrossRef]
28. Garcia-Cela, E.; Kiaitsi, E.; Sulyok, M.; Krska, R.; Medina, A.; Petit Damico, I.; Magan, N. Influence of storage environment on maize grain: CO_2 production, dry matter losses and aflatoxins contamination. *Food Addit. Contam. Part A* **2019**, *36*, 175–185. [CrossRef] [PubMed]
29. Borreani, G.; Tabacco, E.; Schmidt, R.J.; Holmes, B.J.; Muck, R.E. Silage review: Factors affecting dry matter and quality losses in silages. *J. Dairy Sci.* **2018**, *101*, 3952–3979. [CrossRef]
30. Lacey, J.; Hamer, A.; Magan, N. Respiration and dry matter losses in wheat grain under different environmental factors. In *Stored Product Protection 1994*; Highley, E., Wright, E.J., Banks, H.J., Champ, B.R., Eds.; CAB International: Wallingford, UK, 1994; pp. 1007–1013. ISBN 0851989322.
31. Stuper-Szablewska, K.; Kurasiak-Popowska, D.; Nawracała, J.; Perkowski, J. Study of metabolite profiles in winter wheat cultivars induced by *Fusarium* infection. *Cereal Res. Commun.* **2016**, *44*, 572–584. [CrossRef]
32. Yong, R.K.; Cousin, M.A. Detection of moulds producing aflatoxins in maize and peanuts by an immunoassay. *Int. J. Food Microbiol.* **2001**, *65*, 27–38. [CrossRef]
33. Ng, H.-E.; Raj, S.S.A.; Wong, S.H.; Tey, D.; Tan, H.-M. Estimation of fungal growth using the ergosterol assay: A rapid tool in assessing the microbiological status of grains and feeds. *Lett. Appl. Microbiol.* **2008**, *46*, 113–118. [CrossRef]
34. Varga, M.; Bartók, T.; Mesterházy, Á. Determination of ergosterol in *Fusarium*-infected wheat by liquid chromatography-atmospheric pressure photoionization mass spectrometry. *J. Chromatogr. A* **2006**, *1103*, 278–283. [CrossRef]
35. Zhang, H.; Wolf-Hall, C.; Hall, C. Modified microwave-assisted extraction of ergosterol for measuring fungal biomass in grain cultures. *J. Agric. Food Chem.* **2008**, *56*, 11077–11080. [CrossRef] [PubMed]

36. Heidtmann-Bemvenuti, R.; Tralamazza, S.M.; Jorge Ferreira, C.F.; Corrêa, B.; Badiale-Furlong, E. Effect of natural compounds on *Fusarium graminearum* complex. *J. Sci. Food Agric.* **2016**, *96*, 3998–4008. [CrossRef] [PubMed]
37. Ezekiel, C.N.; Odebode, A.C.; Fapohunda, S.O. Zearalenone production by naturally occurring *Fusarium* species on maize, wheat and soybeans from Nigeria. *J. Biol. Environ. Sci.* **2008**, *2*, 77–82.
38. Kerry, R.; Ingram, B.; Garcia-Cela, E.; Magan, N. Spatial analysis of mycotoxins in stored grain to develop more precise management strategies. In *Precision Agriculture*; Wageningen Academic Publishers: Wageningen, The Netherlands, 2019; p. 330.
39. Otten, W.; Gilligan, C.A. Effect of physical conditions on the spatial and temporal dynamics of the soil-borne fungal pathogen Rhizoctonia solani. *New Phytol.* **1998**, *138*, 629–637. [CrossRef]

microorganisms

MDPI

Article

Methodical Considerations and Resistance Evaluation against *Fusarium graminearum* and *F. culmorum* Head Blight in Wheat. Part 3. Susceptibility Window and Resistance Expression

Andrea György, Beata Tóth, Monika Varga and Akos Mesterhazy *

Cereal Research Non-Profit Ltd., 6701 Szeged, Hungary; gyorgyandrea88@gmail.com (A.G.); beata.toth@gabonakutato.hu (B.T.); varga.j.monika@gmail.com (M.V.)

* Correspondence: akos.mesterhazy@gabonakutato.hu

Received: 17 March 2020; Accepted: 22 April 2020; Published: 25 April 2020

Abstract: Flowering is the most favorable host stage for *Fusarium* infection in wheat, which is called the susceptibility window (SW). It is not known how long it takes, how it changes in different resistance classes, nor how stable is the plant reaction in the SW. We have no information, how the traits disease index (DI), *Fusarium*-damaged kernel rate (FDK), and deoxynivalenol (DON) respond within the 16 days period. Seven winter wheat genotypes differing in resistance were tested (2013–2014). Four *Fusarium* isolates were used for inoculation at mid-anthesis, and 4, 8, 11, 13, and 16 days thereafter. The DI was not suitable to determine the length of the SW. In the *Fusarium*-damaged kernels (FDK), a sharp 50% decrease was found after the 8th day. The largest reduction (above 60%) was recorded for DON at each resistance level between the 8th and 11th day. This trait showed the SW most precisely. The SW is reasonably stable in the first 8–9 days. This fits for all resistance classes. The use of four isolates significantly improved the reliability and credit of the testing. The stable eight-day long SW helps to reduce the number of inoculations. The most important trait to determine the SW is the DON reaction and not the visual symptoms.

Keywords: resistance expression; aggressiveness; *F. graminearum*; *F. culmorum*; isolate effect; disease index; *Fusarium*-damaged kernel; deoxynivalenol; susceptibility window; inoculation time and FHB response

1. Introduction

Fusarium head blight (FHB) is one of the most destructive diseases of wheat (*Triticum aestivum* L.) worldwide. Research in the last decades clarifies that the most important toxin-regulating agent is disease resistance [1–4]. Therefore, most of the work belongs to the competence of plant breeding. The artificial inoculations have a larger significance as the natural conditions do not support enough selection work, and this is true also for research. It is long known [5,6] that an increased susceptibility exists during the flowering period, called the susceptibility window (SW). In essence, this is what we know now about this important feature. A possible explanation is that the pollen contains compounds (betaine, choline) stimulating germination of *Fusarium* spores, thus inducing more aggressive disease development [7].

Investigations in inoculation timing have been done in both greenhouse and field experiments. Schroeder and Christensen [8] inoculated in a greenhouse seven spring wheat cultivars at anthesis, milk, and soft dough stages, with spray and single-spikelet inoculation. They concluded that the degree of resistance to both initial infection (type I resistance) and the spread within the spike after infection (type II resistance) could be determined when the inoculation time is known, i.e., artificial inoculation is needed for both. Schroeder and Christensen [8] recorded disease symptoms through evaluation the rate of diseased spikelets and percentages following spray and point inoculation; their data correspond to the disease index. However, they did not find significant differences between resistances to them; the difference for type II resistance (point inoculation) was much lower than *Fusarium*-damaged kernel rate FDK values that were found. Therefore, from our context it is important to include FDK to better understand the role of FDK in the resistance process. Hart et al. [9] spray-inoculated the plants of cultivar Genesee in a greenhouse and seven winter wheat cultivars in a field experiment with a single *F. culmorum* isolate. In the greenhouse experiment, the inoculations were performed at the following developmental stages: $<\frac{1}{4}$ filled, $\frac{1}{4}$–$\frac{1}{2}$ filled, $\frac{1}{2}$–$\frac{3}{4}$ filled, and fully filled. The later inoculations were beyond the flowering time. The early stages were not extensively studied. Inoculation timing in the field ranged from early watery to mid-dough stage. Spikes were covered with plastic bags for varying periods of time following inoculation in both experiments. Deoxynivalenol (DON) production occurred in wheat spikes under favorable moisture conditions, even at late stages of kernel development, while yield reductions were greatest when infections occurred before kernel filling. In a one-year study under a controlled environment, Lacey et al. [10] inoculated one winter wheat cultivar at different time points from spike emergence to harvest using several species of *Fusarium* with varying mist duration after inoculation. Anthesis was the only infection time for which high DON concentrations were observed and both disease severity and DON content sharply decreased for inoculations performed after mid-anthesis. However, significant interactions occurred between infection timing and moisture duration following inoculation, such that DON levels from mid-anthesis infections rose more sharply with increasing postinoculation moisture durations. Del Ponte et al. [11] inoculated a susceptible wheat genotype using a single *F. graminearum* isolate at six stages from mid-anthesis to hard dough in a greenhouse experiment. The percentage of damaged kernels was >94% for inoculations performed between mid-anthesis and late milk stages, but fell to 23% for inoculation at hard dough. The highest DON concentrations were found in samples inoculated at the watery-ripe and early milk stages, but DON was still detected at later stages. In field experiments, Cowger and Arrellano [12] inoculated eight winter wheat cultivars with four *F. graminearum* isolates at mid-anthesis by spray inoculation. The inoculum was a mix of equal proportions of spores of the four isolates. Additional inoculations were made at 10 and 20 days after mid-anthesis in four experimental years. In three of the four years, infections at 10 days after anthesis produced less FDK than infections at anthesis. DON levels were as high from infections at 10 days after anthesis as from infections at anthesis in two years, but lower in two other years. Siou et al. [13] sprayed the highly susceptible winter wheat cultivar Royssac with eight isolates of *F. graminearum*, *F. culmorum* and *F. poae* in a greenhouse experiment. Even the isolate x inoculation time interaction was significant, the authors did not make any comment on this finding. Four dates of inoculation were tested: anther extrusion, 8 days post-anther extrusion, 18 post-anther extrusion (milky kernel development stage) and 28 post-anther extrusion (dough development stage). The highest disease and toxin levels were for inoculations around anthesis, but late infections led to detectable levels of fungus and toxin for the most aggressive isolates. Mesterhazy and Bartók [14] compared fungicide efficacy at full flowering and 10 days thereafter by spraying inoculation. Surprisingly, between disease severity and toxin contamination, no significant difference was found. However, at the first inoculation, the weather was cooler and then later with more rain; there is only weak proof that the window is ten days long. Therefore, the ten days could have only been a random factor.

More isolates in a test are used when aggressiveness of the isolates should be investigated. In serial tests for resistance and phenotyping, only one isolate [15] or a mixture of isolates [10,16,17] are used e.g. one aggressiveness level was applied. Occasionally, the spawn method was also used [18]. In some cases, no source of inoculum is specified [19]. As *F. graminearum* and *F. culmorum* do not have specialized races [20,21] as does, for example, *Puccinia striiformis*, the general conviction is that all inocula are equally good for testing. Two points were missed: (1) there is a variability within species for aggressiveness (the aggressiveness level strongly influences the differentiation of the genotypes) [3,4,22,23], and (2) the aggressiveness of the isolates is not stable, which was proven by tests where more independent isolates were used on different genotypes [2–4,23,24]. As in earlier tests, we worked with four independent isolates, and the isolate role was not highlighted; Therefore in this paper we present details in this matter.

In summary, the papers reviewed confirmed that anthesis is the most susceptible stage for *Fusarium* infection. It also became clear that even late infections can result in high harvest-time DON levels, but the papers do not provide solid information on the length of the SW and its stability within this period, as only two inoculation times represented the early period. For us, it was a problem as to whether or not we have a stability reaction. When we inoculate according the optimal time, we need to inoculate on every second or third day. As weather is seldom stable, every inoculation time might give large differences; the comparison of the data may result in artifacts. This jeopardizes the presentation of the comparable data that makes problems everywhere, in breeding, in genetic analysis, variety registration, etc. When the SW would be stable for a week, it is enough to inoculate once a week and a much higher number of genotypes could be inoculated that will have the same ecological conditions. As we do not have well-supported facts, and it is not known whether the different traits respond similarly or differently, there remained an unsolved problem. We have seen that visual symptoms, FDK, and DON contamination often differ [3,4,22,24,25] Therefore, their role in resistance estimation needed further illumination. The use of one or more inocula was not clarified in the past decades, so we need further research in understanding their role. Therefore, inoculations were planned by using four isolates (two *F. graminearum* and two *F. culmorum*) and seven winter wheat genotypes to provide a deeper understanding of the SW and utilize results in resistance testing at six inoculation dates, from mid-anthesis to 16 days thereafter.

Objectives of this study were: (1) to determine the length of SW in field conditions; (2) to determine the role of visual head symptoms (DI), *Fusarium*-damaged kernels (FDK), and the DON content in forming the length of SW; (3) to describe the role of resistance level in the SW response; and (4) to determine how the results can be used to improve the reliability of the resistance testing.

2. Materials and Methods

2.1. Plant Material

Seven genotypes were selected with differing resistance to FHB, including four winter wheat cultivars from the variety breeding program at Szeged without strong selection for FHB, Hungary, and three lines chosen from our FHB resistance program (Table 1). They had similar flowering times, but differed in FHB resistance level. According to the previous tests, two genotypes proved to be susceptible, one was rated as moderately susceptible, two were rated as moderately resistant, and two were rated as resistant.

Table 1. Genotypes, their abbreviations and resistance classification Szeged, 2013–2014.

Genotype	Resistance Class	Abbreviation
F 569//Ttj/RC103/3/Várkony/4/Ttj/RC103/3/81.60/NB//Kő	R	F569/Kő
F 569//Ttj/RC 103/3/Várkony/4/Ttj/81.F.379	R	F569/81/F379
GK 09.09	MS	GK 9.09
GK Fény	MR	GK Fény
GK Garaboly	S	GK Garaboly
GK Csillag	MR	GK Csillag
GK Futár	S	GK Futár

2.2. Field Experiments for Disease Evaluation

The plant material was evaluated in the nursery of the Cereal Research Nonprofit Ltd. in Szeged, Hungary, (46°14′24″ N, 20°5′39″ E) over two seasons (2012/2013 and 2013/2014). The field experiments were conducted in three replications in a randomized complete block design. A plot (genotype, replicate) consisted of 12 rows, 1.5 m long with a 20 cm row spacing. For the six inoculation times, two row subplots were used. For control, the first inoculation served as the mid-anthesis inoculation (Figure 1). The seven cultivars and three replicates produced 21 plots for an experiment. Each genotype per replications was sown in six (one plot per inoculation date) two-row plots of 1.5 m length in mid-October, using a Wintersteiger Plot Spider planter (Wintersteiger GmbH, Ried, Austria). The width of the plots was approximately 40 cm.

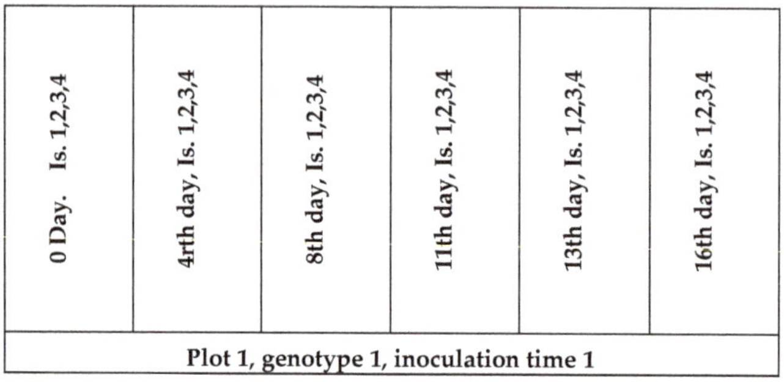

0 Day. Is. 1,2,3,4
4rth day, Is. 1,2,3,4
8th day, Is. 1,2,3,4
11th day, Is. 1,2,3,4
13th day, Is. 1,2,3,4
16th day, Is. 1,2,3,4
Plot 1, genotype 1, inoculation time 1

Figure 1. Map of a plot in the experiment. Isolates: 1, 2, 3, 4, inoculated bunches about 50 cm from each other.

2.3. Inoculum Production and Inoculation Procedure

Fungal suspensions were made in 10 L heat-stable glass flasks filled with 9.4 L liquid Czapek–Dox medium [26]. They were aerated at room temperature for a week. Suspensions were then stored at 4 °C until they were used for inoculation. Inocula contained mycelium and conidia as well. Each genotype was inoculated individually with two *F. culmorum* (F.c. 12375 /1/, F.c. 52.10 /2/) and two *F. graminearum* (F.g. 19.42 /3/, Fg. 13.38 /4/) isolates. The same isolates were used for studies in 2013 and 2014. The cited authors used a different conidium concentration/mL. Schroeder and Christensen [8] applied an inoculum with 5–8 × 10^6 conidia/mL, Hart et al. [9] used 6.6 × 10^4 conidia /mL, Cowger and Arrellano [12] used 1 × 10^5 conidia/mL, and Siou et al. used 2 × 10^4 conidia/mL [13]. This suggests that there is no agreement for which concentration is optimal. The authors did not report the reason for using the given concentration. Theoretically, it should have been some relation with aggressiveness, but none intended to give an explanation. This means that chances were low to find a good solution. However, an aggressiveness test [26] was developed to measure directly the aggressiveness of the given inoculum. This solved the problem and we could test this very important feature. A test needed 10 Petri dishes, 5–5 for two genotypes in seedling stage. Sterile double layer filter paper was placed in 12 cm diameter dishes, 9 mL suspension was uniformly spread on the surface and 25 seeds were placed into the dish in a 5 × 5 binding. Besides the original concentration, 1:1, 1:2, and 1:4 dilutions were also applied, along with a control with sterile distilled water. The rating was made daily from the second to sixth day. The number of healthy germs was evaluated. From the test, the aggressiveness could be directly seen. Therefore, the selection of the inocula was made in this system. As we did not want to relate the aggressiveness of the isolates used to each other, the standardized conidium concentration was not necessary. Even so, the four isolates differed significantly from each other.

Six inoculation dates were tested, the first at anthesis at Feekes growth stage 10.5.1 [27], second at four days, third at eight days, fourth at 11 days, fifth at 13 days, and the sixth 16 days later. Inoculation started on about 10 May. The inoculation was performed with the spray inoculation method [23]. Bunches of 15–25 spikes were sprayed from all sides with a handheld sprayer with the use of 15–20 mL fungal suspension for each sample, as described by Mesterhazy [23,28].

After inoculation, bunches were covered for 48 h in a transparent polyethylene bag, sealed to maintain 100% relative humidity and promote infection. After removing the bags, the plants were loosely bound with a label for identification at half-plant height to allow the leaves to photosynthesize freely.

2.4. Disease Assessment

Evaluation of FHB disease severity started on the 10th day after inoculation and was repeated on 14th, 18th, 22nd, and 26th day, until the control heads started to become yellowish. The percentage of infected spikelets was rated for the whole group of heads. On average, 24 spikelets were in a head. When one spikelet was infected in each head, this was 4%, when only one was in the bunch of 15, then this was 0.3% [28], all percent data mean disease index (DI). This was sensitive enough to provide a comparable data series to FDK and DON data. Their arithmetical mean values across all observations per groups of heads were used as entries for statistical analysis. At maturity (beginning in July), the groups of inoculated heads were harvested manually and stored in paper bags. The samples were threshed using a stationary thresher (Wintersteiger LD 180, Ried, Austria) at low wind, in order to retain the shriveled *Fusarium*-damaged kernels (Mesterházy, 1987, 1995). Chaff was separated using an Ets Plaut-Aubry (41290 Conan-Oucques, France) air separator. FDK was rated visually (estimation of scabby grains with definite tombstone, or chalky-white and rose discoloration as a percentage). All of the work was finished at the end of July.

2.5. Deoxynivalenol Analysis

Of the grains in a group of heads, six grams were separated and milled by a Perten Laboratory Mill (Type: 3310, Perten Instruments, 126 53 Hagersten, Sweden). From this, 1 gram was used for toxin extraction for each replicate of each inoculation date in the case of each isolate with 4 mL of acetonitrile/water (84/16, *v*/*v*) for 2.5 h with a vertical shaker. Following centrifugation (10,000 rpm, 10 min), 2.5 mL of the extract was passed through an activated charcoal/neutral alumina SPE column at a flow rate of 1 mL/min. Thereafter, 1.5 mL of the clear extract was transferred to a vial and evaporated to dryness at 40 °C under vacuum. The residue was dissolved in 500 µL of acetonitrile/water (20/80, *v*/*v*). HPLC separation and quantification were performed on an Agilent 1260 HPLC system (Agilent Technologies, Santa Clara, California, USA), which was equipped with a membrane degasser, a binary pump, a standard autosampler, a thermostatted column compartment, and a diode array detector (DAD). DON was separated on a Zorbax SB-Aq (4.6 × 50 × 3.5 µm) column (Agilent) equipped with a Zorbax SB-Aq guard column (4.6 × 12.5 × 5 µm) thermostatted at 40 °C. The mobile phase A was water, while mobile phase B was acetonitrile. The gradient elution was performed as follows: 0 min, 5% B; 5 min, 15% B; 8 min, 15% B; 10 min, 5% B; 12 min, 5% B. The flow rate was set to 1 mL/min. The injection volume was 5 µL. DON was monitored at 219 nm. This procedure updated the methodology used by Mesterhazy et al. (1999) [3]. The only difference was that the measurements were made by a new Agilent Infinity 1260 HPLC system.

2.6. Statistical Analysis

Correlation analysis was performed using the functions provided by Microsoft Excel 2013's Analysis ToolPak add-in program. The statistical evaluations for four-way ANOVA followed the functions by Sváb (1981) [29] and Weber (1967) [30], with the help of the Microsoft Excel background. The significance between the main effect and the interactions were developed by functions from Weber [30], the *df* values in ANOVA for these are given in Table 2.

Table 2. ANOVAs for disease index, FDK, and DON contamination for the susceptibility window experiment, 2013 and 2014, Szeged.

Source of Variance	df	FHB			FDK			DON		
variance		MS	F	F AxBxCxD	MS	F	F AxBxCxD	MS	F	F AxBxCxD
Genotype A	6	4214.2	223.2 ***	166.3 ***	5524.5	81.07 ***	45.4 ***	738.7	31.1 ***	13.5 ***
Inoculations B	5	680.1	36.0 ***	26.8 ***	9934.2	145.7 ***	81.6 ***	3003.5	126.7 ***	55.2 ***
Isolate C	3	17680.0	936.5 ***	697.9 ***	58206.4	854.2 ***	478.6 ***	15941.6	672.4 ***	293.2 ***
Year D	1	14002.3	741.7 ***	552.7 ***	29487.3	432.7 ***	242.4 ***	12027.7	507.3 ***	221.2 ***
AxB	30	169.0	8.95 ***	6.6 ***	432.4	6.3 ***	3.5 ***	104.0	4.3 ***	1.9 **
AxC	18	400.9	21.2 ***	15.8 ***	774.9	11.3 ***	6.3 ***	363.6	15.3 ***	6.6 ***
AxD	6	740.8	39.2 ***	29.2 ***	1258.9	18.4 ***	10.3 ***	429.4	18.1 ***	7.8 ***
BxC	15	660.6	34.9 ***	26.0 ***	1250.9	18.3 ***	10.2 ***	1194.9	50.4 ***	21.9 ***
BxD	5	604.3	32.0 ***	23.8 ***	3329.4	48.8 ***	27.3 ***	1090.1	45.9 ***	20.0 ***
CxD	3	6059.2	320.9 ***	239.2 ***	20472.8	300.4 ***	168.3 ***	7674.9	323.7 ***	141.1 ***
AxBxC	90	45.9	2.4 ***	1.8 ***	109.6	1.6 ***	0.90ns	63.3	2.6 ***	1.2ns
AxBxD	30	114.6	6.06 ***	4.5 ***	337.0	4.9 ***	2.7 ***	79.4	3.3 ***	1.5 ns
AxCxD	18	80.0	4.2 ***	3.1 ***	249.7	3.6 ***	2.05 ***	269.0	11.3 ***	4.9 ***
BxCxD	15	160.4	8.4 ***	6.3 ***	902.7	13.2 ***	7.42 ***	488.1	20.5 ***	8.9 ***
AxBxCxD	90	25.3	1.34 *	1.0 ns	121.6	1.78 ***	0.99 ns	54.4	2.29 ***	1.0 ns
Within	672	18.9			68.1			23.7		
Total	1007									
Interactions			*p*			*p*			*p*	
A/AB	*df*: 6 /30	24.94	***		12.08	***		7.10	***	
B/AB	*df*: 5 /30	4.02	**		22.97	***		28.88	***	
A/AC	*df*: 6/18	10.51	***		6.74	***		2.03	ns	
C/AC	*df*: 3/18	44.10	***		75.11	***		43.84	***	
B/BC	*df*: 5/15	1.03	ns		7.94	***		2.51	ns	
C/BC	*df*: 3/15	26.76	***		46.53	***		13.34	***	
A/AD	*df*: 6/6	5.69	*		4.39	*		1.720	ns	
D/AD	*df*: 1/6	18.90	***		23.42	***		28.01	***	

*** $p = 0.001$, ** $p = 0.01$, * $p = 0.05$, ns = nonsignificant.

3. Results

The visual data that are generally considered as resistance data did not show any decrease in the 16-day inoculation period (Table 3). At the 8th and 16th day, larger values were found, but values found on the other four days were at the same level. This means that the visual scores did not show any sign of the susceptibility window in the first 16 days, i.e., the whole period acted as a long susceptibility window without signs of decrease, even following the flowering; a decrease was anticipated. The resistance differences between genotypes were highly significant; F569/Kö was the best at 5.18% and GK Futár was the most susceptible at 18.9% disease index. The cultivars reacted rather inconsistently, the most resistant varieties gave the same numbers across the whole period, but others showed variable performance, as shown by the diverging correlation coefficients (Table 3A). On the other hand, the correlations between variety reactions and different inoculation dates were much higher, and only several were not significant at $p = 0.05$. This would indicate a reasonable stability of responses across the different inoculation dates.

Table 3. Susceptibility window for *Fusarium* head blight (FHB) in wheat, disease index data, as the % of infected spikelets, across four independent isolates, 2013–2014, Szeged.

Genotype	Inoculation Days after Mid-Flowering Data DI %						Mean
	0	4	8	11	13	16	
F 569/Kő	3.81	5.76	5.50	5.04	5.58	5.39	5.18
F 569/81.F.379	7.26	4.78	8.52	7.08	3.91	7.40	6.49
GK 09.09	9.43	8.71	10.62	8.30	10.19	14.35	10.27
GK Fény	10.20	12.46	17.29	14.27	14.69	18.15	14.51
GK Garaboly	19.92	16.64	23.09	11.27	11.11	17.98	16.67
GK Csillag	13.32	12.82	15.78	15.47	17.66	26.03	16.85
GK Futár	15.80	16.67	21.66	20.90	16.83	21.49	18.89
Mean	11.39	11.12	14.64	11.76	11.42	15.83	12.69
Limit of significant difference (LSD) 5% variety							0.72
LSD 5% inoculation date							0.67
A/Corr. between cultivars	**F 569/Kő**	**F 569/ 81.F.379**	**GK 09.09**	**GK Fény**	**GK Garaboly**	**GK Csillag**	**GK Futár**
F 569/81.F.379	−0.3721						
GK 09.09	0.1879	0.2855					
GK Fény	0.6138	0.3269	0.7156				
GK Garaboly	−0.2147	0.6670	0.3078	0.1273			
GK Csillag	0.2694	0.1744	0.9295 **	0.7611 *	−0.0281		
GK Futár	0.3364	0.6620	0.4655	0.8466 *	0.1810	0.5522	
B/Corr. between Inoculations	**0**	**4**	**8**	**11**	**13**	**16**	
4	0.9210 **						
8	0.9458 **	0.9745 ***					
11	0.6545	0.8210 *	0.7924 *				
13	0.6178	0.8072 *	0.7287	0.9025 **			
16	0.7372	0.8223 *	0.7736 *	0.8533 *	0.9592 ***		
Mean	0.8791 **	0.9626 ***	0.9392 **	0.9072 **	0.9096 **	0.9351 **	

*** $p = 0.001$, ** $p = 0.01$, * $p = 0.05$.

The FDK data (Table 4A) showed a different picture. Between mid-flowering and eight days after inoculation, except the fourth day (it is significantly lower than the initial and eighth day values), stability was mostly seen, thereafter a sharp, more than 50% reduction was recorded, followed by an additional one-third decrease that remained stable afterwards also by the 16th day. This decrease is highly significant, with limit of significant difference (LSD) 5% at 1.76%. The cultivar differences between means were five-fold, the most resistant showed 5.43% FDK and the most susceptible showed 25.1%. The variation width was near 20%, the LSD 5% value was 1.91, so a roughly ten-fold difference existed between this and the variation width. This meant a rather stable reaction in the first eight days, experimental proof to the presence of the susceptibility window. We should say that the correlations between genotype reactions (Table 4B) were much closer than they were in the FHB disease index, all correlations were significant between $p = 0.05$ and $p = 0.001$. This indicated a similar response of the cultivar reactions in the 16 day inoculation period. The reduction can be seen at all resistance levels. The rate was similar, but the data were much lower in the more resistant genotypes, which explains the closer correlation values. The correlations between the response of the genotypes to different inoculation data were also similar (Table 4C), but in several cases the level was lower, and three correlations were not significant.

Table 4. Susceptibility window for FHB in wheat, FDK data as percentage across fours independent isolates, 2013–2014, Szeged.

Genotype	**Inoculation Days after Mid-Flowering Data: %**						**Mean**
A/Original data	**0**	**4**	**8**	**11**	**13**	**16**	
F 569/Kő	8.96	7.90	8.06	2.88	2.83	1.95	5.43
F 569/81.F.379	18.21	17.00	18.98	7.63	4.13	6.17	12.02
GK 09.09	22.02	12.02	14.00	8.83	9.08	8.04	12.33
GK Csillag	21.86	15.46	20.60	9.27	5.90	5.96	13.18
GK Fény	22.63	21.17	22.31	10.77	5.28	6.65	14.80
GK Futár	28.98	28.09	28.17	13.10	8.82	7.88	19.17
GK Garaboly	41.29	27.04	43.75	16.71	11.29	10.83	25.15
Mean	23.42	18.38	22.27	9.89	6.76	6.78	14.58
LSD 5% genotype							1.90
LSD 5% inoculation							1.76
B/Corr. between Cultivars	**F 569/Kő**	**F 569/81.F.379**	**GK 09.09**	**GK Csillag**	**GK Fény**	**GK Futár**	
F 569/81.F.379	0.9723 ***						
GK 09.09	0.8371 *	0.7689 *					
GK Csillag	0.9624 ***	0.9681 ***	0.8731 *				
GK Fény	0.9742 ***	0.9942 ***	0.7781 *	0.9682 ***			
GK Futár	0.9884 ***	0.9873 ***	0.7802 *	0.9595 ***	0.9945 ***		
GK Garaboly	0.9308 **	0.9471 **	0.8370 *	0.9891 ***	0.9378 **	0.9260 **	
C/Corr. between Inoculations	**0**	**4**	**8**	**11**	**13**		
4	0.8473 *						
8	0.9618 ***	0.8851 **					
11	0.9832 ***	0.9088 **	0.9492 **				
13	0.8960 **	0.6409	0.7512	0.8553 *			
16	0.9412 **	0.7517	0.8409 *	0.9310 **	0.9159 **		

*** $p = 0.001$, ** $p = 0.01$, * $p = 0.05$.

The DON data (Table 5A) for the 0–8th day showed a stable mean DON contamination between 9.23 and 9.81 mg/kg concentration. Most of the cultivars were stable, and the differences between inoculation dates were not large. The decrease of DON contamination was very sharp after the eighth day, only one-third remained from the eighth day amount. This decreased by an additional 50% to the 13th day and a further significant decrease was recorded between the 11th and 13th day. The additional small decrease up to the 16th day was not significant.

Table 5. Susceptibility window for FHB in wheat, DON contamination, mg/kg across four independent isolates, 2013–2014, Szeged.

Genotype	Inoculation, Days after Mid-Flowering, Data: mg/kg						Mean
A/Original Data	**0**	**4**	**8**	**11**	**13**	**16**	
F 569/Kő	2.98	2.68	3.17	1.08	0.92	0.57	1.90
GK 09.09	6.96	7.57	7.10	2.47	2.94	1.21	4.71
F 569/81.F.379	7.06	9.05	9.46	1.94	0.48	1.30	4.88
GK Csillag	10.27	11.01	7.71	2.80	1.28	0.64	5.62
GK Fény	9.84	12.84	12.54	4.62	1.25	1.90	7.16
GK Garaboly	15.36	11.44	15.05	4.53	1.69	1.12	8.20
GK Futár	12.15	13.71	13.66	6.73	2.11	1.01	8.23
Mean	9.23	9.76	9.81	3.45	1.52	1.11	5.81
LSD 5% genotype							1.12
LSD 5% inoculation							1.04
B/Corr. between Cultivars	**F569/Kő**	**GK 09.09**	**F569/81.F.379**	**GK Csillag**	**GK Fény**	**GK Garaboly**	
GK 09.09	0.9740 ***						
F 569/81.F.379	0.9590 ***	0.9516 ***					
GK Csillag	0.9315 **	0.9619 ***	0.9239 **				
GK Fény	0.9497 **	0.9478 **	0.9913 ***	0.9362 **			
GK Garaboly	0.9894 ****	0.9389 **	0.9334 **	0.9184 **	0.9285 **		
GK Futár	0.9579 ***	0.9499 **	0.9591 ***	0.9439 **	0.9829 ***	0.9488 **	
C/Corr. between Inoculations	**0**	**4**	**8**	**11**	**13**	**16**	
4	0.8372 *						
8	0.9035 **	0.8760 **					
11	0.7830 *	0.8575 **	0.8551 *				
13	0.2636	0.1840	0.1688	0.3784			
16	0.2300	0.4884	0.5361	0.3456	0.0534		

**** $p = 0.0001$, *** $p = 0.001$, ** $p = 0.01$, * $p = 0.05$.

Therefore, counting by FDK and FDK, the susceptibility window lasted for the same eight days for all resistance classes. Also, different cultivar reactions can be recognized. The most resistant F569/Kő had the lowest DON at about 3 mg/kg, and this could decrease the DON level below the EU limit of 1.25 mg/kg. The case of the GK Fény variety is important, since in 2010 the large FHB epidemic damaged it only moderately in contrast to GK Garaboly, which had the highest DON contamination at flowering. We observed a similar phenomenon with GK Csillag; its resistance was better than GK Fény. Attention should be paid to the fact that in normal testing only the first inoculation was made and the others were not existing. The data show also that to see the resistance behavior for delicate and highly important cultivars, the 11th- and 16th-day inoculation is suggested. By this way we could detect cultivars like GK Csillag, that was rather susceptible at mid-anthesis and could be discarded based on disease index, but in later inoculation dates it gave the second place following the most resistant genotype. Correlations between genotype reactions were very close, as it was for FDK, but at a higher level (Table 5B). Therefore, cultivars across inoculation times behaved similarly. We should also consider that good correlations existed for the DON data only between the first four inoculation dates, during the later inoculation periods no significant correlations were found (Table 5C). This means that the resistance relationships do not automatically follow this range later on, so therefore unexpected phenomena may occur. Only one example, GK Csillag, had 10 ppm DON content at mid-flowering and 0.64 ppm on the last inoculation. F569/Kő started with 2.98 and finished with 0.57 mg/kg, practically the same value. The DON data in graphical form (Figure 2) shows the genotype reaction during the six inoculation periods. The DON regulation seemed to be more complicated than usually expected.

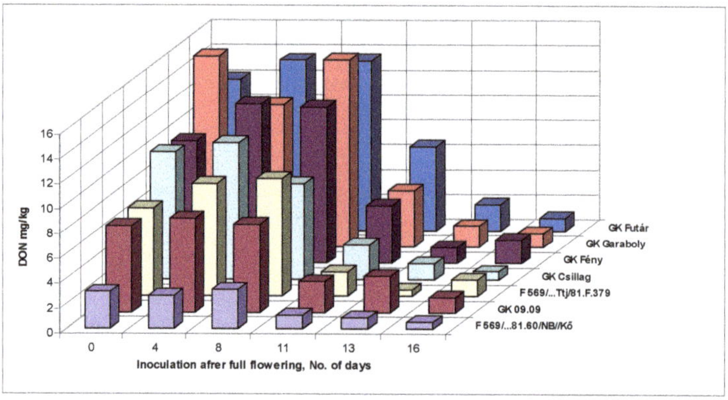

Figure 2. Susceptibility window against FHB in wheat, DON contamination of the wheat genotypes at different inoculation dates, Szeged, 2013 and 2014. LSD 5% genotypes: 1.12; inoculation days: 1.04.

The general means of the isolate data showed rather large differences for DI, but for all isolates the genotype differences were significant (Figure 3). Fg 19.42 isolate was the most aggressive; the others displayed only 40–50% of its value. The FDK data showed larger isolate differences than DI, the resistance differentiation less expressed for the three less aggressive isolate (Figure 4). We had the largest difference between isolates for DON compared to DI and FDK (Figure 5); the Fg 19.42 isolate kept its excellent performance, the Fc 12375 showed yet significant differences, and the two least aggressive isolates did not show a proper differentiation, as they showed for DI and FDK. It seems that even at medium aggressiveness in DI or FDK can cause a problem in DON response.

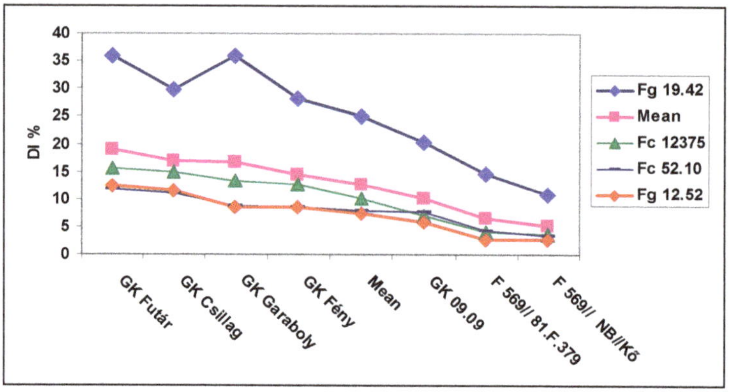

Figure 3. Susceptibility window in wheat, disease index (DI) % general means for the four isolates in the seven genotypes differing in resistance, 2013 and 2014, Szeged. LSD 5% between any data: 2.01.

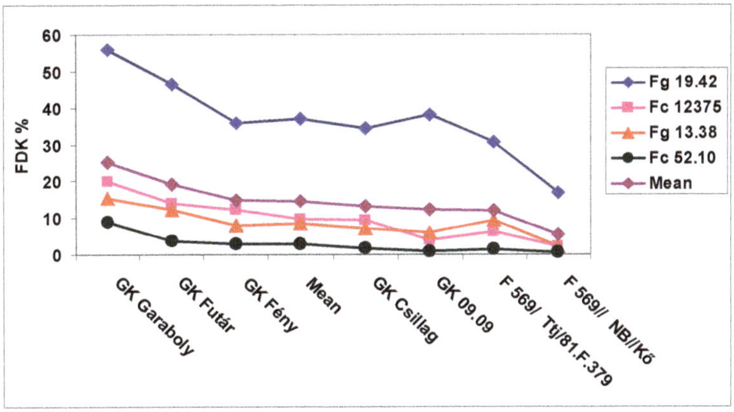

Figure 4. Susceptibility window in wheat, FDK % for the general means of the four isolates in the seven genotypes differing in resistance, 2013 and 2014, Szeged. LSD 5% between any data: 3.81.

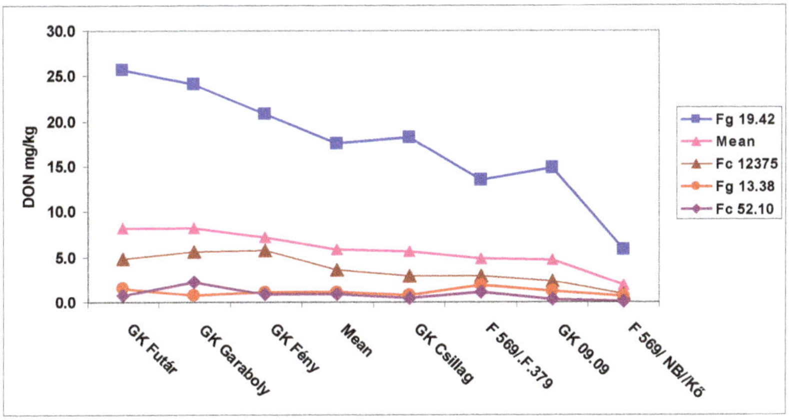

Figure 5. Susceptibility window in wheat, DON mg/kg values for the general means of the four isolates in the seven genotypes differing in resistance, 2013 and 2014, Szeged. LSD 5% between any data: 2.25.

Looking at the three traits across cultivars, we observed a clear picture of how the FHB data did not show any sign of decreasing during the 16-day period. However, the FDK and DON showed much similarity (Figure 6). The correlation for disease index/FDK was r = −0.1639, DI/DON was r = −0.2347, and FDK/DON was highly significant at r = 0.9673 with $p = 0.01$. This is again an argument that FDK is a more important trait to characterize resistance level than DI.

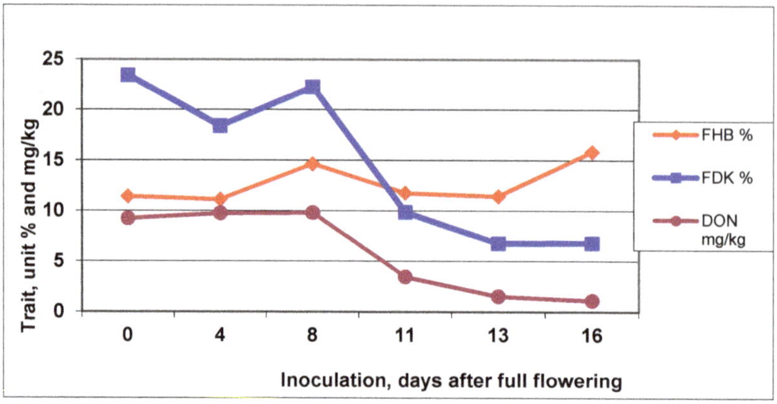

Figure 6. FHB disease index, FDK and DON means across genotypes at six inoculation time points during 2013 and 2014, Szeged

The general means for the genotypes and traits showed several tendencies (Figure 7). The most resistant F569/Kő showed the lowest values for all three traits. The F569/81.F.379 had an FDK value twice as large compared to the FHB than the most resistant genotype. As we have the mean values of six inoculations, the resistance data are more reliable than only with one inoculation at full flowering stage. Thus, this two-fold difference does not seem to have occurred due to random chance. It is remarkable that GK Csillag was for visual symptoms among the more susceptible cultivars, but one of the best in counting for DON.

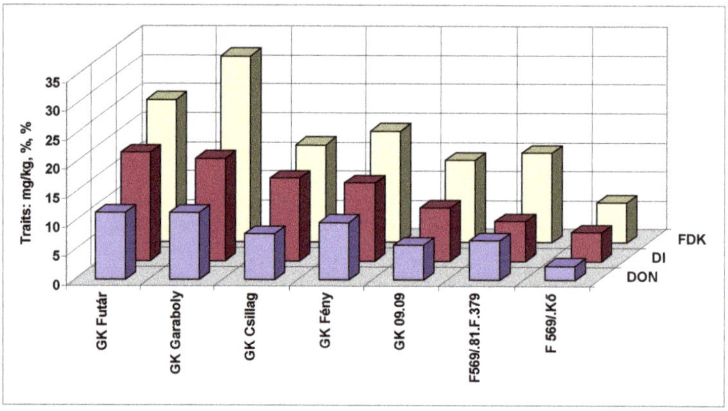

Figure 7. Susceptibility window against FHB in wheat showing the general pattern for genotypes and traits FHB (DI) %, FDK %, and DON mg/kg, 2013 and 2014, Szeged

The ANOVA of the three traits (Table 2) showed highly significant main effects. As the four-way interaction was significant in all tests, we also calculated the F-value for the AxBxCxD interaction, but basically the result remained very similar to what we found for F-values against the "Within." Beyond this, for us especially, the two-way interactions were important, which contained the main effect Genotype (A) and the other three main effects: Inoculations, Isolates, and Year (AxB, AxC, AxD). When the main effect is dominant above the two-way interaction, it means that the genotype performance is stable under the different inoculation times. Thus, in different inoculation times, even we found a significant influence, but with a modifying effect only and the probability of a very variable performance at these conditions was very low. When the main effect does not differ significantly from the interaction, it means that the stability of the main effect is poor, and we may count it with differing responses. It is also important whether the traits responded similarly or differently. There were examples for both. The Genotype (A) main effect was dominant over the AxB interaction (MS = 24.94); its level is shown by asterisks. This dominance was very strong at FHB, medium for FDK, and lowest for DON, so the traits behaved differently, even the level of significance was $p = 0.001$. For the inoculation time (B), this was opposite to AxB; for FHB data this was just above the limit of significance (as between inoculations the difference was small), but it was very high for FDK and DON, where the last inoculations gave only small DON content compared to the first three inoculation dates. The Genotype (A) main effect to AxC interaction was highly significant and the tendency corresponded to the fact we found with the Genotype main effect tested against the AxB interaction. The Isolate (C) effect was similarly dominant against AxC interaction for all traits, indicating that the main factor was the Isolate effect, being rather stable at different traits. Regarding the Inoculation date (B) interaction against the BxC interaction, we did not see significant differences except FDK; for FHB and DON the result was neutral. However, the Isolate (C) effect was highly significant for all traits against BxC interaction, but the smallest number was found for DON, and even it was highly significant. The Genotype effect against AxD was significant at $p = 0.05$ for FHB and FDK, but no difference with DON, indicating no special interaction between Genotype and Year. The Year effect was highly significant for all traits against AxD, showing a strong dominance of the year over the interaction, i.e., the stability of the Genotype performance in different years.

The degrees of freedom for the interactions were the same as those listed in the main ANOVA (Table 2). We counted the means for two years for all cultivars and isolates separately for each trait and then calculated the Pearson correlation coefficients for all isolates between traits (Table 6). All correlations were significant for isolate 3. None were significant for isolate 4. Two were significant for isolate 1 and only one for isolate 2. Furthermore, all were highly significant for the mean reactions. One general conclusion is certain, namely, the different isolates do not respond the same way. Therefore, the use of a single isolate can be risky. Here we should remark that any of the four isolates can be one in a paper where only one inoculum was used. For a publication where only one was available for phenotyping, a large variation might occur, and the genetic meaning of the data is instable. This is valid for mapping work, but also for resistance evaluation of the genotypes in breeding or cultivar registration.

Table 6. Susceptibility window against FHB in wheat, 2013 and 2014. Correlation between different traits and the corresponding trait values. For orientation, the general means of the data are given.

Isolates	Correlations			Means for Isolates		
	DI/FDK	DI/DON	FDK/DON	FHB DI%	FDK%	DON mg/kg
1 F.c. 12375/1	0.774 *	0.732	0.904 **	10.19	9.80	3.62
2 F.c. 52.10 /2	0.427	0.173	0.916 **	7.48	2.91	0.88
3 F.g. 19.42	0.863 *	0.959 ***	0.899 **	25.13	36.93	17.60
4 F.g. 13.38/4	0.499	−0.062	0.262	7.96	8.69	1.16
Mean	0.771 *	0.869 *	0.921 **	14.58	12.69	5.81

*** $p = 0.001$, ** $p = 0.01$, * $p = 0.05$; DI = disease index; F.c.: *F. culmorum*; F.g.: *F. graminearum*.

4. Discussion

4.1. Susceptibility Window

Despite the fact that the susceptibility window or the flowering indicated susceptibility is well known in the *Fusarium* literature, Selby and Manns [31], as cited by Stack [32] and Atanasoff [5], mentioned it as a susceptible stage, gave no details were given. Several papers analyzed the whole period between flowering and early ripening, e.g., del Ponte et al. [11], with six inoculations from flowering to late stages of kernel development, but it was not detailed for the early SW period. The largest toxin DON contamination was found at the watery milk stage with 98 mg/kg, the smallest was only 1.2 mg/kg at the end of the vegetation period. For the disease index, here the 16-day long testing period was rather stable, with nearly no significant difference between the inoculation day's means. This means that on this basis the identification of the SW was not possible. For FDK and DON, a sharp decrease was found in all genotypes between the 8th and 11th day. This would mean that late palea and lemma inoculations were successful, but its spreading to the developing grains was inhibited by an unknown effect. We should realize that the SW could be identified only by the FDK and DON data. The stability for DON was the most useful in the first eight days. As DON is the ultimate trait most needed, this is good news for the artificial inoculation. This means that during the first eight days we can pool the genotypes flowering in the first six days, for another we can pool those that flower between the 7th and 12th days, and the rest can be inoculated when the latest flower. As inoculations should not be made on every second or third day, they can be pooled to one day. The background weather noise in the data will be smaller and the data will be more comparable. In the last years we used two or maximum three inoculation dates [24] and a scientific control test series was needed, how this modification influences the exactness of the experi9ment5ation. The conclusion is that two inoculations were suitable when the season is warmer; in cooler weather three inoculations were needed. It might happen that ecological conditions for the two inoculation dates were similar thereafter. When the variation in the first and second group have similar means, similar minimum and maximum values, then the whole data basis can be pooled. If not, for a resistance test where earlier, middle flowering or late controls are included, this makes no problem in selection, but creates a problem when comparing the resistance of the different flowering groups. This pooled inoculation time can also help in fungicide tests, where cultivars having a difference of 4–5 days in flowering can be inoculated on the same day, when similar flowering-time genotypes could not be applied.

4.2. Resistance Expression

Resistance differences were large and mostly similar across the six inoculation dates. As we had in regular tests only one inoculation time, we cannot see variation in responses, but basic differences between years and isolates were not found. The data showed that differences were the largest for the early inoculations, later the differences become smaller. Resistance is measured so the behaviors of the cultivars for the visual reaction (D) can be evaluated. This corresponds to the general assumption that the resistance should be evaluated by the disease symptoms of the heads. A great number of papers using this approach prove the strength of this idea clearly. It has long been clear that the fine regulation of FDK and DON might follow different patterns. First, we recognized that they are not absolutely interdependent [3,23], indicating that even these three traits may correlate well, a part of the genotypes react differently and this allowed us to describe specific resistance traits to FDK and DON. Mesterhazy et al. [4] showed that about 80% of the decrease in DON is a consequence of the higher resistance, but other resistance-independent genetic regulation(s) exists. For this reason, it is not surprising that the same FHB values and different FDK and DON values can be seen (Figure 3). As the data show the means of six inoculation times, the supporting power of the data is larger than it would be for only one single isolate. As in visual symptoms, no clear tendency exists, and we can only state that differences in resistance exist, corresponding with earlier data. However, for FDK, five genotypes had stable performance during the first three inoculations, but GK Garaboly and GK Csillag produced larger differences. We do not think this happened by random chance. From the fourth inoculation the three more resistant genotypes did not show large variation, but afterwards a continuous decrease was seen. The DON data again showed a differing picture. Between the first four inoculations we had significant correlations, but the correlations for DON between the first four and the last two inoculations did not show a significant relationship. For example, GK Csillag had 10 mg/kg toxin content for the first inoculation, but afterwards had only 0.64 mg/kg for the sixth inoculation. This is nearly the same value (0.57 mg/kg) that was found for the most resistant genotype F569/Kő, which had only 2.98 mg/kg DON for the first inoculation. Such behavior of GK Csillag was hypothesized earlier, but the first scientific proof was produced in this paper. Such tests may contribute to better descriptions of the resistance behavior of highly important cultivars for FDK and DON, even in a regular screening it is not enough to present data to understand and explain the "irrational" behavior of the given cultivar. Such a trait may have great economic importance; it is not only a scientific result. Even GK Csillag and F569/.Kö had nearly the same DON contamination on the 16th day, the F569/.Kő resists much better in the early, highly susceptible phase. Csillag and F569/.Kő can be protected effectively by fungicides with lower DON content than the 1.25 mg/kg EU limit. With higher susceptibility without the extra ability of GK Csillag to reduce DON, the effect of fungicide efficacy can be problematic.

We have recognized that very young, strongly differentiating tissues are more susceptible, as seen in young roots during germination [26,33], and this paper together with many others, starting from Atanasoff [5], indicate a high susceptibility in the flowering time. Older tissues are less sensitive as demonstrated by del Ponte et al. [11], Cowger and Arrellano [12], and Siou et al. [13]. This study showed also that the very young head tissues are more severely infected than the later inoculated ones. However, the lemma and palea keep their susceptibility in the first 16 days, so their infection was continuously at a high level without expressed change in any direction. It seemed that the grains differed from this, since after a week (here eight days) a significant decrease in the infection severity was found by a mean 50% reduction. Thus, the data support our earlier finding [23] that FDK is regulated not only by the visual symptoms, but also by other mechanisms not known in detail. The toxin contamination decreased most intensively, seeming that aging was the most important role here. However, beside the physiological and epidemiological responses, genetic effects can also be discovered. To continue these investigations, breeders can have more profound information about their varieties, and when growers know about is, they can request such cultivars.

It seems that resistance expression is much more complex than generally supposed. Without mapping FDK and DON, a reliable picture of the genetic background of the genotypes is not possible. The differences might be explained not only by sometimes poor methodology or environment, but genetic effects also play a significant role.

4.3. Advantage of Using More Isolates

Most papers use a single inoculum for resistance tests, most often mixtures from different isolates. As the *Fusarium* head blight pathogens do not have specialized races [20,23,34,35] such as yellow rust, this seems reasonable. However, the experimental data gave variable results for the different isolates [34]. It seems that there is a general assumption that the aggressiveness of an isolate is stable. However, this cannot be tested when we have only one single inoculum. It is known that between years we have large differences for the same isolate. When we have more parallel isolates, the interest is in the ranking of the isolates, as this shows really the changing aggressiveness [3,4,21]. The source of the differences is the differing aggressiveness of the individual inocula, either in the case of pure isolates or mixtures. The correlations and isolate specific means showed this trend clearly (Table 6). The data also showed remarkably that every isolate has a different pattern to cause head symptoms, FDK or DON. As we do not have stability in aggressiveness across years and isolates, we do not have the same infection patterns for the different isolates and as a consequence we receive highly variable correlations between traits in different isolates. The risk is high, working with one inoculum to receive less reliable results. For research and phenotyping, we need exact data as far as possible. This is the reason that we use four independent inocula in such tests. Ecologically, the results are comparable as they were measured under the same ecological conditions, so an environment/epidemic severity interaction within a specific year is not the case. However, 2–4 years of study is necessary because between years and other sources of variance we have mostly significant interactions. There is an additional practical necessity. When projects or contract work are done, the probability is high to get low infection pressure and the test can be less successful than it should be. For this, a high infection pressure is very important. The chance is very low that not all isolates will give the accepted results in a year. When one or two fail, the test can be yet successful. With a single isolate, however, the year should be repeated, if it is possible at all.

The possible exact phenotyping is the key factor in breeding, genetic studies, and all branches of research where interactions between plant and any other influencing agents are studied. Less reliable resistance data may lead to under- or overestimation of resistance, may cause QTL artifacts in research, etc. To improve the quality of the work, the new approaches may help further both theoretical genetic research and practical breeding.

4.4. Resistance Testing Aspects

The pooling of inoculation dates makes it possible for treating a larger population in an inoculation and genotype differences can be identified better. The aggressiveness of the inocula for large scale tests should be checked before use. In this way, the use of inocula with standardized conidium concentration with low or no aggressiveness can be avoided. We need less inoculation times; the results will be more comparable. As resistance expression differs for different isolates, the use of more isolates gives more reliable data on resistance level and its reliability, therefore representing a higher phenotyping quality where it is needed. All published reports stress the need for low DON contamination. As DON contamination does not correlate well in about 20% of the genotypes where DON overproduction and relative DON reduction [4] occurs, its testing should be included. Therefore, serial testing for DON in an advanced stadium of breeding is important to discard toxin overproducers and highly susceptible genotypes during breeding and variety registration tests [24].

Author Contributions: Conceptualization, A.M. and A.G.; methodology, A.M., A.G.; M.V., and B.T.; investigation, A.G., B.T., and M.V.; data curation, A.G. and A.M.; writing—original draft preparation, A.M. and A.G.; writing—review and editing, A.M. and A.G.; supervision, A.M.; project administration, B.T. and A.M. All authors have read and agreed to the published version of the manuscript.

Funding: MycoRed EU FP7 (KBBE-2007-2-5-05), GOP-1.1.1-11-2012-0159 projects (supported by Hungarian and EU sources) and TUDFO/51757/2019-ITM (2019-2023). The authors also acknowledge the kind help of Dr. Matyas Cserhati in correcting the English of the manuscript.

Conflicts of Interest: The authors declare no conflict of interest.

Abbreviations

SW	susceptibility window
DI	disease index
DON	deoxynivalenol
FHB	*Fusarium* head blight
LSD	Limit of significant difference

References

1. Bai, G.H.; Plattner, R.; Desjardins, A.; Kolb, F.; McIntosh, R. Resistance to *Fusarium* head blight and deoxynivalenol accumulation in wheat. *Plant Breed.* **2001**, *120*, 1–6. [CrossRef]
2. Mesterhazy, A. Breeding for resistance against *Fusarium* head blight in wheat. Chapter 13. In *Mycotoxin Reduction in Grain Chains*; Leslie, J.F., Logrieco, A.F., Eds.; Wiley Blackwell: Hoboken, NJ, USA, 2014; pp. 189–208.
3. Mesterhazy, A.; Bartok, T.; Mirocha, C.M.; Komoroczy, R. Nature of resistance of wheat to *Fusarium* head blight and deoxynivalenol contamination and their consequences for breeding. *Plant Breed.* **1999**, *118*, 97–110. [CrossRef]
4. Mesterhazy, A.; Lehoczki-Krsjak, S.; Varga, M.; Szabo-Hever, A.; Toth, B.; Lemmens, M. Breeding for fhb resistance via *Fusarium* damaged kernels and deoxynivalenol accumulation as well as inoculation methods in winter wheat. *Agric. Sci.* **2015**, *6*, 970–1002. [CrossRef]
5. Atanasoff, D. *Fusarium* blight (scab) of wheat and other cereals. *J. Agric. Res.* **1920**, *20*, 1–32.
6. Parry, D.W.; Jenkinson, P.; McLeod, L. *Fusarium* ear blight (scab) in small grain cereals—A review. *Plant Pathol.* **1995**, *44*, 207–238. [CrossRef]
7. Strange, R.N.; Smith, H. A fungal growth stimulant in anthers which predisposes wheat to attack by *Fusarium graminearum*. *Physiol. Plant Pathol.* **1971**, *1*, 141–150. [CrossRef]

8. Schroeder, H.W.; Christensen, J.J. Factors affecting resistance of wheat to scab by *Gibberella zeae*. *Phytopathology* **1963**, *53*, 831–838.
9. Hart, L.P.; Pestka, J.J.; Liu, M.T. Effect of kernel development and wet periods on production of deoxynivalenol in wheat infected with *Gibberella zeae*. *Phytopathology* **1984**, *74*, 1415–1418. [CrossRef]
10. Lacey, J.; Bateman, G.L.; Mirocha, C.J. Effects of infection time and moisture on development of spike blight and deoxynivalenol production by *Fusarium* spp. in wheat. *Ann. Appl. Biol.* **1999**, *134*, 277–283. [CrossRef]
11. Del Ponte, E.M.; Fernandes, J.M.C.; Bergstrom, G.C. Influence of growth stage on *Fusarium* head blight and deoxynivalenol production in wheat. *J. Phytopathol.* **2007**, *155*, 577–581. [CrossRef]
12. Cowger, C.; Arellano, C. Plump kernels with high deoxynivalenol linked to late *Gibberella zeae* infection and marginal disease conditions in winter wheat. *Phytopathology* **2010**, *100*, 719–728. [CrossRef] [PubMed]
13. Siou, D.; Gélisse, S.; Laval, V.; Repinçay, C.; Canalès, R.; Suffert, F.; Lannou, C. Effect of wheat spike infection timing on *Fusarium* head blight development and mycotoxin accumulation. *Plant Pathol.* **2013**, *63*, 390–399. [CrossRef]
14. Mesterhazy, A.; Bartok, T. Control of *Fusarium* head blight of wheat by fungicide and its effect in the toxin contamination of the grains. *Pflanzenschutz Nachrichten Bayer* **1996**, *49*, 187–205.
15. Soresi, D.; Carrera, A.; Echenique, V.; Garbus, I. Identification of genes induced by *Fusarium graminearum* inoculation in the resistant durum wheat line Langdon(Dic-3A)10 and the susceptible parental line. *Langdon Microbiol. Res.* **2015**, *177*, 53–66. [CrossRef] [PubMed]
16. Andersen, K.F.; Madden, L.V.; Paul, P.A. *Fusarium* head blight development and deoxynivalenol accumulation in wheat as influenced by post-anthesis moisture patterns. *Phytopathology* **2015**, *105*, 210–219. [CrossRef] [PubMed]
17. Zwart, R.S.; Muylle, H.; van Bockstaele, E.; Roldan-Ruiz, I. Evaluation of genetic diversity of *Fusarium* head blight resistance in European winter wheat. *Theor. Appl. Genet.* **2008**, *117*, 813–828. [CrossRef]
18. Kubo, K.; Kawada, N.; Nakajima, T.; Hirayae, K.; Fujita, M. Field evaluation of resistance to kernel infection and mycotoxin accumulation caused by *Fusarium* head blight in western Japanese wheat (*Triticum aestivum* L.) cultivars. *Euphytica* **2014**, *200*, 81–93. [CrossRef]
19. Ji, F.; Wu, J.; Zhao, H.; Xu, J.; Shi, J. Relationship of deoxynivalenol content in grain, chaff, and straw with *Fusarium* head blight severity in wheat varieties with various levels of resistance. *Toxins* **2015**, *7*, 728–742. [CrossRef]
20. VanEeuwijk, F.A.; Mesterhazy, A.; Kling, C.I.; Ruckenbauer, P.; Saur, L.; Bürstmayr, H.; Lemmens, M.; Maurin, M.; Snijders, C.H.A. Assessing non-specificity of resistance in wheat to head blight caused by inoculation with European strains of *Fusarium culmorum, F. graminearum* and *F. nivale*, using a multiplicative model for interaction. *Theor. Appl. Genet.* **1995**, *90*, 221–228. [CrossRef]
21. Mesterhazy, A.; Bartok, T.; Kászonyi, G.; Varga, M.; Tóth, B.; Varga, J. Common resistance to different *Fusarium* spp. causing *Fusarium* head blight in wheat. *Eur. J. Plant Pathol.* **2005**, *112*, 267–281. [CrossRef]
22. Booth, C. *The Genus Fusarium*; Commonwealth Mycological Institute: Kew, Surrey, UK, 1971.
23. Mesterhazy, A. Types and components of resistance to *Fusarium* head blight of wheat. *Plant Breed.* **1995**, *114*, 377–386. [CrossRef]
24. Mesterházy, Á.; Varga, M.; György, A.; Lehoczki-Krsjak, S.; Tóth, B. The role of adapted and non-adapted resistance sources in breeding resistance of winter wheat to *Fusarium* head blight and deoxynivalenol contamination. *World Mycotoxin J.* **2018**, *11*, 539–557. [CrossRef]
25. Mesterhazy, A.; Varga, M.; Tóth, B.; Kótai, C.; Bartók, T.; Véha, A.; Ács, K.; Vágvölgyi, C.; Lehoczki-Krsjak, S. Reduction of deoxynivalenol (DON) contamination by improved fungicide use in wheat. Part 1. Dependence on epidemic severity and resistance level in small plot tests with artificial inoculation. *Eur. J. Plant Pathol.* **2018**, *151*, 39–55. [CrossRef]
26. Mesterhazy, A. Effect of seed production area on the seedling resistance of wheat to *Fusarium* seedling blight. *Agronomie* **1985**, *5*, 491–497. [CrossRef]
27. Large, E.C. Growth stages in cereals. *Plant Pathol.* **1954**, *3*, 128–129. [CrossRef]
28. Mesterhazy, A. Selection of head blight resistant wheats through improved seedling resistance. *Plant Breed.* **1987**, *98*, 25–36. [CrossRef]

29. Sváb, J. *Biometriai Módszerek a Kutatásban (Biomertrical Methods in Research)*, 3rd ed.; Agricultural Publishing House: Ho Chi Minh City, Vietnam, 1981; p. 557. (In Hungarian)
30. Weber, E. *Grundriss der Biologischen Statistik (Fundaments of the Biological Statistics)*; VEB Fisher Verlag: Jena, Germany; Berlin, Germany, 1967.
31. Selby, A.D.; Manns, T.F. Studies in diseases of cereals and grasses. II. The fungus of wheat scab as a seed and seedling parasite. *Ohio Agric. Exp. Sta. Bull.* **1909**, *203*, 212–236.
32. Stack, R.W. History of *Fusarium* head blight with emphasis on North America. In *Fusarium Head Blight of Wheat and Barley*; Leonard, K.J., Bushnell, W.R., Eds.; APS Press: St. Paul, MN, USA, 2003; pp. 1–34.
33. Mesterhazy, A. Reaction of winter wheat varieties to four *Fusarium* species. *Phytopathol. Z. J. Phytopathol.* **1977**, *90*, 104–112. [CrossRef]
34. Mesterhazy, A. Breeding wheat for *Fusarium* head blight resistance in Europe. In *Fusarium Head Blight of Wheat and Barley*; Leonard, K., Bushnell, W., Eds.; APS Press: St. Paul, MN, USA, 2003; pp. 211–240.
35. Miedaner, T.; Perkowski, J. Correlations among *Fusarium culmorum* head blight resistance, fungal colonization and mycotoxin contents in winter rye. *Plant Breed.* **1996**, *115*, 347–351. [CrossRef]

microorganisms

MDPI

Article

Multiple Fungal Metabolites Including Mycotoxins in Naturally Infected and *Fusarium*-Inoculated Wheat Samples

Valentina Spanic [1,*], Zorana Katanic [2], Michael Sulyok [3], Rudolf Krska [3,4], Katalin Puskas [5], Gyula Vida [5], Georg Drezner [1] and Bojan Šarkanj [6]

1 Agricultural Institute Osijek, Juzno predgradje 17, 31000 Osijek, Croatia; georg.drezner@poljinos.hr
2 Department of Biology, Josip Juraj Strossmayer University of Osijek, Cara Hadrijana 8a, 31000 Osijek, Croatia; zorana.katanic@biologija.unios.hr
3 Institute of Bioanalytics and Agro-Metabolomics, Department of Agrobiotechnology (IFA-Tulln), University of Natural Resources and Life Sciences Vienna (BOKU), Konrad Lorenzstr. 20, 3430 Tulln, Austria; michael.sulyok@boku.ac.at (M.S.); rudolf.krska@boku.ac.at (R.K.)
4 Institute for Global Food Security, School of Biological Sciences, Queen's University Belfast, University Road, Belfast BT7 1NN, Northern Ireland, UK
5 Agricultural Institute, Centre for Agricultural Research, Brunszvik u. 2, 2462 Martonvásár, Hungary; puskas.katalin@agrar.mta.hu (K.P.); vida.gyula@agrar.mta.hu (G.V.)
6 Department of Food Technology, University Centre Koprivnica, University North, Trg dr. Žarka Dolinara 1, 48000 Koprivnica, Croatia; bsarkanj@unin.hr
* Correspondence: valentina.spanic@poljinos.hr; Tel.: +385-31-515-563

Received: 21 March 2020; Accepted: 14 April 2020; Published: 17 April 2020

Abstract: In this study, the occurrence of multiple fungal metabolites including mycotoxins was determined in four different winter wheat varieties in a field experiment in Croatia. One group was naturally infected, while the second group was inoculated with a *Fusarium graminearum* and *F. culmorum* mixture to simulate a worst-case infection scenario. Data on the multiple fungal metabolites including mycotoxins were acquired with liquid chromatography with mass spectrometry (LC-MS/MS) multi-(myco)toxin method. In total, 36 different fungal metabolites were quantified in this study: the *Fusarium* mycotoxins deoxynivalenol (DON), DON-3-glucoside (D3G), 3-acetyldeoxynivalenol (3-ADON), culmorin (CULM), 15-hydroxyculmorin, 5-hydroxyculmorin, aurofusarin, rubrofusarin, enniatin (Enn) A, Enn A1, Enn B, Enn B1, Enn B2, Enn B3, fumonisin B1, fumonisin B2, chrysogin, zearalenone (ZEN), moniliformin (MON), nivalenol (NIV), siccanol, equisetin, beauvericin (BEA), and antibiotic Y; the *Alternaria* mycotoxins alternariol, alternariolmethylether, altersetin, infectopyron, tentoxin, tenuazonic acid; the *Aspergillus* mycotoxin kojic acid; unspecific metabolites butenolid, brevianamid F, cyclo(L-Pro-L-Tyr), cyclo(L-Pro-L-Val), and tryptophol. The most abundant mycotoxins in the inoculated and naturally contaminated samples, respectively, were found to occur at the following average concentrations: DON (19,122/1504 μg/kg), CULM (6109/1010 μg/kg), 15-hydroxyculmorin (56,022/1301 μg/kg), 5-hydroxyculmorin (21,219/863 μg/kg), aurofusarin (43,496/1266 μg/kg). Compared to naturally-infected samples, *Fusarium* inoculations at the flowering stage increased the concentrations of all *Fusarium* mycotoxins, except enniatins and siccanol in Ficko, the *Aspergillus* metabolite kojic acid, the *Alternaria* mycotoxin altersetin, and unspecific metabolites brevianamid F, butenolid, cyclo(L-Pro-L-Tyr), and cyclo(L-Pro-L-Val). In contrast to these findings, because of possible antagonistic actions, *Fusarium* inoculation decreased the concentrations of the *Alternaria* toxins alternariol, alternariolmethylether, infectopyron, tentoxin, tenuazonic acid, as well as the concentration of the nonspecific metabolite tryptophol.

Keywords: fusarium; LC-MS/MS; mycotoxin; occurrence; wheat

1. Introduction

Wheat is the basic staple food and its global consumption is about 66 kg/per capita worldwide [1]. Various diseases can affect the heads of wheat, and severe infection can result in decreased grain yield and quality. Furthermore, wheat and its products can be contaminated with mycotoxins produced by different fungi that can be found in the field and/or postharvest [2]. Among the most important risks associated with cereal consumption are mycotoxins, heavy metals, pesticide residues, and alkaloids. It is very important to monitor mycotoxins in all stages of wheat production from the field to end-use quality usage. Fusarium head blight (FHB), caused by several *Fusarium* species, mainly *Fusarium graminearum*, *F. culmorum*, and *F. avenaceum*, is a devastating disease of wheat, associated with mycotoxin contamination with a significant threat to animal and human health. The most important mycotoxins in wheat are mainly *Fusarium* toxins, such as deoxynivalenol (DON), with its acetylated forms (15 acetyl-deoxynivalenol or 15AcDON, and 3 acetyl-deoxynivalenol, or 3AcDON), zearalenone (ZEN), nivalenol (NIV), fumonisins (FB), T-2, and HT-2 toxins and its modified forms [2–4]. Beside *Fusarium* spp., species from the genera *Alternaria*, *Penicillium*, and *Aspergillus* are critical in the maintenance of food safety and can cause mycotoxin contaminations of cereals [5,6]. For example, *Aspergillus flavus* can infect wheat grain in the field but also contaminates stored grains when temperature and water activity are favorable [7].

The most common mycotoxin in wheat and wheat-based products in the European Union (EU) is the well-known type-B trichothecene, DON [8]. Moreover, a very limited number of mycotoxins are subject to legislation and regular monitoring. As far as cereals are concerned, aflatoxins, fumonisins, DON, zearalenone (ZEN), and ochratoxin A (OTA) are those most often analyzed [9]. For these reasons, the other mycotoxins, which until now have not received detailed scientific attention, are commonly indicated as 'novel' or 'emerging' mycotoxins [10,11]. Therefore, there is an urgent need to acquire data on the presence and diffusion of these emerging mycotoxins in field crops, in relationship to different climatic conditions, in order to perform proper risk characterization, risk assessment, and afterwards propose maximum limits in the food chain. The European Union has established maximum levels for DON [12], where unprocessed wheat, cereal flour, bread and wheat-based foods for infants and young children must not contain more than 1250, 750, 500, and 200 μg/kg of DON, respectively. DON is known to cause food refusal, vomiting, and depressed immune function, resulting in poor weight gain [13]. Culmorins (CULMs) are tricyclic sesquiterpene diols that can be produced by *F. culmorum*, *F. graminearum*, and *F. venenatum* [14]. Nevertheless, CULM is considered an "emerging mycotoxin" whose synthesis and toxicology will be of greater interest for food safety consideration in the future [11]. It was also confirmed that CULM suppresses one of the important steps of DON detoxification—glucuronidation [15] and therefore can increase DON's toxic effect if it co-occurs with DON. Genes for CULM and trichothecene production co-occur in other fungal species closely related to *F. graminearum* [16]. Nivalenol (NIV) to some extent co-occurs with DON. Although NIV is less toxic to plants compared to DON, it has more severe toxic effects in animals and humans [17]. ZEN is often co-produced with DON by *Fusarium* spp. such as *F. graminearum* [18]. Moreover, recent studies revealed an increased presence of modified *Fusarium* mycotoxins and so-called emerging mycotoxins, particularly enniatins (Enns), beauvericin (BEA), and moniliformin (MON) [19,20] which are far less investigated. Enns can be produced by several fungal species including *Fusarium* spp., and the enniatin analogs enniatin A (EnnA), enniatin A1 (EnnA1), enniatin B (EnnB), enniatin B1 (EnnB1), and enniatin B2 (EnnB2) are reported to be the most prevalent ones in cereals in Europe [21,22]. Beauvericin (BEA) can co-occur with Enns, since they can be produced by the same *Fusarium* species [21]. Moniliformin (MON) often was found in *Fusarium*-damaged durum wheat grains due to *F. avenaceum* infection [23,24]. The European Union maximum limits for other *Fusarium* toxins (ZEN and fumonisins B1 and B2, and T-2 and HT-2 toxins) in cereals and cereal-based products have been established by Commission Regulation [12]. Different reference points were established for mycotoxins as follows: the benchmark dose lower confidence limit of 10% (BMDL10) extra risk for aflatoxin B1 at 170 ng/kg per body weight (bw) per day; tolerable daily intake (TDI) of 2 μg/kg bw per day for fumonisins B1,

B2, and B3; tolerable weekly intake (TWI) of 0.1 μg/kg bw per week for OTA; BMDL05 of 200 μg/kg bw per day for MON; and 90 μg/kg bw per day of BEA was established as a concentration with a low risk (for genotoxic carcinogens such as aflatoxins) or no risk (for other mycotoxins) [25]. The control of *Fusarium* fungal infection in the field by growing resistant cultivars could be the most efficient method to control FHB [16] and mitigating mycotoxin accumulation in the end-use products. To reduce the risk of *Fusarium* contamination, the application of preventive agricultural practices is also important, such as crop selection, rotation, tillage, irrigation, and the proper use of fungicides, as partial control of FHB [26,27]. Furthermore, mycotoxin production by mycotoxigenic fungal species is dependent on water activity, temperature, and CO_2 levels [28,29].

In addition to these better-known compounds, other secondary metabolites produced by *Fusarium* species may be detected and investigated using multi-analytic methods developed in recent years based on liquid chromatography coupled to mass spectrometry [30,31]. Modified and emerging mycotoxins which cover DON derivatives (DON-3-glucoside, acetyl-DONs, nor-DONs and deepoxy-DON), nivalenol, T-2 and HT-2 toxins, enns, BEA, moniliformin, and fumonisins are not regulated by EU law. Furthermore, multiple fungal metabolites including mycotoxins are frequently observed [32,33]. This is a topic of great concern, as co-contaminated samples might still exert adverse health effects due to additive/synergistic interactions of the mycotoxins.

This research summarizes the occurrence of the mycotoxins/metabolites produced by *Fusarium*, *Alternaria,* and *Aspergillus*, as well as several unspecific metabolites in naturally-infected and *Fusarium*-inoculated wheat samples. To our best knowledge, the majority of mycotoxins/metabolites analyzed in this study are poorly characterized [6], and their occurrence and concentration in commercially produced wheat varieties are mainly unexplored.

2. Materials and Methods

2.1. Plant Material and Field Trial

The entire field experiment was conducted from October 2018 to July 2019 at Osijek (45°32′ N, 18°44′ E), Croatia. The soil type is a eutric cambisol. The average annual precipitation in the growing season is 531.3 mm, and the average annual temperature is 10.9 °C. Plots consisted of eight row plots with a 2.5 m length and a 1.08 m width at a sowing rate of 330 seeds/m^2. The weed control was conducted with the herbicide Sekator (100 g/L amidosulfuron and 25 g/L iodosulfuronmethyl-sodium) at wheat tillering (GS 31). A total of 170 kg N/ha was applied to the plots as a granular ammonium nitrate fertilizer and was split between GS 23 and 45. The experiment was conducted with two experimental replications in two treatments (trials): naturally-infected and *Fusarium*-inoculated. In each treatment, the same four winter varieties were used (Ficko and Pepeljuga originated from the Agricultural Institute Osijek, Croatia; Mv Karizma and Mv Kolompos originated from the Agricultural Institute in Martonvasar, Hungary). The grains were taken by harvesting the whole plot with a Wintersteiger cereal plot combine-harvester. The harvested grains were mixed thoroughly from two replications in each treatment, and 100 g grain samples were taken from each plot to analyze the mycotoxin content.

2.2. Inoculum Production and Inoculation Procedure

The *Fusarium* species used in this experiment were two the most prevalent casual agents of FHB: *F. graminearum* strain (PIO 31), previously isolated from winter wheat collected in East Croatia and *F. culmorum* strain (IFA 104), obtained from IFA, Tulln. Conidial inoculum of the *F. graminearum* was produced in mung bean medium [34], while *F. culmorum* spores were produced by a mixture of wheat and oat grains [35]. Conidial concentrations of both species were determined using a hemocytometer and were set to 1×10^5 mL^{-1}. The spore suspensions were set to a concentration so that a single bottle of one strain contained a sufficient amount of suspension (>900 mL), which could be diluted in 100 L of water immediately before inoculation (100 mL/m^2). One treatment was grown according to standard agronomical practice with no usage of fungicide and without misting treatment, while another treatment was subjected to two inoculation events using a tractor-back sprayer with *Fusarium* spp. at the time of flowering (Zadok's scale 65) [36]. Misting was provided by spraying with a tractor back-sprayer on several occasions.

2.3. Mycotoxin LC-MS/MS Analyses

Determination of mycotoxins with the LC-MS/MS method in wheat grains and wheat malt was performed as previously described [31]: In brief, 5.00 g of ground wheat was extracted with 20 mL of extraction solvent composed of acetonitrile (AcN):water (W):acetic acid (HAC) = 79:20:1 (*v:v:v*) on a rotary shaker (GFL 3017, GFL; Burgwedel, Germany) for 90 min at room temperature in a horizontal position. After extraction, 500 μL of the extract was diluted with 500 μL of dilution solvent composed of AcN:W:HAC = 20:79:1 in vials. Finally, 5 μL was injected into an LC-MS/MS system composed of a QTrap 5500 MS/MS (Sciex, Foster City, CA, USA) coupled with an Agilent 1290 series UHPLC system (Agilent Technologies, Waldbronn, Germany). The separation of analytes was performed on a Gemini C18 column (150 × 4.6 mm i.d., 5 μm particle size) with a 4 × 3 mm precolumn with the same characteristics (Phenomenex, Torrance, CA, USA). The eluents used were composed of methanol (MeOH):W:HAC = 10:89:1 (*v:v:v*) as eluent A, and MeOH:W:HAC = 97:2:1 (*v:v:v*) as eluent B. The analysis was performed on the fully validated method described in detail by Sulyok et al. (2020) for measurement of 500+ mycotoxins and other secondary metabolites.

2.4. Statistical Analysis

The data were evaluated for the distribution by the Shapiro–Wilk W-test, and the homoscedasticity was determined by Levene's test. Since the data did not show a normal distribution, the exact comparison between the two groups (naturally-infected and inoculated) was tested by the Mann–Whitney U test. The comparison of the differences in the data distribution between different tested varieties was performed by Kruskal–Wallis ANOVA. Statistical tests and boxplot graphs were performed in Statistica 13.1. (TIBCO Software Inc., Palo Alto, CA, USA), shown in the supplementary file. Graphs were prepared by using the statistical software GraphPad Prism 5.0 (GraphPad Software Inc., San Diego, CA, USA).

3. Results

3.1. *Fusarium* Metabolites/Mycotoxins in *Fusarium*-Inoculated and Naturally-Infected Samples

Deoxynivalenol (DON), DON-3-glucoside (D3G), 3-acetyldeoxynivalenol (3-ADON), culmorin (CULM), 15-hydroxyculmorin, 5-hydroxyculmorin, aurofusarin, rubrofusarin, enniatin (Enn) A1, B, B1 (Figure 1a–k) were found in all tested samples, although sometimes in ppb concentrations, while enniatin (Enn) B2 was present in at least half of the tested samples (Figure 1) Chrysogin was also found in all tested samples (Figure 1m), while zearalenone (ZEN), moniliformin (MON), nivalenol (NIV), siccanol, and equisetin were present in some of the tested samples (Figure 1n–r).

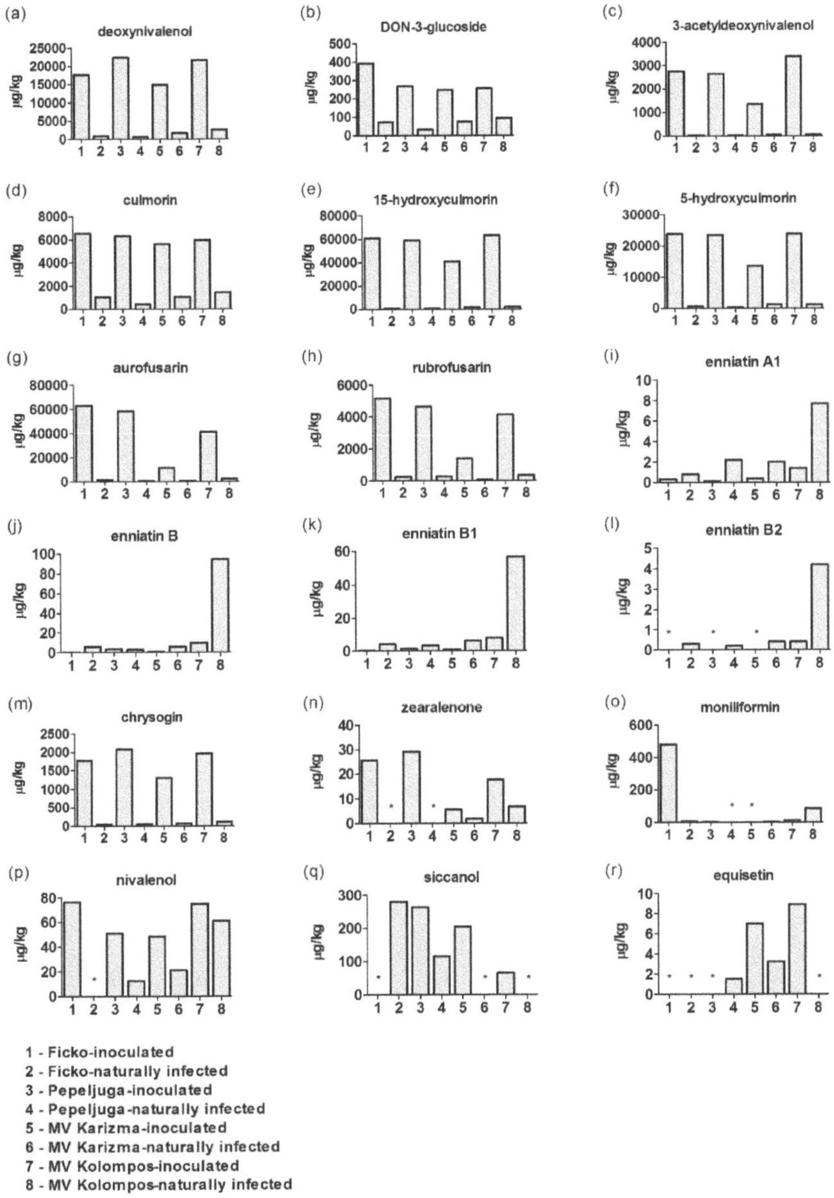

Figure 1. *Fusarium* mycotoxins/metabolites in *Fusarium*-inoculated and naturally-infected samples: deoxynivalenol (**a**), DON-3-glucoside (**b**), 3-acetyldeoxynivalenol (**c**), culmorin (**d**), 15-hydroxyculmorin (**e**), 5-hydroxyculmorin (**f**), aurofusarin (**g**), rubrofusarin (**h**), enniatin A1 (**i**), enniatin B (**j**), enniatin B1 (**k**), enniatin B2 (**l**), chrysogin (**m**), zearalenone (**n**), moniliformin (**o**), nivalenol (**p**), siccanol (**q**), and equisetin (**r**). The asterisk (*) indicates < LOD.

The concentrations of *Fusarium* mycotoxins DON, CULM, 3-ADON, 15-hydroxyculmorin, D3G, and 5-hydroxyculmorin were significantly elevated in the inoculated samples compared to naturally-infected samples of all tested winter wheat varieties (Supplementary Figure S1, Figure 1a–f).

For other detected *Fusarium* metabolites, i.e., aurofusarin, rubrofusarin, and chrysogin, significant increases also occurred in inoculated samples compared to those that were naturally infected (Supplementary Figure S2, Figure 2c,d,f). The levels of DON measured in four FHB-inoculated samples were 17,586 μg/kg in Ficko, 22,337 μg/kg in Pepeljuga, 14,946 μg/kg in Mv Karizma, and 21,622 μg/kg in Mv Kolompos. Although FHB-inoculated samples were far more contaminated with DON, naturally-infected samples also contained higher levels of DON, as well as several other *Fusarium* mycotoxins, than expected (Figure 1a). Moreover, in Mv Karizma and Mv Kolompos, recorded DON concentrations were 1764 μg/kg and 2682 μg/kg, respectively, which is above the maximal allowed concentration for human consumption. As expected, D3G and 3-ADON were found in all samples that contained DON (Figure 1b,c). The highest concentration of D3G was recorded in Ficko in the inoculated samples (393 μg/kg), while the lowest concentration was in naturally-infected samples of Pepeljuga (31 μg/kg). 3-ADON concentration ranged from 9 μg/kg in naturally-infected samples to 3399 μg/kg in *Fusarium*-inoculated samples. FHB-inoculated samples of Mv Karizma had at least 2 times lower concentrations of 3-ADON compared to inoculated samples of other varieties.

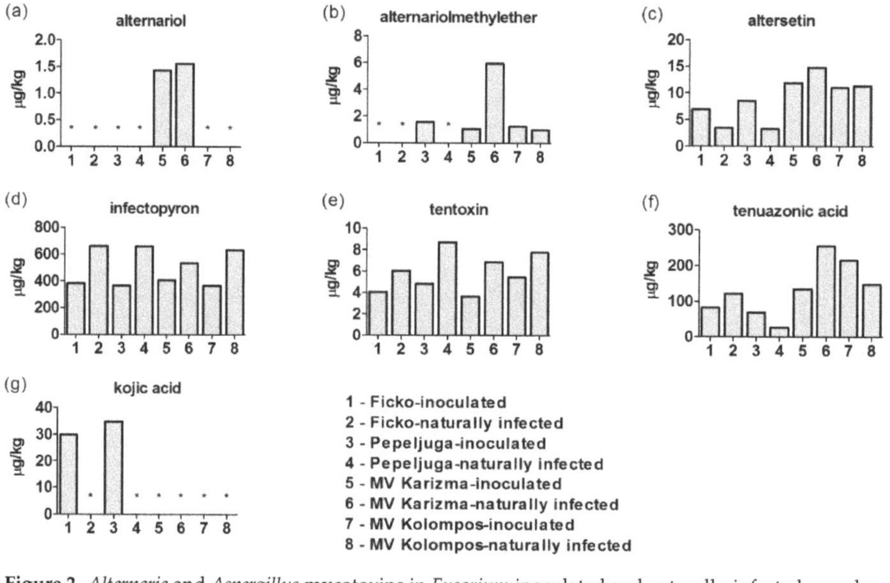

Figure 2. *Alternaria* and *Aspergillus* mycotoxins in *Fusarium*-inoculated and naturally-infected samples: alternariol (**a**), alternariolmethylether (**b**), altersetin (**c**), infectopyron (**d**), tentoxin (**e**), tenuazonic acid (**f**), and kojic acid (**g**). The asterisk (*) indicates < LOD.

The highest concentration of CULM (6551 μg/kg) was recorded in FHB-inoculated samples of Ficko, and the lowest was obtained in naturally-infected samples of Pepeljuga (443 μg/kg) (Figure 1d). The highest concentrations of 15-hydroxyculmorin and 5-hydroxyculmorin were found in FHB-inoculated samples of Mv Kolompos—63,329 and 23,864 μg/kg, respectively—similar to the levels detected in inoculated samples of Ficko and Pepeljuga, but lower levels were detected in inoculated samples of Mv Karizma (Figure 1e,f).

In this study, pigments were also detected; aurofusarin was detected in the range of 735 to 63,098 μg/kg and rubrofusarin from 65 to 5136 μg/kg. Although aurofusarin and rubrofusarin levels in all inoculated samples increased compared to naturally-infected samples, inoculated samples of Mv Karizma had at least three times lower concentrations of these mycotoxins compared to other varieties (Figure 1g,h).

The highest level of enniatins A1, B, B1, and B2 (8, 94, 57, and 4 μg/kg, respectively) were found in naturally-infected samples of Mv Kolompos. Overall, these mycotoxins were shown to be present in lower concentrations in FHB-inoculated samples compared to naturally-infected samples or were not detected at all in the inoculated samples (Figure 1i–l). The various enniatins were ranked as follows, in descending order of incidence and mean concentration: enniatin B > enniatin B1 > enniatin A1 > enniatin B2 > enniatin A > enniatin B3. Total enniatin levels varied from traces (0.02 μg/kg, enniantin A) to 94 μg/kg (enniantin B).

Chrysogin, in the inoculated samples compared to those that were naturally infected, decreased by 97, 97, 94, and 94%, respectively, in Ficko, Pepeljuga, Mv Karizma, and Mv Kolompos (Figure 1m).

All inoculated and two naturally-infected samples were contaminated with ZEN, and the highest concentration of this mycotoxin was detected in FHB-inoculated samples (26 μg/kg). No ZEN was detected in naturally-infected samples of Ficko and Pepeljuga, and all detected concentrations were below legal limits set in the EU by the European Commission regulation 1881/2006 (Figure 1n).

In the case of MON, the concentration in the inoculated sample of Ficko was 480 μg/kg, which is a 99% increase in relation to naturally-infected grain, while other varieties were far less contaminated with MON, and its concentration in other tested samples ranged from 3 (Pepeljuga, inoculated sample) to 86 μg/kg (Mv Kolompos, inoculated sample) (Figure 1o).

Nivalenol (NIV) was found in seven out of eight samples. Here, the inoculated samples had increased NIV levels compared to the naturally-infected grains, although the difference between naturally-infected and inoculated samples was the least pronounced for Mv Kolompos, in which high values of NIV were detected in both treatments (Figure 1p). The differences between treatments were not statistically significantly different (data not shown).

Inoculated sample of Ficko and naturally-infected samples of Mv Karizma and Mv Kolompos had no siccanol. The highest amount was recorded in naturally-infected samples of Ficko (300 μg/kg) (Figure 1q). The presence of siccanol is not specific for *Fusarium* strains used in this research, and such a high concentration in naturally-infected sample likely came from natural co-contamination by other *Fusarium* species.

Equisetin was detected only in samples of Mv Karizma, both FHB-inoculated and naturally-infected, as well as in naturally-infected samples of Pepeljuga and inoculated samples of Mv Kolompos (Figure 1r).

In addition to the mycotoxins presented in Figure 1, beauvericin, enniatin A, fumonisin B1, fumonisin B2, and antibiotic Y were also detected in the same samples, while enniatin B3, which was tested, was not detected in any of the analyzed samples. Concentrations of beauvericin and enniatin A were very low: 0.5 μg/kg of beauvericin was detected in inoculated samples of Ficko and enniatin A was detected in naturally-infected samples of Pepeljuga, Mv Karizma, and Mv Kolompos at concentrations of 0.5 μg/kg, 0.1 μg/kg, and 0.2 μg/kg, respectively. The inoculated sample of Pepeljuga was the only variety in which fumonisin B1 and B2 were detected at concentrations of 54 μg/kg and 15 μg/kg, respectively, while antibiotic Y was only detected in inoculated samples of Mv Karizma at a concentration of 73 μg/kg. The presence of those metabolites was not specific for *Fusarium* strains used in this research and probably resulted from natural co-contamination by other *Fusarium* species. T2 and HT2 toxins were absent in those tested wheat varieties in both treatments.

3.2. *Alternaria and Aspergillus Metabolites/Mycotoxins in Fusarium-Inoculated and Naturally-Infected Samples*

Concentrations of six *Alternaria* and *Aspergillus* metabolites/mycotoxins in *Fusarium*-inoculated and naturally-infected samples of four wheat varieties are presented in Figure 2. Alternariol was detected only in variety Mv Karizma, in both inoculated and naturally-infected samples (Figure 2a). The highest concentration of alternariolmethylether was detected in naturally-infected Mv Karizma (6 μg/kg), while naturally-infected Pepeljuga and Ficko in both treatments had no alternariolmethylether (Figure 2b). Altersetin, infectopyron, tentoxin, and tenuazonic acid were detected in all varieties in both treatments

(Figure 2c–f). In naturally-infected samples, compared to inoculated samples, the mean infectopyron and tentoxin concentrations were significantly increased (Supplementary Figure S2, Figure 2a,b). The *Aspergillus* metabolite kojic acid was found only in FHB-inoculated samples of Ficko and Pepeljuga (Figure 2g).

3.3. Other Metabolites in Fusarium-Inoculated and Naturally-Infected Samples

Concentrations of additional five unspecific metabolites in *Fusarium*-inoculated and naturally-infected samples are presented in Figure 3. Brevianamid F was found in both treatments in Ficko, and in naturally-infected samples of Mv Karizma and Mv Kolompos (Figure 3a). Butenolid was found in all samples with increased concentrations in inoculated samples compared to naturally-infected samples (Figure 3b) (Supplementary Figure S2, Figure 2e). Cyclo(L-Pro-L-Tyr) and cyclo(L-Pro-L-Val) were increased in inoculated samples (Figure 3c,d), while tryptophol had higher concentrations in naturally-infected samples (Figure 3e).

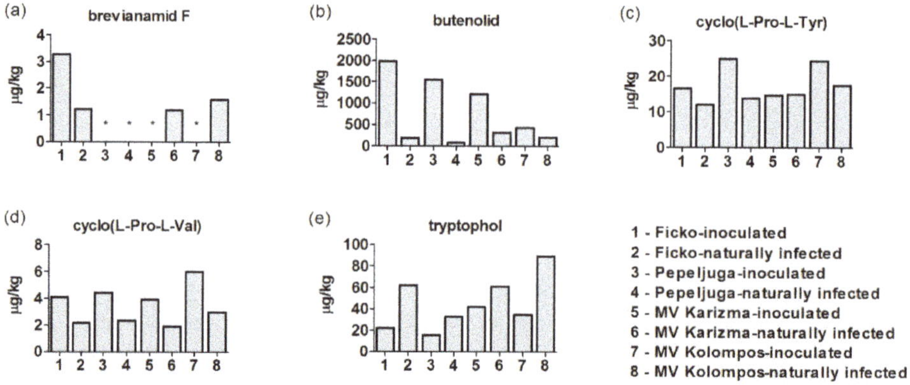

Figure 3. Other metabolites in *Fusarium*-inoculated and naturally-infected samples: brevianamid F (**a**), butenolid (**b**), cyclo(L-Pro-L-Tyr) (**c**), cyclo(L-Pro-L-Val) (**d**), and tryptophol (**e**). The asterisk (*) indicates < LOD.

There were no observed statistical differences between different wheat varieties tested by Kruskal–Wallis ANOVA (data not shown).

4. Discussion

This study presents the first report on the occurrence of 36 fungal metabolites in the wheat grain of four different wheat varieties from Croatia and Hungary and their concentrations after artificial inoculation with *Fusarium* spp. compared to non-inoculated field-grown wheat plants.

The major mycotoxins occurring in wheat, at levels of potential concern for human and animal health, are *Fusarium* mycotoxins [2]. *F. graminearum* occurs worldwide as well as in Croatia [37] and is the most important producer of deoxynivalenol (DON), type B trichothecene. The most frequently detected mycotoxins in wheat grain are DON, FB, ZEN, produced by *Fusarium* species, and aflatoxins (AFs) and ochratoxin A (OTA) produced by *Aspergillus* and *Penicillium* species, respectively [38]. In temperate areas, DON is the most prevalent mycotoxin in wheat [39]. Field survey reports clearly indicate that the mycotoxins most frequently produced in cereal head blight by *F. graminearum* and *F. culmorum* in all European countries are DON, 3-acetyldeoxynivalenol (3-ADON), and zearalenone (ZEN) [40].

Several studies reported a high incidence of multi-mycotoxin contamination in cereals and agricultural commodities [41]. The current investigation showed that in 2019, wheat samples in eastern Croatia were co-contaminated by 24 *Fusarium* mycotoxins/metabolites. A study carried out in Italy showed that at least 80% of wheat samples were contaminated with one mycotoxin, while two mycotoxins were found in 27% of contaminated samples; 38% of the analyzed samples were contaminated with three or more mycotoxins [42].

4.1. Deoxynivalenol (DON), DON-3-Glucoside (D3G), and 3-Acetyldeoxynivalenol (3-ADON)

As regards the contamination levels, the obtained results showed that mycotoxins such as DON, D3G, and CULM, which are mainly produced by *F. culmorum* and *F. graminearum*, were the most abundant mycotoxins in the environmental conditions of the considered field experiments. Average levels of DON contamination did not exceed risk threshold levels for Ficko and Pepeljuga in naturally-infected samples, but as the content range was very wide, all *Fusarium*-inoculated and two naturally-infected (Mv Karizma and Mv Kolompos) samples exceeded the maximum levels for DON contamination. The mean and maximum DON concentrations of naturally-infected wheat samples are summarized in Figure 1 and are lower than the maximum DON concentrations reported in the European wheat summary—a concentration as high as 10,000 μg/kg DON was reported in the European results [8].

Wheat samples had a high incidence of 3-ADON, which was observed in inoculated wheat samples. The modified mycotoxin, D3G was detected and quantified in both inoculated and naturally-infected samples. D3G, one of the several masked mycotoxins, is a phase II plant metabolite of the *Fusarium* mycotoxin DON [43], which can be hydrolyzed in the digestive tract of mammals, thus contributing to the total dietary DON exposure of individuals [20].

4.2. Culmorin (CULM), 15-Hydroxyculmorin, and 5-Hydroxyculmorin

The average value of one of the "emerging mycotoxins" culmorin (CULM) in investigated inoculated samples was 6109 μg/kg, and in naturally-infected samples, this value was 1010 μg/kg. In Norway, the concentration in naturally-infected wheat was lower at median concentrations of 100 μg/kg [44]. This concentration was much higher in durum wheat in central Italy, where it was found that *F. graminearum* produced CULM at concentrations that were very high (2500–14,000 μg/kg) [45]. It was found that naturally-contaminated Norwegian wheat, barley, and oat samples with high DON concentrations also contained CULM and hydroxyculmorins at relatively high levels [46]. In addition, CULM and various hydroxy-culmorins (5- and 15-hydroxy-culmorin) were present in the same concentration ranges as DON in naturally-contaminated grain. This was also concluded by other researchers where levels were typically positively correlated with the amount of DON [44,47]. In Croatian wheat samples in the brewing industry, DON ranged from 595 μg/kg [48] to 2723 μg/kg in wheat malt [49], while concentrations in barley were considerably lower (up to 17.6 μg/kg) [50].

In the current research along with CULM, 15-hydroxyculmorin and 5-hydroxyculmorin also occurred. According to one study [51] CULM, 15-hydroxyculmorin, 5-hydroxyculmorin, and 15-hydroxyculmoron were detected after inoculation with *F. graminearum*, showing enhanced DON toxicity to insects, impacting both growth and mortality, although according to some researchers, CULM and DON can have a synergistic effect on toxicity [52]. Recent findings indicate that CULM can suppress the activity of uridine diphosphate glucosyltransferases (UGTs) that catalyze the glycosylation of DON into the less toxic DON 3-glucoside (D3G) [15]. Compared to DON, in the current research, D3G was found in lower concentrations. All varieties had higher CULM production in inoculated samples compared to those that were naturally infected, which can result in synergistic phytotoxic effects of DON and CULM, when CULM is present at a higher concentration than DON [16], which was not the case in our research.

4.3. Aurofusarin–Rubrofusarin

A pioneer study identified aurofusarin, rubrofusarin, and their derivatives among the pigments [53]. It was concluded that all aurofusarin-producing organisms also have the potential to produce rubrofusarin because the latter pigment is an intermediate of the aurofusarin biosynthetic pathway [54]. This was confirmed in this research, where all samples with aurofusarin contained rubrofusarin. Similar concentrations of aurofusarin in this research were detected as in a study where researchers detected aurofusarin at 10,400–14,0000 μg/kg in Italian samples of durum wheat [54]. Mutants with the absence of the pigment aurofusarin seem to produce an increased amount of ZEN [55]. In contrast, in the current research, samples (Ficko, Pepeljuga, and Mv Karizma in the naturally-infected treatment) with the lowest production of aurofusarin showed the lowest production of ZEN.

4.4. *Enniantins (Enns)*

Mycotoxin contamination by emerging *Fusarium* mycotoxins, such as beauvericin and enniatins, represents a problem of global concern, especially in northern Europe [19,56]. The various enniatins were ranked as follows, in descending order of incidence and mean concentration: enniatin B > enniatin B1 > enniatin A1 > enniatin B2 > enniatin A > enniatin B3. Similarly, Enn analogue concentrations displayed the following gradient (EnnB1 > EnnB2 > EnnA1 > EnnB > EnnA > EnnB3) [45]. In a French study done with wheat, durum wheat, triticale, and barley, enniatin B was the toxin present in the largest amounts in all crops, followed by enniatin B1, enniatin A1, and enniatin A [57]. In other research, the various enniatins were ranked as follows, in descending order of incidence and mean concentration: enniatin B > enniatin B1 > enniatin A1 > enniatin A [19,58]. Enniatin B (trace to 4.8 μg/g) was detected in 12 samples, enniatin B1 (trace to 1.9 μg/g) was detected in eight samples, and enniatin A1 (trace to 6.9 μg/g) was detected in 10 samples [59]. In the current research, enniatin concentrations varied between varieties but were the highest in Mv Kolompos and lower in other varieties. It was expected that more enniatins were presented in naturally-infected samples due to the fact that *Fusarium* species used for inoculation do not produce enniatins.

4.5. *Fumonisin*

To date, only the fumonisins FB1 and FB2 appear to be toxicologically significant. The occurrence of FB1 in cereals, primarily maize, has been associated with serious outbreaks of leukoencephalomalacia (LEM) in horses and pulmonary oedema in pigs. However, *F. proliferatum* appears as a source of fumonisins in wheat grain, and fumonisins might, at very low levels (in the ppb range), be frequently present in wheat [60]. Fumonisins have been reported to cause diseases in humans and animals after consumption of contaminated food and feed, especially esophageal cancer [61]. In the current research, the concentrations of fumonisin B1 and B2 were in accordance with other studies where the natural occurrence of fumonisin B1 (FB1) ranged from 15–155 μg/kg, fumonisin B2 (FB2) ranged from 12–86 μg/kg, and fumonisin B3 (FB3) ranged from 13–64 μg/kg [62]. In other research, concentrations of FB1 were higher, ranging from 958 to 4906 μg/kg [63].

4.6. *Zearalenone (ZEN), Moniliformin (MON), and Beauvericin (BEA)*

F. graminearum is among the main species producing ZEN in Bulgarian wheat where more than half of the samples tested were contaminated with ZEN [64]. In the current research, we used a strain of *F. gramienarum*, so it was expected that higher concentrations of ZEN would occur in the inoculated samples, but as confirmed in previous research, *F. graminearum* prefers lower temperatures for the production of ZEN [65].

Moniliformin (MON) had higher concentrations in inoculated samples of Ficko than in durum wheat, which contained 45.1 μg kg^{-1} of MON in the 2013/2014 season and 171.7 μg kg^{-1} of MON in the 2012/2013 season [66].

In the current research, only one sample contained beauvericin (BEA) in trace amounts. Opposite to that in previous Italian research, BEA was detected (0.64 to 3.5 μg/g) in all investigated samples [59]. Also, BEA was detected at concentrations <10 μg/kg in field samples from Norway and Finland [19]. Although MON and BEA were not produced by the *Fusarium* species tested, they occurred in small amounts due to natural infection by other *Fusarium* species in the field.

4.7. Other Fusarium Metabolites

No T-2 toxin or HT-2 toxins were reported in the current investigated wheat samples compared to the notable percentages of European contaminated wheat samples [8]. Equisetin occurred sporadically in some samples due to natural infection by some other *Fusarium* species that are capable of producing it [67]. Besides *Aspergillus*, the main producers of antibiotic Y are *F. avenaceum* and *F. acuminatum*, but as *F. avenaceum* occurred in lower abundance, it was expected that antibiotic Y occurred only in one sample [68].

4.8. Alternaria, Aspergillus Mycotoxins and Other Metabolites

In the current research, besides *Fusarium* mycotoxins/metabolites, *Alternaria* and *Aspergillus* mycotoxins and some other metabolites were detected. This is in accordance with previous research where in field experiments, mycotoxins produced by several different *Fusarium* spp. were detected together with traces of those produced by *Alternaria* spp. [69]. Commonly, mycotoxinogenic fungi are divided into two groups: preharvest (mainly *Fusarium* species) and postharvest (mainly *Aspergillus* and *Penicillium* species) fungi.

The results for most *Alternaria* mycotoxins are in accordance with research where contamination levels in grains were <100 µg/kg and maximum concentrations were <1000 µg/kg [70]. However, the maximum observed tenuazonic acid (TeA) contamination level in wheat was 4224 µg/kg, which is a much higher concentration than that in the current research, i.e., 125 µg/kg in inoculated samples and 138 µg/kg in naturally-infected samples. The higher presence of *Alternaria* mycotoxins in naturally infected samples is expected, since the *Alternaria* spp. did not have to compete with high levels of used *Fusarium* species. Therefore, higher levels of *Alternaria* mycotoxins were documented. It is worth noting that there were statistically significantly higher levels of tentoxin and infectopyron in naturally-infected samples compared to inoculated samples ($p = 0.03$).

Kojic acid, which is produced by *Aspergillus* and *Penicillium* species [71], was observed in two inoculated samples in the current research.

5. Conclusions

This study detected the presence of a broad range of metabolites in wheat, and future attention should be paid not only to the mycotoxins addressed by regulations, but also to emerging and modified mycotoxins/metabolites. The results showed:

- the presence of 36 wheat mycotoxins/metabolites under field conditions in Croatia under *Fusarium* inoculation and in a natural infection environment in the harvest year 2019;
- the most abundant fungal metabolites were DON, CULM, 15-hydroxyculmorin, 5-hydroxyculmorin, and aurofusarin;
- an increase in DON in inoculated (treated) compared to naturally-infected samples;
- the absence of T2 and HT2 toxins in wheat produced commercially;
- decreased values of Enns and Fumonisins in the inoculated samples—these are produced by other *Fusarium* species that probably out-competed the *F. culmorum*/*F. graminearum* in the inoculated samples.

More wheat varieties included in the research could prove to be a better step towards identifying the mycotoxins in wheat in Croatia, which will be examined in our next investigation.

Supplementary Materials: The following are available online at http://www.mdpi.com/2076-2607/8/4/578/s1, Figure S1: The boxplot diagrams for the statistically significant differences between two treatments (inoculated and naturally-infected) for the major *Fusarium graminearum* and *F. culmorum* metabolites. Figure S2: The boxplot diagrams for the statistically significant differences between two treatments (inoculated and naturally-infected) for the other detected metabolites.

Author Contributions: Conceptualization, investigation, writing—original draft preparation, project administration V.S.; writing—review and editing, software Z.K.; writing—review and editing, investigation K.P.; resources G.V. and G.D.; writing—review and editing, methodology, M.S. and R.K.; writing—review and editing, formal analysis, B.Š.; All authors have read and agreed to the published version of the manuscript.

Funding: This research was funded by the European Union, who provided the EUROPEAN REGIONAL DEVELOPMENT FUND, grant number KK.01.1.1.04.0067, and the bilateral corporation Hungary–Croatia.

Conflicts of Interest: The authors declare no conflict of interest.

References

1. Food and Agriculture Organization UN (FAO). FAOSTAT, Statistic Division, Database 2014. Available online: http://faostat.fao.org (accessed on 10 January 2020).
2. Cheli, F.; Pinotti, L.; Rossi, L.; Dell'Orto, V. Effect of milling procedures on mycotoxin distribution in wheat fractions: A review. *LWT Food Sci. Technol.* **2013**, *54*, 307–314. [CrossRef]
3. Bottalico, A. Fusarium diseases of cereals: Species complex and related mycotoxin profiles, in Europe. *J. Plant Pathol.* **1998**, *80*, 85–103. [CrossRef]
4. Rodrigues, I.; Naehrer, K. A three-year survey on the worldwide occurrence of mycotoxins in feedstuffs and feed. *Toxins* **2012**, *4*, 663–675. [CrossRef] [PubMed]
5. Pitt, J.; Wild, C.; Baan, R.; Gelderblom, W.; Miller, J.; Riley, R.; Wu, F. Fungi producing significant mycotoxins. In *Improving Public Health Through Mycotoxin Control*; Pitt, J., Wild, C., Baan, R., Gelderblom, W., Miller, J., Riley, R., Wu, F., Eds.; IARC Scientific Publication: Lyon, France, 2012; Volume 158, pp. 1–30.
6. Bryła, M.; Waśkiewicz, A.; Ksieniewicz-Woźniak, E.; Szymczyk, K.; Jędrzejczak, R. Modified fusarium mycotoxins in cereals and their products—Metabolism, occurrence, and toxicity: An updated review. *Molecules* **2018**, *23*, 963. [CrossRef]
7. Campbell, K.W.; White, D.G. Evaluation of corn genotypes for resistance to Aspergillus ear rot, kernel infection, and aflatoxin production. *Plant Dis.* **1995**, *79*, 1039–1045. [CrossRef]
8. Stanciu, O.; Banc, R.; Cozma, A.; Filip, L.; Miere, D.; Mañes, J.; Loghin, F. Occurrence of Fusarium mycotoxins in wheat from Europe—A review. *Acta Univ. Cibiniensis Ser. E Food Technol.* **2015**, *19*, 35–60. [CrossRef]
9. Binder, E.M. Managing the risk of mycotoxins in modern feed production. *Anim. Feed Sci. Technol.* **2007**, *133*, 149–166. [CrossRef]
10. Streit, E.; Naehrer, K.; Rodrigues, I.; Schatzmayr, G. Mycotoxin occurrence in feed and feed raw materials worldwide: Long-Term analysis with special focus on Europe and Asia. *J. Sci. Food Agric.* **2013**, *93*, 2892–2899. [CrossRef]
11. Gruber-Dorninger, C.; Novak, B.; Nagl, V.; Berthiller, F. Emerging mycotoxins: Beyond traditionally determined food contaminants. *J. Agric. Food Chem.* **2017**, *65*, 7052–7070. [CrossRef]
12. European Union (EU). *Commission Regulation (EC) No 1881/2006 of 19 December 2006 Setting Maximum Levels for Certain Contaminants in Foodstuffs*; Consolidated Version from 28.07.2017 Including Amendments; European Union (EU): Brussels, Belgium, 2017; Available online: http://eur-lex.europa.eu/legal-content/EN/TXT/?uri=CELEX:02006R1881-20170728 (accessed on 2 November 2019).
13. World Health Organization. Available online: https://www.who.int/foodsafety/publications/monographs/en/ (accessed on 25 November 2019).
14. Langseth, W.; Ghebremeskel, M.; Kosiak, B.; Kolsaker, P.; Miller, J.D. Production of culmorin compounds and other secondary metabolites by *Fusarium culmorum* and *F. graminearum* strains isolated from Norwegian cereals. *Mycopathologia* **2001**, *152*, 23–34. [CrossRef]
15. Woelflingseder, L.; Warth, B.; Vierheilig, I.; Schwartz-Zimmermann, H.; Hametner, C.; Nagl, V.; Novak, B.; Šarkanj, Š.; Berthiller, F.; Adam, G.; et al. The Fusarium metabolite culmorin suppresses the In Vitro glucuronidation of deoxynivalenol. *Arch. Toxicol.* **2019**, *93*, 1729–1743. [CrossRef] [PubMed]
16. Wipfler, R.; McCormick, S.P.; Proctor, R.H.; Teresi, J.M.; Hao, G.; Ward, T.J.; Alexander, H.J.; Vaughan, M.M. Synergistic phytotoxic effects of Culmorin and trichothecene mycotoxins. *Toxins* **2019**, *11*, 555. [CrossRef] [PubMed]
17. Cheat, S.; Gerez, J.R.; Cognié, J.; Alassane-Kpembi, I.; Bracarense, A.P.; Raymond-Letron, I.; Oswald, I.P.; Kolf-Clauw, M. Nivalenol has a greater impact than deoxynivalenol on pig jejunum mucosa In Vitro on explants and In Vivo on intestinal loops. *Toxins* **2015**, *7*, 1945–1961. [CrossRef] [PubMed]

18. Tančić, S.; Stanković, S.; Lević, J.; Krnjaja, V. Correlation of deoxynivalenol and zearalenone production by Fusarium species originating from wheat and maize grain. *Pestic. Phytomed.* **2015**, *30*, 99–105. [CrossRef]
19. Jestoi, M. Emerging Fusarium mycotoxins fusaproliferin, beauvericin, enniatins, and moniliformin: A review. *Crit. Rev. Food Sci. Nutr.* **2008**, *48*, 21–49. [CrossRef] [PubMed]
20. Berthiller, F.; Krska, R.; Domig, K.J.; Kneifel, W.; Juge, N.; Schuhmacher, R.; Adam, G. Hydrolytic fate of deoxynivalenol-3-glucoside during digestion. *Toxicol. Lett.* **2011**, *206*, 264–267. [CrossRef]
21. Liuzzi, V.; Mirabelli, V.; Cimmarusti, M.; Haidukowski, M.; Leslie, J.; Logrieco, A.; Caliandro, R.; Fanelli, F.; Mulè, G. Enniatin and beauvericin biosynthesis in Fusarium species: Production profiles and structural determinant prediction. *Toxins* **2017**, *9*, 45. [CrossRef]
22. Prosperini, A.; Berrada, H.; Ruiz, M.J.; Caloni, F.; Coccini, T.; Spicer, L.J.; Perego, M.C.; Lafranconi, A. A Review of the mycotoxin enniatin B. *Front. Public Health* **2017**, *5*, 304. [CrossRef]
23. Schütt, F.; Nirenberg, H.I.; Demi, G. Moniliformin production in the genus Fusarium. *Mycotoxin Res.* **1998**, *14*, 35–40. [CrossRef]
24. Tittlemier, S.A.; Roscoe, M.; Trelka, R.; Patrick, S.K.; Bamforth, J.M.; Gräfenhan, T.; Schlichting, L.; Fu, B.X. Fate of moniliformin during milling of Canadian durum wheat, processing, and cooking of spaghetti. *Can. J. Plant Sci.* **2014**, *94*, 555–563. [CrossRef]
25. Ojuri, O.T.; Ezekiel, C.N.; Sulyok, M.; Ezeokoli, O.T.; Oyedele, O.A.; Ayeni, K.I.; Eskola, M.K.; Šarkanj, B.; Hajšlová, J.; Adeleke, R.A.; et al. Assessing the mycotoxicological risk from consumption of complementary foods by infants and young children in Nigeria. *Food Chem. Toxicol.* **2018**, *121*, 37–50. [CrossRef] [PubMed]
26. Dill-Macky, R.; Jones, R.K. The effect of previous crop residues and tillage on Fusarium head blight of wheat. *Plant Dis.* **2000**, *84*, 71–76. [CrossRef] [PubMed]
27. Mesterházy, A.; Tóth, B.; Varga, M.; Bartók, T.; Szabó-Hevér, A.; Farády, L.; Lehoczki-Krsjak, S. Role of fungicides, application of nozzle types, and the resistance level of wheat varieties in the control of Fusarium head blight and Deoxynivalenol. *Toxins* **2011**, *3*, 1453–1483. [CrossRef] [PubMed]
28. Medina, A.; Schmidt-Heydt, M.; Rodríguez, A.; Parra, R.; Geisen, R.; Magan, N. Impacts of environmental stress on growth, secondary metabolite biosynthetic gene clusters and metabolite production of xerotolerant/xerophilic fungi. *Curr. Genet.* **2015**, *61*, 325–334. [CrossRef] [PubMed]
29. Medina, A.; Akbar, A.; Baazeem, A.; Rodriguez, A.; Magan, N. Climate change, food security and mycotoxins: Do we know enough? *Fungal Biol. Rev.* **2017**, *31*, 143–154. [CrossRef]
30. Pereira, V.L.; Fernandes, J.O.; Cunha, S.C. Mycotoxins in cereals and related foodstuffs: A review on occurrence and recent methods of analysis. *Trends Food Sci. Technol.* **2014**, *36*, 96–136. [CrossRef]
31. Sulyok, M.; Stadler, D.; Steiner, D.; Krska, R. Validation of an LC-MS/MS-based dilute-and-shoot approach for the quantification of > 500 mycotoxins and other secondary metabolites in food crops: Challenges and solutions. *Anal. Bioanal. Chem.* **2020**, 2607–2620. [CrossRef]
32. Ezekiel, C.N.; Sulyok, M.; Ogara, I.M.; Abia, W.A.; Warth, B.; Šarkanj, B.; Turner, P.C.; Krska, R. Mycotoxins in uncooked and plate-ready household food from rural northern Nigeria. *Food Chem. Toxicol.* **2019**, *128*, 171–179. [CrossRef]
33. Blandino, M.; Scarpino, V.; Sulyok, M.; Krska, R.; Reyneri, A. Effect of agronomic programmes with different susceptibility to deoxynivalenol risk on emerging contamination in winter wheat. *Eur. J. Agron.* **2017**, *85*, 12–24. [CrossRef]
34. Lemmens, M.; Haim, K.; Lew, H.; Ruckenbauer, P. The effect of nitrogen fertilization on Fusarium head blight development and deoxynivalenol contamination in wheat. *Phytopathology* **2004**, *152*, 1–8. [CrossRef]
35. Snijders, C.H.A.; Van Eeuwijk, F.A. Genotype X strain interactions for resistance to Fusarium head blight caused by *Fusarium culmorum* in winter wheat. *Theor. Appl. Genet.* **1991**, *81*, 239–244. [CrossRef]
36. Zadoks, J.C.; Chang, T.T.; Konzac, F.C. A decimal code for the growth stages of cereals. *Weed Res.* **1974**, *14*, 415–421. [CrossRef]
37. Spanic, V.; Lemmens, M.; Drezner, G. Morphological and molecular identification of Fusarium species associated with head blight on wheat in East Croatia. *Eur. J. Plant Pathol.* **2010**, *128*, 511–516. [CrossRef]
38. Priyanka, S.R.; Venkataramana, M.; Balakrishna, K.; Murali, H.S.; Batra, H.V. Development and evaluation of a multiplex PCR assay for simultaneous detection of major mycotoxigenic fungi from cereals. *J. Food Sci. Technol.* **2015**, *52*, 486–492. [CrossRef]
39. Mishra, S.; Ansari, K.M.; Dwivedi, P.D.; Pandey, H.P.; Das, M. Occurrence of deoxynivalenol in cereals and exposure risk assessment in Indian population. *Food Control* **2013**, *30*, 549–555. [CrossRef]

40. Bottalico, A.; Perrone, G. Toxigenic Fusarium species and mycotoxins associated with head blight in small-grain cereals in Europe. *Eur. J. Plant Pathol.* **2002**, *108*, 611–624. [CrossRef]
41. Streit, E.; Schatzmayr, G.; Tassis, P.; Tzika, E.; Marin, D.; Taranu, I.; Tabuc, C.; Nicolau, A.; Aprodu, I.; Puel, O.; et al. Current situation of mycotoxin contamination and co-occurrence in animal feed—Focus on Europe. *Toxins* **2012**, *4*, 788–809. [CrossRef]
42. Alkadri, D.; Rubert, J.; Prodi, A.; Pisi, A.; Mañes, J.; Soler, C. Natural co-occurrence of mycotoxins in wheat grains from Italy and Syria. *Food Chem.* **2014**, *157*, 111–118. [CrossRef]
43. Kovač, M.; Šubarić, D.; Bulaić, M.; Kovač, T.; Šarkanj, B. Yesterday masked, today modified: What do mycotoxins bring next? *Arch. Ind. Hig. Toxicol.* **2018**, *69*, 196–214. [CrossRef]
44. Uhlig, S.; Eriksen, G.; Hofgaard, I.; Krska, R.; Beltrán, E.; Sulyok, M. Faces of a changing climate: Semi-Quantitative multi-mycotoxin analysis of grain grown in exceptional climatic conditions in Norway. *Toxins* **2013**, *5*, 1682–1697. [CrossRef]
45. Beccari, G.; Colasante, V.; Tini, F.; Senatore, M.T.; Prodi, A.; Sulyok, M.; Covarelli, L. Causal agents of fusarium head blight of durum wheat (*Triticum durum* desf.) in central Italy and their In Vitro biosynthesis of secondary metabolites. *Food Microbiol.* **2018**, *70*, 17–27. [CrossRef]
46. Ghebremeskel, M.; Langseth, W. The occurrence of culmorin and hydroxy-culmorins in cereals. *Mycopathologia* **2000**, *152*, 103–108. [CrossRef]
47. Khaneghah, A.M.; Kamani, M.H.; Fakhri, Y.; Coppa, C.F.S.C.; de Oliveira, C.A.F.; Sant'Ana, A.S. Changes in masked forms of deoxynivalenol and their co-occurrence with culmorin in cereal-based products: A systematic review and meta-analysis. *Food Chem.* **2019**, *294*, 587–596. [CrossRef]
48. Mastanjević, K.; Šarkanj, B.; Rudolf, K.; Sulyok, M.; Warth, B.; Mastanjević, K.; Šantek, B.; Krstanović, V. From malt to wheat beer: A comprehensive multi-toxin screening, transfer assessment and its influence on basic fermentation parameters. *Food Chem.* **2018**, *254*, 115–121. [CrossRef]
49. Mastanjević, K.; Šarkanj, B.; Warth, B.; Krska, R.; Sulyok, M.; Mastanjević, K.; Šantek, B.; Krstanović, V. *Fusarium culmorum* multi-toxin screening in malting and brewing by-products. *LWT* **2018**, *98*, 642–645. [CrossRef]
50. Habschied, K.; Krska, R.; Sulyok, M.; Lukinac, J.; Jukić, M.; Šarkanj, B.; Krstanović, V.; Mastanjević, K. The influence of steeping water change during malting on the multi-toxin content in Malt. *Foods* **2019**, *8*, 478. [CrossRef]
51. McCormick, S.P.; Alexander, N.J.; Harris, L.J. CLM1 of *Fusarium graminearum* encodes a longiborneol synthase required for culmorin production. *Appl. Environ. Microbiol.* **2010**, *76*, 136–141. [CrossRef]
52. Rotter, R.G.; Trenholm, H.L.; Prelusky, D.B.; Hartin, K.E.; Thompson, B.K.; Miller, J.D. A preliminary examination of potential interactions between deoxynivalenol (DON) and other selected Fusarium metabolites in growing pigs. *Can. J. Anim. Sci.* **1992**, *72*, 107–116. [CrossRef]
53. Ashley, J.N.; Hobbs, B.C.; Raistrick, H. Studies in the biochemistry of micro-organisms: The crystalline colouring matters of *Fusarium culmorum* (wg Smith) sacc. and related forms. *Biochem. J.* **1937**, *31*, 385. [CrossRef]
54. Frandsen, R.J.; Schutt, C.; Lund, B.W.; Staerk, D.; Nielsen, J.; Olsson, S.; Giese, H. Two novel classes of enzymes are required for the biosynthesis of aurofusarin in *Fusarium graminearum*. *J. Biol. Chem.* **2011**, *286*, 10419–10428. [CrossRef]
55. Malz, S.; Grell, M.N.; Thrane, C.; Maier, F.J.; Rosager, P.; Felk, A.; Albertsen, K.S.; Salomon, S.; Bohn, L.; Schäfer, W. Identification of a gene cluster responsible for the biosynthesis of aurofusarin in the *Fusarium graminearum* species complex. *Fungal Genet. Biol.* **2005**, *42*, 420–433. [CrossRef]
56. Mortensen, A.; Granby, K.; Eriksen, F.D.; Cederberg, T.L.; Friis-Wandall, S.; Simonsen, Y.; Broesbøl-Jensen, B.; Bonnichsen, R. Levels and risk assessment of chemical contaminants in byproducts for animal feed in Denmark. *J. Environ. Sci. Health B* **2014**, *49*, 797–810. [CrossRef]
57. Orlando, B.; Grignon, G.; Vitry, C.; Kashefifard, K.; Valade, R. Fusarium species and enniatin mycotoxins in wheat, durum wheat, triticale and barley harvested in France. *Mycotoxin Res.* **2019**, *35*, 69–380. [CrossRef]
58. Stanciu, O.; Juan, C.; Miere, D.; Loghin, F.; Mañes, J. Occurrence and co-occurrence of Fusarium mycotoxins in wheat grains and wheat flour from Romania. *Food Control* **2017**, *73*, 147–155. [CrossRef]
59. Logrieco, A.; Rizzo, A.; Ferracane, R.; Ritieni, A. Occurrence of Beauvericin and Enniatins in wheat affected by *Fusarium avenaceum* head blight. *Appl. Environ. Microbiol.* **2002**, *68*, 82–85. [CrossRef]

60. Amato, B.; Pfohl, K.; Tonti, S.; Nipoti, P.; Dastjerdi, R.; Pisi, A.; Karlovsky, P.; Prodi, A. *Fusarium proliferatum* and fumonisin B1 co-occur with Fusarium species causing Fusarium head blight in durum wheat in Italy. *J. Appl. Bot. Food Qual.* **2017**, *88*, 228–233. [CrossRef]
61. Reddy, K.R.N.; Salleh, B.; Saad, B.; Abbas, H.K.; Abel, C.A.; Shier, W.T. An overview of mycotoxin contamination in foods and its implications for human health. *Toxin Rev.* **2010**, *29*, 3–26. [CrossRef]
62. Chehri, K.; Jahromi, S.T.; Reddy, K.R.N.; Abbasi, S.; Salleh, B. Occurrence of *Fusarium* spp. and Fumonisins in stored wheat grains marketed in Iran. *Toxins* **2010**, *2*, 2816–2823. [CrossRef]
63. Da Rocha Lemos Mendes, G.; Alves dos Reis, T.; Corrêa, B.; Badiale-Furlong, E. Mycobiota and occurrence of Fumonisin B1 in wheat harvested in Southern Brazil. *Cienc. Rural* **2015**, *45*. [CrossRef]
64. Beev, G.; Denev, S.; Bakalova, D. Zearalenone—Producing activity of *Fusarium graminearum* and *Fusarium oxysporum* isolated from Bulgarian wheat. *Bulg. J. Agric. Sci.* **2013**, *19*, 255–259.
65. Habschied, K.; Šarkanj, B.; Klapec, T.; Krstanović, V. Distribution of zearalenone in malted barley fractions dependent on *Fusarium graminearum* growing conditions. *Food Chem.* **2011**, *129*, 329–332. [CrossRef]
66. Gorczyca, A.; Oleksy, A.; Gala-Czekaj, D.; Urbaniak, M.; Laskowska, M.; Waśkiewicz, A.; Stępień, L. Fusarium head blight incidence and mycotoxin accumulation in three durum wheat cultivars in relation to sowing date and density. *Sci. Nat.* **2018**, *105*, 2. [CrossRef]
67. Venkatesh, N.; Keller, N.P. Mycotoxins in conversation with bacteria and fungi. *Front. Microbiol.* **2019**, *10*, 403. [CrossRef]
68. Golinski, P.; Kostecki, M.; Lasocka, I.; Wisniewska, H.; Chelkowski, J.; Kaczmarek, Z. Moniliformin accumulation and other effects of *Fusarium avenaceum* (Fr.) Sacc. on kernels of winter wheat cultivars. *J. Phytopathol.* **1996**, *144*, 459–499. [CrossRef]
69. Scarpino, V.; Reyneri, A.; Sulyok, M.; Krska, R.; Blandino, M. Effect of fungicide application to control Fusarium head blight and 20 Fusarium and Alternaria mycotoxins in winter wheat (*Triticum aestivum* L.). *World Mycotoxin J.* **2015**, *8*, 499–510. [CrossRef]
70. Fraeyman, S.; Croubels, S.; Devreese, M.; Antonissen, G. Emerging Fusarium and Alternaria mycotoxins: Occurrence, toxicity and toxicokinetics. *Toxins* **2017**, *9*, 228. [CrossRef]
71. Bentley, R. From miso, sake and shoyu to cosmetics: A century of science for kojic acid. *Nat. Prod. Rep.* **2006**, *23*, 1046–1062. [CrossRef]

microorganisms

MDPI

Article

Mycotoxins in Flanders' Fields: Occurrence and Correlations with *Fusarium* Species in Whole-Plant Harvested Maize

Jonas Vandicke [1,*], Katrien De Visschere [2], Siska Croubels [3], Sarah De Saeger [4], Kris Audenaert [1] and Geert Haesaert [1,*]

1 Department of Plants and Crops, Faculty of Bioscience Engineering, Ghent University, Valentin Vaerwyckweg 1, 9000 Ghent, Belgium; Kris.Audenaert@UGent.be
2 Biosciences and Food Sciences Department, Faculty Science and Technology, University College Ghent, Research Station HoGent-UGent, Diepestraat 1, 9820 Bottelare, Belgium; katrien.devisschere@hogent.be
3 Department of Pharmacology, Toxicology and Biochemistry, Faculty of Veterinary Medicine, Ghent University, Salisburylaan 133, 9820 Merelbeke, Belgium; Siska.Croubels@ugent.be
4 Department of Bio-analysis, Faculty of Pharmaceutical Sciences, Ghent University, Ottergemsesteenweg 460, 9000 Ghent, Belgium; sarah.desaeger@ugent.be
* Correspondence: Jonas.Vandicke@UGent.be (J.V.); Geert.Haesaert@UGent.be (G.H.)

Received: 18 October 2019; Accepted: 15 November 2019; Published: 18 November 2019

Abstract: Mycotoxins are well-known contaminants of several food- and feedstuffs, including silage maize for dairy cattle. Climate change and year-to-year variations in climatic conditions may cause a shift in the fungal populations infecting maize, and therefore alter the mycotoxin load. In this research, 257 maize samples were taken from fields across Flanders, Belgium, over the course of three years (2016–2018) and analyzed for 22 different mycotoxins using a multi-mycotoxin liquid chromatography-tandem mass spectrometry (LC-MS/MS) method. DNA of *Fusarium graminearum*, *F. culmorum* and *F. verticillioides* was quantified using the quantitative polymerase chain reaction (qPCR). Multi-mycotoxin contamination occurred frequently, with 47% of samples containing five or more mycotoxins. Nivalenol (NIV) was the most prevalent mycotoxin, being present in 99% of the samples, followed by deoxynivalenol (DON) in 86% and zearalenone (ZEN) in 50% of the samples. Fumonisins (FUMs) were found in only 2% of the samples in the wet, cold year of 2016, but in 61% in the extremely hot and dry year of 2018. Positive correlations were found between DON and NIV and between *F. graminearum* and *F. culmorum*, among others. FUM concentrations were not correlated with any other mycotoxin, nor with any *Fusarium* sp., except *F. verticillioides*. These results show that changing weather conditions can influence fungal populations and the corresponding mycotoxin contamination of maize significantly, and that multi-mycotoxin contamination increases the risk of mycotoxicosis in dairy cattle.

Keywords: Maize; mycotoxins; *Fusarium*; monitoring; forage; silage; maize ear rot; nivalenol; fumonisins

1. Introduction

Ensiling forage crops is a common way of ensuring a continuous and stable supply of feed throughout the year in dairy husbandry. These silages, mostly grass or maize [1], represent 50–80% of the diet of dairy cows during the winter [2]. Especially in North-Western Europe, fodder maize cultivation for on-farm use is an essential part of dairy husbandry [3]. In the region of Flanders, Belgium, more than 127,000 ha of silage maize was grown in 2018, making it the second most grown crop behind pasture [4].

Maize silages can be contaminated with mycotoxins, secondary metabolites produced by a variety of moldy fungi. Mycotoxins can cause several acute and chronic toxic effects to humans and animals when ingested.

In general, ruminants are less sensitive to mycotoxins than monogastrics due to the ability of the ruminal flora to degrade several mycotoxins to less toxic substances [5]. However, not all mycotoxins are degraded in this way. Some can be converted to molecules with a higher toxicity level (e.g., zearalenone (ZEN) to α-zearalenol (α-ZEL)), while others are not even converted at all (e.g., fumonisins (FUMs)) [6–9]. Moreover, if both the rumen microflora and the pH are not stable (e.g., in calves, high-yielding cows or animals in the transition period), mycotoxin metabolism is reduced [10]. Therefore, dairy cattle are susceptible to mycotoxic effects as well, including gastroenteritis, reduced feed intake and reduced fertility [10–12], leading to economic losses [13].

Mycotoxins can be produced in the field (preharvest) as well as in the silage (postharvest). In preharvest field conditions within temperate regions, mycotoxins are mainly produced by *Fusarium* spp., causing maize stem and ear rot [1,14–16]. Two main types of maize ear rot can be distinguished: red ear rot (or *Gibberella* ear rot), primarily caused by *F. graminearum*, *F. culmorum* and *F. poae*, and pink ear rot (or *Fusarium* ear rot), primarily caused by *F. verticillioides, F. proliferatum* and *F. subglutinans* [17–20]. The distribution and prevalence of these *Fusarium* spp. is dependent upon geography and climate. In Europe, the most isolated *Fusarium* species are *F. graminearum* and *F. culmorum*, dominantly in the North, and *F. verticillioides*, mostly found in the South [19,21–23]. However, maize ear rot is always caused by a *Fusarium* complex, rather than by a single species [16,19,24]. Different *Fusarium* spp. interact with each other, leading to possible synergistic effects for infection, although reports have been contradictory [25–29]. Different pathways can be used by *Fusarium* spp. to infect maize plants, and while some *Fusarium* spp. prefer a primary infection via the silks (e.g., *F. graminearum*), others use a systemic transmission from root to kernel, or co-occur as a secondary infection when insects damage the kernels (e.g., *F. verticillioides*) [30–34]. This makes the prevention and control of *Fusarium* spp. in maize very difficult and complex.

During infection, *Fusarium* spp. can produce a variety of mycotoxins. Some of the most well-known *Fusarium* mycotoxins include deoxynivalenol (DON), causing reduced feed intake and diarrhea; zearalenone (ZEN), causing fertility problems; and the fumonisins (FUMs), causing liver and kidney injuries. Other important *Fusarium* mycotoxins include nivalenol (NIV), T-2 toxin (T2), diacetoxyscirpenol (DAS) and enniatins (ENN), among others [11,12,35,36]. Mycotoxins produced by other fungal species, such as aflatoxins produced by *Aspergillus* spp., are rarely found in temperate climates [37–40]. However, climate change may influence the geographical spread of mycotoxin-producing fungi in Europe, causing more tropical fungi, such as *Aspergillus flavus* and *Fusarium verticillioides*, to migrate northward [23,41–46].

Fusarium spp. cannot survive postharvest silage conditions if the silage is firmly pressed and sealed hermetically, but *Fusarium* mycotoxins are stable molecules that may remain unchanged during the silage process [24,47–50]. If a silage is not pressed and sealed correctly and oxygen remains present, *Fusarium* spores may germinate and colonize the maize silage and produce additional mycotoxins [51–53]. Furthermore, other fungal species such as *Penicillium* spp. and *Aspergillus* spp. are well adapted to the silage conditions and may produce additional mycotoxins [47,54–57].

This cocktail of mycotoxins, coming from different fungal species, has led to the observation that almost every maize or maize silage sample is contaminated with at least one mycotoxin, and often multiple. Numerous surveys have been conducted in many regions in the world [22,24,37–40,48,54,58–62]. However, most of these surveys were conducted on samples of maize ears, rather than on the entire plant. Some reports state that the ear can be a representative sample for the entire plant [38], although the fungal species composition and mycotoxin concentrations can differ [16]. Furthermore, most surveys focused on a selection of mycotoxins, rather than screening the entire mycotoxin load. Severe multi-mycotoxin contaminations could hence be overlooked.

The European Union (EU) has set a maximum level for aflatoxin B1 (AFB1) [63] and guidance values for DON, ZEN, ochratoxin A (OTA), fumonisin B1 (FB1) and B2 (FB2), T-2 toxin (T2) and HT-2 toxin (HT2) in several food- and feedstuffs, including maize [64,65]. No recommendations have been formed on lesser researched mycotoxins like NIV, modified mycotoxins like 3- or 15-acetyldeoxynivalenol (3-ADON and 15-ADON), or emerging mycotoxins like enniatins (ENN) [66,67].

Neither do these guidance values take into account any possible synergistic effects of multi-mycotoxin contamination [40,62,68–72]. As a result, one cannot assess whether a particular feed sample is safe, based on these guidance values alone [2]. A better strategy to safeguard livestock health would be to avoid fungal infection and the production of mycotoxins in the first place.

The aim of this research was to investigate the natural mycotoxin load in harvested maize plants intended for silage in the Northwestern European region of Flanders over the course of three years, and link these concentrations to the presence of certain mycotoxigenic *Fusarium* species. A total of 257 samples were taken from harvested maize fields across Flanders during 2016–2018. Samples were analyzed for 22 different mycotoxins using a multi-mycotoxin liquid chromatography-tandem mass spectrometry (LC-MS/MS) method. Then, using a quantitative polymerase chain reaction (qPCR), the DNA of three of the most prevalent *Fusarium* spp. in Flanders, namely *F. graminearum*, *F. culmorum* and *F. verticillioides* [21,73,74], was quantified in the same maize samples. With these data, we were able to quantify the mycotoxin load of silage maize fields in practice, compare mycotoxin occurrence between different years and weather conditions, and identify correlations between these mycotoxins and the corresponding *Fusarium* spp.

2. Materials and Methods

2.1. Maize Sampling

A total of 106 dairy farmers across Flanders were contacted to participate in this study from 2016 till 2018. The selected maize fields were scattered throughout Flanders, grown on different soils and in different micro climates, and the number of selected fields was proportional to the intensity of maize production in that region. Data regarding daily temperature, rainfall, relative humidity and radiation for each growing season were obtained from 17 weather stations spread across Flanders (Figure 1). A few months before harvest, each farmer received a plastic bag, a label and a manual, in which the sampling technique was explained. Sampling was done by taking at least 10 samples per trailer of harvested maize. These samples were mixed and a subsample of ca. 1 kg was put into a plastic bag. The bag was then sealed airtight and stored in a freezer until it was collected by the researchers. After sample collection, a subsample of ca. 5 g was taken for quantitative polymerase chain reaction (qPCR) analysis and stored in a freezer at −20 °C until further analysis; the remaining sample was dried in an airstream of 65 °C for four days. The dried maize sample was then milled in a 0.5 mm sieve, and stored until further mycotoxin analysis.

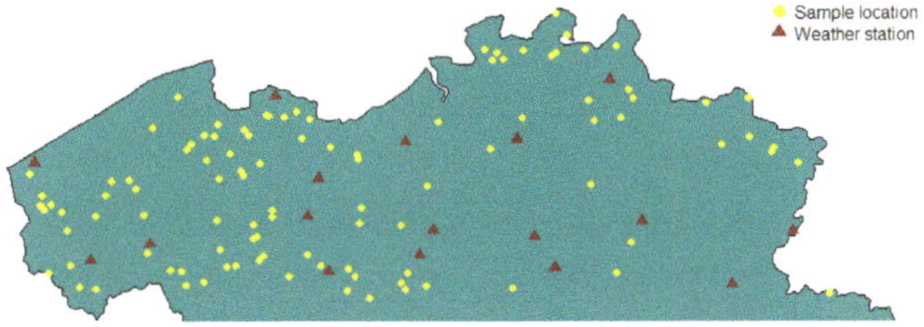

Figure 1. Location of the 106 dairy farms and 17 weather stations in Flanders, Belgium.

2.2. Reagents and Chemicals for LC-MS/MS

Methanol (LC-MS grade), glacial acetic acid (LC-MS grade), and analytical grade acetonitrile were purchased from Biosolve B.V. (Valkenswaard, The Netherlands). Analytical grade acetic acid and ammonium acetate were obtained from Merck (Darmstadt, Germany), while analytical grade n-hexane and methanol were purchased from VWR International (Zaventem, Belgium). Water was purified using a Milli-Q Gradient System (Millipore, Brussels, Belgium).

Certified mycotoxin standard solutions in acetonitrile of OTA (10 μg/mL), aflatoxin mix (AFB1, AFB2, AFG1 and AFG2) (20 μg/mL), fumonisin mix (FB1 and FB2) (50 μg/mL), sterigmatocystin (STERIG) (50 μg/mL), DON (100 μg/mL), deepoxy-deoxynivalenol (DOM) (50 μg/mL), ZEN (100 μg/mL), NIV (100 μg/mL), neosolaniol (NEO) (100 μg/mL), T2 (100 μg/mL), 3-ADON (100 μg/mL), DAS (100 μg/mL), 15-ADON (100 μg/mL), and F-X (100 μg/mL) were obtained from Romer Labs (Tulln an der Donau, Austria). Alternariol (AOH), alternariol monomethylether (AME), zearalanone (ZAN) and enniatin B (ENN B) were purchased from Sigma-Aldrich (Bornem, Belgium), fumonisin B3 was obtained from Promec unit (Tygerberg, South Africa), and roquefortine C (ROQ-C) was purchased from Alexis Biochemicals (Enzo Life Sciences BVBA, Zandhoven, Belgium). Stock solutions were prepared in acetonitrile/water (50/50 *v*/*v*) for FB3 (1 mg/ml), methanol/dimethylformamide (60/40 *v*/*v*) for AOH and AME (1 mg/ml), and methanol for ROQ-C and ZAN (1 mg/ml). All stock solutions were stored at −20 °C, except for the stock solution of FB3 (4 °C). Working solutions were prepared by diluting the stock solutions in methanol, and were stored at −20 °C for three months. Three standard mixture working solutions were prepared in methanol. The first contained the mycotoxins AFB1, AFB2, AFG1 and AFG2 (2 ng/μL); OTA (5 ng/μL); ZEN and T2 (10 ng/μL); and DON, FB1 and FB2 (40 ng/μL). The second contained DAS (0.5 ng/μL); ROQ-C (1 ng/μL); 15-ADON (2.5 ng/μL); 3-ADON and STERIG (5 ng/μL); NEO and AOH (10 ng/μL); NIV, F-X and AME (20 ng/μL); and FB3 (25 ng/μL). The third contained ENN B (10 ng/μL). The standard mixtures were stored at −20 °C and used for a maximum of three months.

2.3. Sample Preparation for LC-MS/MS

Twenty-two mycotoxins were extracted from the samples according to the methodology described by Monbaliu et al. [75]. Five grams of dried maize sample was spiked with internal standards ZAN and DOM at a concentration of 200 and 250 μg/kg, resp. The spiked sample was kept in the dark for 15 min and extracted with 20 ml of extraction solvent (acetonitrile/water/acetic acid (79/20/1, *v*/*v*/*v*)), and then agitated on a vertical shaker for 1 h. After centrifuging for 15 min at 3300 g, the supernatant was passed through a preconditioned C18 solid phase extraction (SPE) column (Alltech, Lokeren, Belgium). The eluate was diluted to 25 ml with extraction solvent and defatted with 10 ml n-hexane. In order to recover all 22 mycotoxins, two different clean-up pathways were followed. In the first pathway, 10 ml of extract was diluted with 20 ml acetonitrile/acetic acid (99/1 *v*/*v*), passed through a Multisep®226, AflaZon+ multifunctional column from Romer Labs (Tulln, Austria) and washed with 5 ml acetonitrile/acetic acid (99/1 *v*/*v*). For the second pathway, 10 ml extract was filtered using a Whatman glass microfilter (VWR International, Zaventem, Belgium). Two milliliters of this filtered extract was combined with the MultiSep 226 eluate from the first pathway and evaporated to dryness. The residue was then redissolved into 150 μL of mobile phase (water/methanol/acetic acid (57.2/41.8/1, *v*/*v*/*v*)) and 5 mM ammonium acetate. Lastly, the solution was centrifuged for 5 min at 14,000 ×*g* using ultra free-MC centrifuge filters (Millipore, Bedford, MA, USA).

2.4. Mycotoxin Analysis by LC-MS/MS

The samples were analyzed using a micromass Quattro Premier XE triple quadrupole mass spectrometer coupled with a Waters Acquity UPLC system (Waters, Milford, MA, USA). Data processing was done using the Masslynx™ (4.1 version) and Quanlynx® (4.1 version) software (Micromass, Manchester, UK). The analytical column used was a Symmetry C18, 5 μm, 2.1 × 150 mm, with a guard column of the same material (3.5 μm, 10 mm × 2.1 mm) (Waters, Zellik, Belgium) kept at room temperature. The injection volume was 10 μL. Capillary voltage was set at 3.2 kV with a source block temperature and desolvation temperature of 120 and 400 °C, resp. Liquid chromatography conditions and MS parameters were followed as described by Monbaliu et al. [75].

2.5. LC-MS/MS Quality Control

To compensate for matrix effects and losses during extraction and cleanup, DOM (a structural analogue of DON) and ZAN (a structural analogue of ZEN) were used as internal standards. For each mycotoxin, five blank samples were spiked at five concentration levels. A cutoff (CO) level was established for every mycotoxin. The CO levels were based on the current regulatory levels, if available [64,65,76]; else, the CO level was chosen arbitrarily. The decision limit CCα was defined as the concentration at the y-intercept plus 2.33 times the standard deviation (SD) of the within lab reproducibility (α = 1%). The apparent recovery was calculated by dividing the observed value from the calibration plot by the spiked level. Linearity was tested graphically using a scatter plot, and the linear regression model was evaluated using a lack-of-fit test.

LOD and LOQ were estimated for each separate mycotoxin using the blank samples spiked at five different concentrations, which provided a signal-to-noise (S/N) ratio of 3 and 10, resp., in accordance to the definitions set by the International Union of Pure and Applied Chemistry (IUPAC). The interday repeatability was calculated using the relative standard deviation (RSD) at the spiked concentration levels.

2.6. qPCR Analysis

A quantitative PCR (qPCR) assay was used to quantify the total *F. graminearum*, *F. verticillioides* and *F. culmorum* DNA content in the maize samples from 2017 and 2018. In 2016, no samples for qPCR were taken. These three species were selected based on the known fungal species composition in temperate climates and in Belgium in particular [21,35,73,74], and to cover most *Fusarium* producers of mycotoxins that were included in the LC-MS/MS analysis [15,35,77]. Each subsample (5 g) was crushed with liquid nitrogen using a pestle and mortar and approx. 150 mg (the exact amount was weighted) was transferred to a 1.5 ml Eppendorf tube for DNA extraction. DNA was extracted from harvested maize samples using a CTAB method modified for use with fungi [78]. The total amount of DNA was quantified with a Quantus fluorometer (Promega, Leiden, The Netherlands), and stored at −20 °C. Then qPCR analysis was performed. The qPCR mix consisted of 6.25 µL of GoTaq®qPCR Master Mix (Promega, Leiden, The Netherlands), the corresponding primers (0.625 µL primer, 5 µM), 2 µL of DNA, 0.208 µL CXR reference dye (Promega, Leiden, The Netherlands), and watered to 12 µL. The used primers were FgramB379 forward (CCATTCCCTGGGCGCT), FgramB411 reverse (CCTATTGACAGGTGGTTAGTGACTGG), FculC561 forward (CACCGTCATTGGTATGTTGTCACT), FculC614 reverse (CGGGAGCGTCTGATAGTCG), Fver356 forward (CGTTTCTGCCCTCTCCCA), and Fver412 reverse (TGCTTGACACGTGACGATGA) [79]. The qPCR analysis was performed using a CFX96 system (Bio-Rad, Temse, Belgium), including the following thermal settings: 95 °C for 3 min; 40 cycles of 95 °C for 10 s, and 60 °C for 30 s, followed by dissociation curve analysis at 65 to 95 °C.

2.7. Statistical Analysis

The Pearson correlation coefficient was used to detect relations between different mycotoxins, between mycotoxins and fungal DNA, and between different *Fusarium* spp. at a significance level of $p = 0.05$. For calculation of the correlation coefficients, four outliers were discarded in the *F. verticillioides* DNA data and one in the *F. graminearum* DNA data. All statistical analyses were conducted using the R software package (R Core Team, Vienna, Austria) version 3.4.3 [80].

3. Results

3.1. Mycotoxin Levels in Harvested Maize Samples in 2016–2018

Incidence, mean, median and maximum concentrations, and the numbers of samples exceeding the European regulations can be found in Table 1; Complete results per sample can be found in supplementary Table S1. NIV was the most prevalent mycotoxin, being present in 99.2% of all samples between 2016 and 2018. DON was present in all samples in 2017, but only in 64.7% of the samples in 2018. Over the three years, DON and its derivates 3-ADON and 15-ADON (described together as DON+) were the second most prevalent mycotoxins. ZEN's highest incidence was in 2016, with 64.8% of the samples contaminated, while only 40.7% and 42.4% of the samples were contaminated in 2017 and 2018, resp. FB1, FB2 and FB3 incidence rose considerably from 2016 till 2018, with a total fumonisin incidence (described as FUM) of only 2.5% in 2016, to 19.8% in 2017 and 61.2% in 2018. AOH, AME, DAS, FX, T2, STERIG and ROQ-C were detected sporadically and never reached incidences higher than 11.0%. NEO, AFB1, AFB2, AFG1, AFG2 and OTA were never detected.

Mean concentration of NIV rose from 650.7 µg/kg in 2016, to 719.0 µg/kg in 2017 and 881.9 µg/kg in 2018. The highest mean concentration of DON was found in 2017 (557.5 µg/kg), while the lowest concentration was found in 2018 (186.5 µg/kg). Concentrations for NIV and DON went as high as 6776.3 µg/kg and 5322.5 µg/kg, resp. These concentrations were detected in the same sample from a maize field in 2017. This sample contained the highest total mycotoxin load of all years, with a total mycotoxin concentration of 13,747.6 µg/kg. Mean concentrations of fumonisin (FUM) rose simultaneously with its incidence, from 1.3 µg/kg in 2016 to 327.0 µg/kg in 2018. The average total mycotoxin load in a maize sample from 2016 till 2018 was 1692.0 µg/kg. Over the three years, 2.3% and 7.8% of the maize samples exceeded the EU guidance values for DON and ZEN, resp. No samples exceeded the guidance values for FB1, FB2, FB3 or T2, nor the maximum level for AFB1.

A vast majority of the maize samples was contaminated with more than one mycotoxin. Only one out of the 257 samples analyzed over the course of three years contained no mycotoxins, while 46.7% of all samples contained five or more mycotoxins. The median load was four mycotoxins per sample. When comparing the multi-mycotoxin contamination of each year (Figure 2), it is clear that the most diversely contaminated maize samples were found in 2017 and 2018. In 2018, two samples even contained 10 different mycotoxins. In 2016, the maximum number of detected mycotoxins in one sample was seven. However, 62.6% of the samples in 2016 contained five or more different mycotoxins, leading to the highest median mycotoxin load per sample (five mycotoxins per sample, compared to four mycotoxins per sample for 2017 and 2018, resp.).

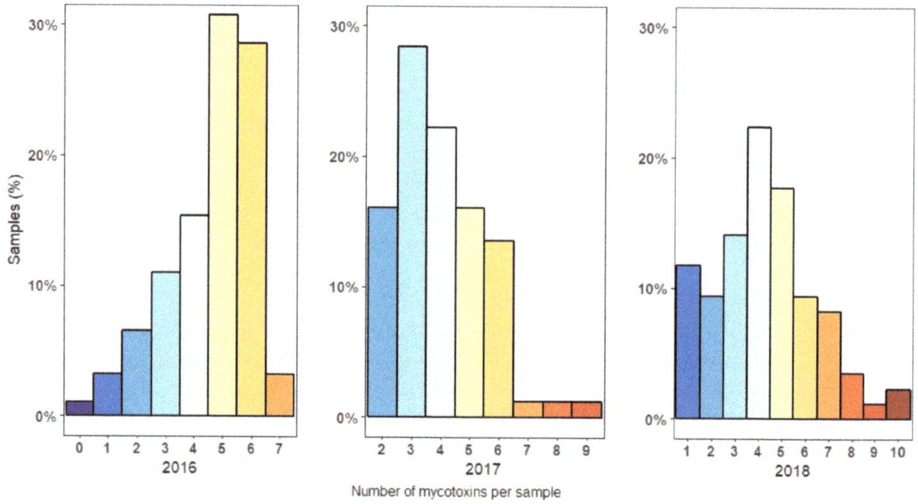

Figure 2. The relative number of maize samples contaminated with a certain number of different mycotoxins for 2016, 2017 and 2018.

Table 1. Mycotoxin contamination detected in maize samples at harvest in Flanders, Belgium, from 2016 till 2018.

	Positive Samples (%)				Mean Concentration ᵃ (μg/kg)				Median Concentration (μg/kg)				Max. Concentration (μg/kg)				Samples Exceeding EU Recommendation (%)ᵇ			
	2016	2017	2018	2016–2018	2016	2017	2018	2016–2018	2016	2017	2018	2016–2018	2016	2017	2018	2016-2018	2016	2017	2018	2016–2018
n samples	91	81	85	257	91	81	85	257	91	81	85	257	91	81	95	257	91	81	95	257
NIV	98.9	100	98.8	99.2	650.7	719.0	881.9	748.7	527.5	460.6	782.1	587.1	2368.2	6776.3	2409.5	6776.3				
DON	92.3	100	64.7	85.6	449.0	557.5	186.5	396.4	263.1	337.4	121.3	215.3	2777.4	5322.4	2110.5	5322.4	2.2	3.7	1.0	2.3
3-ADON	78.0	29.6	15.3	42.0	53.5	36.3	23.3	38.1	43.2	0	0	0	297.0	380.3	1046.8	1046.8				
15-ADON	64.8	51.3	12.9	43.4	95.0	81.2	15.2	64.5	71.3	17.7	0	0	819.3	769.2	248.6	819.3				
DON+ᶜ	95.6	100	64.7	86.8	597.5	675.0	225.0	498.7	376.9	406.7	130.3	261.5	3050.1	6471.9	2110.5	6471.9				
ZEN	64.8	40.7	42.4	49.8	100.5	158.5	175.5	159.7	71.2	0	0	0	1085.6	1411.9	2791.6	2791.6	1.1	8.6	12.6	7.8
ENN B	42.9	18.5	45.9	36.2	133.1	77.7	180.3	149.5	56.2	27.5	70.7	46.2	1375.1	1041.9	1984.9	1984.9				
AOH	n.d.	3.7	9.4	4.3	n.d.	1.4	6.5	2.6	n.d.	0	0	0	n.d.	49.1	208.6	208.6				
AME	2.2	3.7	10.6	5.4	0.8	11.5	19.8	10.5	0	0	0	0	49.1	370.6	452.8	452.8				
FB1ᵈ	2.5	19.8	61.2	28.6	1.5	61.1	247.4	106.5	0	0	54.0	0	70.2	1362.9	4414.9	4414.9	0	0	0	0
FB2ᵈ	n.d.	4.9	24.7	10.2	n.d.	9.0	61.6	24.4	n.d.	0	0	0	n.d.	412.6	1427.4	1427.4	0	0	0	0
FB3ᵈ	n.d.	7.4	18.8	9.0	n.d.	3.4	18.0	7.4	n.d.	0	0	0	n.d.	90.5	451.2	451.2				
FUMᶜ	2.5	19.8	61.2	28.6	1.3	73.6	327.0	131.8	0	0	58.7	0	70.2	1782.8	6293.5	6293.5				
DAS	11.0	8.6	5.9	8.6	0.3	0.3	0.4	0.4	0	0	0	0	6.1	10.3	14.9	14.9				
FX	n.d.	7.4	2.4	3.1	n.d.	14.2	2.7	5.4	n.d.	0	0	0	n.d.	223.6	161.6	223.6				
T2	1.1	n.d.	8.2	3.1	0.2	n.d.	6.2	2.1	0	n.d.	0	0	16.8	n.d.	121.6	121.6	0	0	0	0
STERIG	1.1	n.d.	1.2	0.8	0.2	n.d.	2.6	0.9	0	n.d.	0	0	15.1	n.d.	73.3	204.8				
ROQ-Cᵈ	n.d.	2.5	2.9	1.7	n.d.	0.6	0.6	0.4	n.d.	0	0	0	n.d.	30.4	24.6	30.4				
TOTALᶜ	98.9	100	100	100	1485.1	1729.9	1877.4	1692.0	1309.6	1088.2	1596.1	1309.6	4153.4	13747.6	8309.0	13747.6				

n.d.: Not detected. a: Arithmetic mean. b: EU regulations: 2000 μg/kg for DON (complementary and complete feedstuffs for calves (< 4 months)); 500 μg/kg for ZEN complementary and complete feedstuffs for calves and dairy cattle; 20,000 μg/kg for FB1+FB2 (calves (< 4 months)); 250 μg/kg for T2 (compound feed) (European Commission, 2006, 2013). c: DON+ = the sum of the incidence/concentrations of DON, 3-ADON and 15-ADON; FUM = the sum of the incidence/concentrations of FB1, FB2 and FB3; TOTAL = The sum of the incidence/concentrations of all detected mycotoxins. d: In 2016, only 79 samples were analyzed for FB1, FB2 and FB3. In 2018, only 68 samples were analyzed for ROQ-C.

3.2. Correlations between Different Mycotoxins

A heat map with correlations between different mycotoxins for 2016–2018 is shown in Figure 3. NIV was significantly correlated with DON (r = 0.38, $p < 0.001$) and its derivates 3-ADON (r = 0.22, $p < 0.001$) and 15-ADON (r = 0.28, $p < 0.001$). NIV was also significantly correlated with ZEN and ENN B, although the correlations were rather weak (r = 0.21, $p < 0.001$ and r = 0.12, $p = 0.0496$, resp.). Other correlations were non-existent or not significant. When splitting the data per year, similar results were obtained, however some differences occurred (Figures A1–A3). For instance, the correlation between NIV and DON was the strongest in 2017 (r = 0.65, $p < 0.001$), but was not significant in 2016 and 2018 (r = 0.08, $p = 0.455$ and r = 0.21, $p = 0.058$, resp.). Furthermore, ZEN and ENN B were significantly correlated in 2016 and 2017 (r = 0.35, $p < 0.001$ and r = 0.37, $p < 0.001$, resp.). FUMs were not correlated with any other mycotoxin in any year, except with 15-ADON in 2018 (r = 0.31, $p = 0.004$).

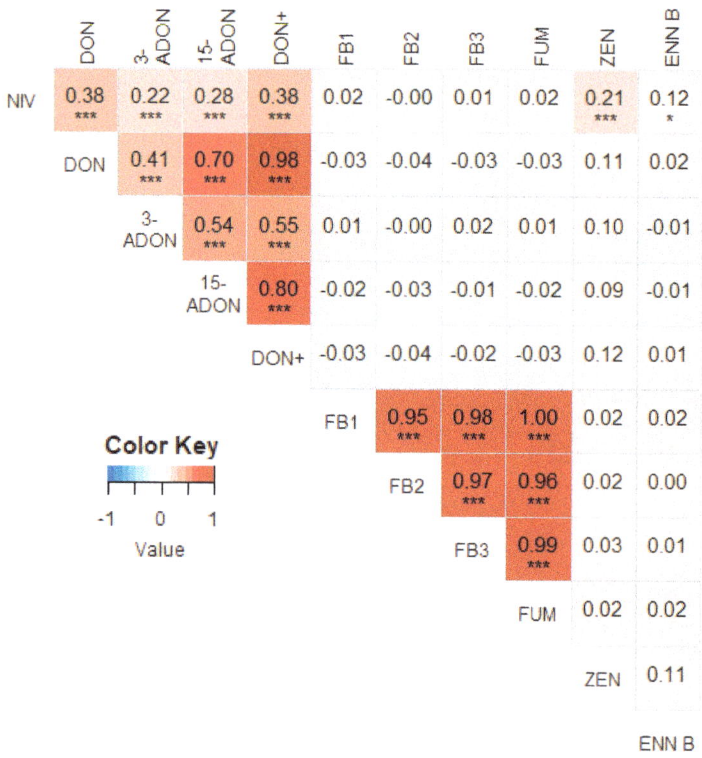

Figure 3. Heat map based on the pairwise Pearson correlation coefficients between the measured mycotoxin concentrations from 2016–2018. A darker blue color indicates a stronger negative correlation, a darker red color indicates a stronger positive correlation. Significant correlations are indicated with asterisks (* $p < 0.05$, *** $p < 0.01$). DON+ = the sum of the concentrations of DON, 3-ADON and 15-ADON. FUM = the sum of the concentrations of FB1, FB2 and FB3.

3.3. Fusarium spp. DNA in Maize Samples in 2017-2018

Incidence of *F. graminearum*, *F. verticillioides*, *F. culmorum* and *Fusarium* spp. in maize samples in 2017, 2018 and both years combined, is shown in Figure 4; Complete results per sample can be found in supplementary Table S1. In 2017, every maize sample was contaminated with at least one *Fusarium* sp., while in 2018, 36% of the samples were free of *Fusarium* spp. DNA. In both years, *F. verticillioides* was detected most often, with a prevalence of 99% in 2017 and 54% in 2018. *F. graminearum* and *F. culmorum* were detected in 90% and 85% of the maize samples in 2017, and 43% and 51% in 2018, resp.

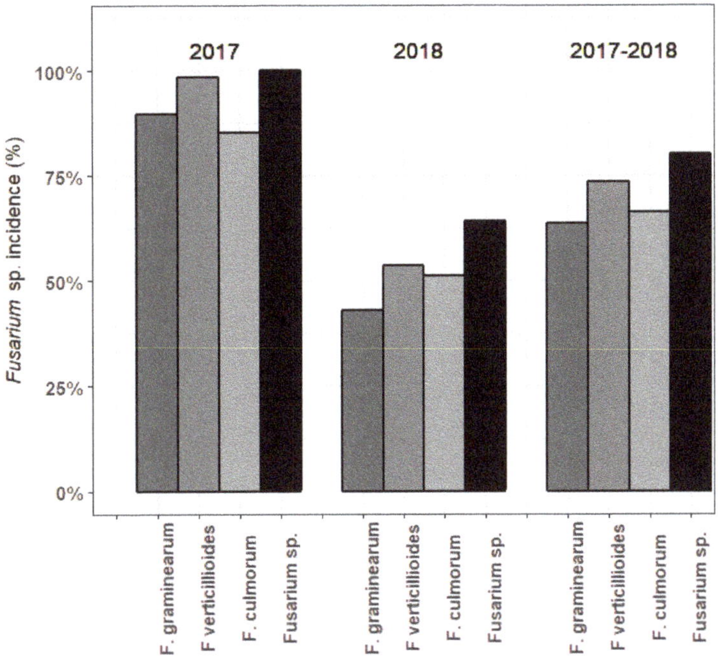

Figure 4. Incidence of *F. graminearum*, *F. verticillioides*, *F. culmorum* and *Fusarium* spp. in general, in samples of harvested maize in 2017, 2018 and in both years combined.

3.4. Correlations between Mycotoxin Concentrations and Fusarium spp. DNA

Using qPCR analysis, we could calculate correlations between mycotoxin concentrations and *Fusarium* spp. DNA on maize fields, and interspecies correlations between *Fusarium* species (Figure 5). Rather weak but significant correlations were found between the amount of *F. graminearum* DNA and *F. culmorum* DNA ($r = 0.21$, $p = 0.009$) and *F. verticillioides* DNA and *F. culmorum* DNA ($r = 0.19$, $p = 0.024$). A strong significant correlation was found between DON+ and *F. graminearum* DNA ($r = 0.53$, $p < 0.001$). Both *F. graminearum* and *F. culmorum* were significantly correlated with higher concentrations of NIV ($r = 0.35$, $p < 0.001$ and $r = 0.36$, $p < 0.001$, resp.). Furthermore, *F. verticillioides* DNA was positively correlated with FUM ($r = 0.20$, $p < 0.016$), but an even stronger correlation was found between FUM and *F. graminearum* ($r = 0.27$, $p < 0.001$). However, the latter correlation is based primarily on one data point with a high concentration of FUM and a high *F. graminearum* DNA content. When removed from the dataset, the resulting correlation is no longer significant ($r = -0.03$, $p = 0.707$). Similarly, when removing two outliers from the dataset with a very high FUM content, the correlation between *F. verticillioides* DNA and FUM becomes more profound ($r = 0.45$, $p < 0.001$). Lastly, when eliminating one outlier from the *F. culmorum* and *F. verticillioides* data, the correlation becomes non-existent ($r = -0.06$, $p = 0.482$) (See Figure A4). Other correlations were less dependent upon outliers. Splitting the data per year yields similar results as the combined data (Figures A5 and A6).

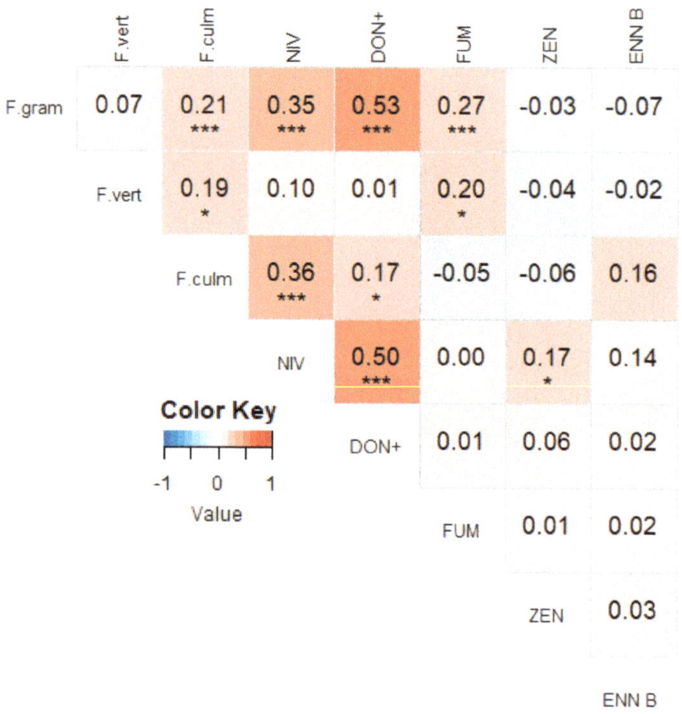

Figure 5. Heat map based on the pairwise Pearson correlation coefficients between measured mycotoxin concentrations and DNA of *F. graminearum*, *F. verticillioides* and *F. culmorum* from 2017–2018. A darker blue color indicates a stronger negative correlation, a darker red color indicates a stronger positive correlation. Significant correlations are indicated with asterisks (* $p < 0.05$, *** $p < 0.01$). DON+ = the sum of the concentrations of DON, 3-ADON and 15-ADON. FUM = the sum of the concentrations of FB1, FB2 and FB3.

4. Discussion

In our survey, NIV was the most prevalent mycotoxin. Only one out of 257 samples was free of NIV. Concentrations went as high as 6776.3 µg/kg. DON and its derivates 3-ADON and 15-ADON, described together as DON+, were present in 86.8% of the samples. ZEN was found in 49.8% of the samples. Binder et al. [37] took samples of several feedstuffs in Europe and Asia. For maize in Europe, they found DON in 81% of the samples, ZEN in 63%, FUM in 56% and AFB1 in 21%. NIV was not tested. Eckard et al. [24] sampled 20 fields of silage maize in Switzerland for one year. They found DON in every sample, with concentrations up to 2990 µg/kg. ZEN was found in 79% and NIV in 42% of the samples. Goertz et al. [22] sampled maize ears in Germany for two years, and found that incidence and concentrations differed between years. ZEN, NIV and DON and its derivates were detected more frequently and in higher concentrations in a temperate year than in a hot and warm year, while FUMs were only detected in the latter. Van Asselt et al. [38] found that only a quarter of the sampled maize ears in the Netherlands were contaminated with mycotoxins, but 84% of those contained NIV, with concentrations up to 1671 µg/kg. Kosicki et al. [39] found ZEN, DON and NIV in 92%, 89% and 77% of Polish maize ear samples between 2011 and 2014. FUMs were detected in 58% of the samples.

Overall, the results of our survey of Flemish maize are in line with previous research, although the overwhelming incidence and concentrations of NIV have not been described before. NIV is often overlooked when analyzing for mycotoxin contamination.

NIV-producing populations of *F. graminearum* and *F. culmorum* are emerging however [81–83], possibly due to the increased use of wheat in a rotation with maize [44]. Furthermore, some reports state NIV may be even more toxic than DON and other trichothecene mycotoxins [84–87]. More than 24% of the samples in our survey contained NIV concentrations higher than 1000 µg/kg, and 12 samples (4.7%) even exceeded 2000 µg/kg, the EU guidance value for DON. Our research shows that NIV is present in nearly every maize field in Flanders, and often in high concentrations. This mycotoxin should therefore always be included in analyses, especially in Central and North Europe.

Multi-mycotoxin contamination was very common in our survey. Only one sample contained none of the 22 analyzed mycotoxins. 46.7% of maize samples were contaminated with five or more mycotoxins, and two samples in 2018 were even contaminated with 10 different mycotoxins. Schollenberger et al. [62] found up to 12 trichothecenes in one sample, Drejer Storm et al. [48] found up to seven mycotoxins in one sample, and Streit et al. [67] found that up to 69 secondary metabolites including mycotoxins may co-occur in one sample. This multi-mycotoxin contamination is not covered in the current EU regulations. In our survey, 2% of the samples exceeded the EU guidance value for DON, and 7.8% for ZEN. None of the samples exceeded the guidance values or maximum levels for FUM, AFB1 or T2. However, other mycotoxins (e.g., NIV) are not included in the EU regulations, and as mentioned before, multi-mycotoxin contamination and possible synergistic effects are also not included. The sample with the highest overall mycotoxin concentration in our survey contained seven different mycotoxins, i.e. NIV, DON, FX, 3-ADON, 15-ADON, DAS and ZEN, with a total mycotoxin load of 13,474.6 µg/kg. This sample only exceeded the EU regulation for DON, being 2000 µg/kg. But one could assume that its toxicity will be far higher than that of a sample containing only DON in a concentration above 2000 µg/kg. Synergistic or additive toxic effects of a combined mycotoxin contamination have been demonstrated in previous literature, especially with mycotoxins that share a similar chemical structure or are produced by the same fungal species [68,70–72,88]. The EU regulations should therefore be re-evaluated and expanded in the future to account for multi-mycotoxin contamination [2].

As expected, the concentrations of DON and its derivates 3-ADON and 15-ADON were strongly positively correlated [66]. Similarly, the fumonisins FB1, FB2 and FB3, were strongly correlated. Other significant positive correlations were found between NIV; and DON (and its derivates), ZEN and ENN B. NIV is known to be primarily produced by *F. culmorum* in temperate regions, while DON is mainly produced by *F. graminearum* [35,77,89,90]. Since DON and NIV are positively correlated, the amount of *F. graminearum* and *F. culmorum* DNA were expected to be positively correlated as well. This was indeed the case, although the correlation was not particularly strong ($r = 0.21$).

FUMs were not correlated with any other mycotoxin. Likewise, the main fumonisin producing *Fusarium* species, *F. verticillioides*, was not correlated with *F. graminearum*. The correlation between *F. verticillioides* and *F. culmorum* was significant ($r = 0.19$), despite the absence of a correlation between NIV and FUM. However, as explained earlier in the Results section, some outliers may have skewed the data. In this case, omitting one outlier from the dataset effaced the corresponding correlation between *F. verticillioides* and *F. culmorum* ($r = -0.06$). Similarly, removing one outlier made the unexpected correlation between *F. graminearum* and FUM non-significant ($r = -0.03$), and removing two outliers made the expected but rather weak correlation between *F. verticillioides* and FUM more profound ($r = 0.45$). Other correlations were stable and less dependent upon outliers. With these adjustments, we could conclude that the main fumonisin-producer *F. verticillioides* is positively correlated with FUM; the main NIV-producer *F. culmorum* is correlated with NIV; and the main DON-producer *F. graminearum* is correlated with DON+. The latter relation could be demonstrated anecdotally, because the sample with the highest DON+ concentration also had the highest levels of *F. graminearum* DNA. These results are in accordance with the previous literature [22,24,83]. Other researchers have found a correlation between DON and ZEN [39,40,61,91], which was not the case in our survey, except in 2017 (Figure A2).

Fusarium graminearum and *F. culmorum* share a positive correlation, meaning that they can co-exist and produce mycotoxins on the same plant. On the other hand, *F. verticillioides* is not correlated with *F. graminearum* nor with *F. culmorum*.

This could be caused by differing optimal growing conditions, since *F. verticillioides* prefers warm temperatures and dry conditions, while *F. graminearum* and *F. culmorum* both prefer colder and wetter conditions [19,23]. Moreover, the co-occurrence of different fungal species on the same plant may have a significant impact on fungal development and mycotoxin production [92–94]. Indeed, most plant diseases are caused by a complex of species rather than by a single species, which may lead to synergistic effects [95]. Previous research has shown that *F. graminearum* and *F. verticillioides* may co-occur and produce mycotoxins on the same plant when infected artificially, but the type of interactions may differ depending on the weather conditions [25–27,29,96]. FUM production is mainly reduced when *F. graminearum* and *F. verticillioides* are co-inoculated, whereas DON production is increased; ZEN production is not affected [26]. When co-inoculated with *Aspergillus parasiticus*, ZEN and DON production by *F. graminearum* is not infected, while AFB1 production by *A. parasiticus* is significantly reduced [97]. Furthermore, a high amount of fungal inoculum does not necessarily lead to higher mycotoxin concentrations [98]. These effects of fungal co-occurrence may explain why *F. graminearum* and *F. verticillioides* are not correlated in our survey, and why certain expected correlations between fungal species and/or mycotoxins have not been observed.

There was a clear year-to-year difference in the observed mycotoxin incidences and concentrations and the presence of *Fusarium* spp. DNA, related to changes in the weather conditions. This has been observed multiple times in past literature [22,40,58,99]. A summary of the weather conditions of each year (2016–2018) can be found in Table 2. 2016 was a year with high precipitation, especially in June, and a high relative humidity (RH). 2017 had less rainfall and less radiation, but similar temperatures. 2018 was an extremely dry year, with only 241 mm of precipitation during the growing season, and the highest temperatures ever recorded in Belgium, up to 41.8 °C on the 25th of July [100]. These extreme, dry and warm temperatures led to a number of different observations: More diversely-contaminated samples, but a lower median mycotoxin load per sample; A reduction of the incidence and concentrations of DON and its derivates; more samples that were highly contaminated with ZEN, and thereby exceeded EU guidance values; more incidence of *Alternaria* mycotoxins AOH and AME; and most remarkably, a strong increase in the incidence and concentrations of FUMs. 61.2% of maize samples were contaminated with FUMs in 2018, with a concentration of up to 6293.5 μg/kg, versus 19.8% in 2017 and only 2.5% in 2016. Contrastingly, the incidence of *F. verticillioides* did not rise, but was lower compared to 2017 (99% and 54%, resp.). Since mycotoxin production is influenced by temperature and water levels [23,26,101,102], the specific growing conditions in 2018 could have reduced *F. verticillioides* infection but induced FUM production. In general, less maize samples were contaminated with *Fusarium* spp. in 2018 compared to 2017 (100% and 64%, resp.), with *F. verticillioides* being the most prevalent species in both years. Scauflaire et al. [21] found that in maize ears and stalks in Wallonia, Belgium, *F. graminearum* was the predominant species, while *F. verticillioides* occurred only sporadically. The same conclusions were drawn in Switzerland [61] and the UK [44]. The dissimilar results of our survey compared to these studies could be explained by the abnormal weather conditions in Belgium in 2017 and 2018, causing a shift in the fungal populations. *F. verticillioides* infection and, correspondingly, FUM production is higher in warm and dry years [22,23,40,103]. Many maize fields in 2018 were of very low quality and were harvested with little to no cobs developed, possibly explaining the lower general incidence of *Fusarium* spp. in that year. Furthermore, *Fusarium* spp. generally infect a plant in a species complex [19]. Only three *Fusarium* spp. were included in our qPCR analysis. It is possible that other species were present as well, and produced mycotoxins of their own. In the previous literature, 11 to 23 different *Fusarium* species were isolated from maize fields in Belgium [21], the UK [44], Switzerland [24,61], Germany [22] and the Netherlands [38]. Possibly, infections by *F. poae* (NIV, DAS), *F. avenaceum* (ENN B), *F. proliferatum* (FUM), *F. crookwellense* (NIV, ZEN) or other *Fusarium* species occurring in Belgium [21,74] could explain the incidence of certain related mycotoxins [19,104].

Table 2. Weather parameters during the 2016–2018 maize growing seasons in Flanders, Belgium. The growing season start date is based on the first maize field in our database being sown; the end date is based on the last maize field being harvested. Mean and range values are based on daily weather measurements from 17 weather stations across Flanders (Figure 1).

Year	Growing Season	Rainfall (mm)		Relative Humidity (%)		Average Temperature (°C)		Total Daily Radiation (W/m^2)	
		Mean	Range (Min. - Max.)	Mean	Range (Min. - Max.)	Mean	Range (Min. - Max.)	Mean	Range (Min. - Max.)
2016	20.04–26.10 (189 days)	423	283–610	80.5	72.8–86.3	15.7	15.1–16.7	3839	3697–4024
2017	10.04–28.10 (201 days)	344	186–541	78.6	70.3–85.3	15.7	15.1–16.2	3389	2600–3902
2018	19.04–14.10 (178 days)	241	155–335	74.2	69.6–84.7	17.5	16.0–18.3	4294	4109–4710

5. Conclusions

In conclusion, this 3-year study has demonstrated the shifting mycotoxin load in silage maize fields at harvest due to changing weather conditions, possibly induced by climate change. Fumonisins, produced by *F. verticillioides*, which is more prevalent in tropical climates, were detected sporadically in Flanders in wet and cold years, but were found far more frequent during dry and hot years. Nivalenol was found in all but one of the samples, across all three years, making it the most stable and widespread mycotoxin. Concentrations went as high as 6776 µg/kg. Aflatoxins were not found, but *Aspergillus* spp. grow at similar conditions as *F. verticillioides*, so these mycotoxins should not be overlooked in future surveys. In order to monitor the effect of climate change on these changing weather conditions and on subsequent mycotoxin production, a yearly sampling should be continued.

The next step will be to identify the underlying cultivation, environmental and climatic factors that influence mycotoxin contamination in the field, and to create a prediction model for farmers based on these data. Ultimately, this research could help reduce mycotoxin contamination in silage maize and reduce mycotoxicosis in dairy cattle.

Supplementary Materials: The following are available online at http://www.mdpi.com/2076-2607/7/11/571/s1, Table S1: Results of the LC-MS/MS and qPCR analysis of 257 maize samples in Flanders in 2016-2018.

Author Contributions: Conceptualization, G.H., K.A. and S.C.; formal analysis, J.V.; investigation, J.V. and K.D.V.; methodology, S.D.S.; resources, S.D.S. and G.H.; data curation, J.V.; Writing—Original draft preparation, J.V.; Writing—Review and editing, K.A. and G.H.; visualization, J.V.; supervision, K.A. and G.H.; project administration, G.H.; funding acquisition, G.H. and S.C.

Funding: This research is part of the project 'Ontwikkeling van een beslissingsondersteunend adviessysteem voor een betere beheersing van mycotoxines in maïskuilen', funded by Flanders Innovation and Entrepreneurship (VLAIO) (140971).

Acknowledgments: The authors kindly thank the dairy farmers who participated in this study. They also thank Sofie Landschoot and Xiangrong Chen (UGent, Faculty of Bioscience Engineering) for help with the qPCR analysis, and Christ'l Detavernier, Marthe De Boevre and Mario Van de Velde (UGent, Faculty of Pharmaceutical Sciences) for help with the LC-MS/MS analysis.

Conflicts of Interest: The authors declare no conflict of interest. The funders had no role in the design of the study; in the collection, analyses, or interpretation of data; in the writing of the manuscript, or in the decision to publish the results.

Appendix A

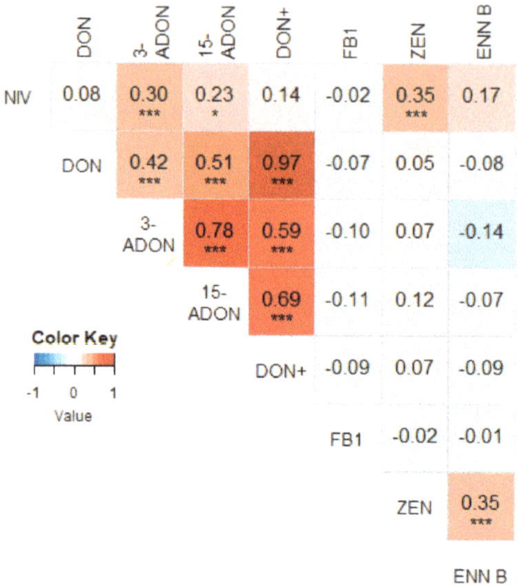

Figure A1. Heat map based on the pairwise Pearson correlation coefficients between the measured mycotoxin concentrations in 2016. A darker blue color indicates a stronger negative correlation, a darker red color indicates a stronger positive correlation. Significant correlations are indicated with asterisks (* $p < 0.05$, *** $p < 0.01$). DON+ = the sum of the concentrations of DON, 3-ADON and 15-ADON.

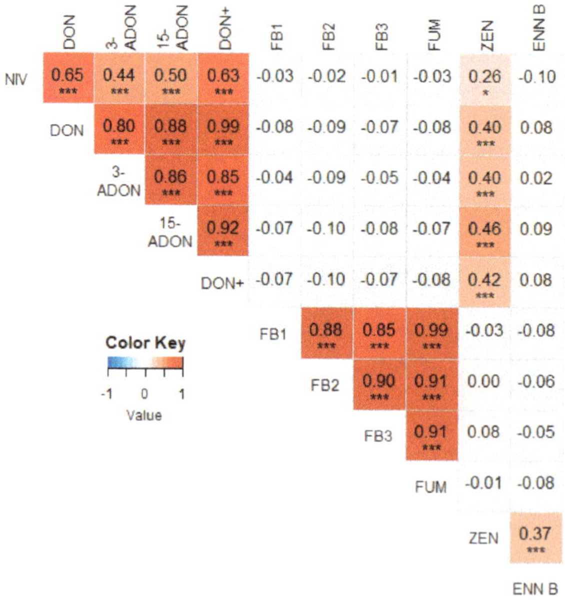

Figure A2. Heat map based on the pairwise Pearson correlation coefficients between the measured mycotoxin concentrations in 2017. A darker blue color indicates a stronger negative correlation, a darker red color indicates a stronger positive correlation. Significant correlations are indicated with asterisks (* $p < 0.05$, *** $p < 0.01$). DON+ = the sum of the concentrations of DON, 3-ADON and 15-ADON.

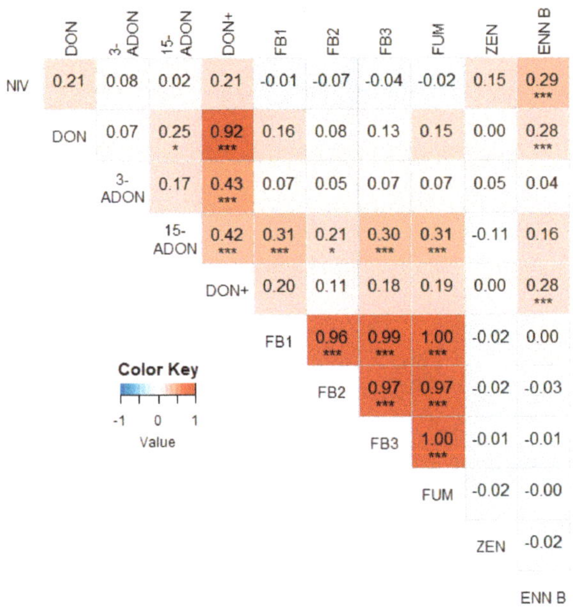

Figure A3. Heat map based on the pairwise Pearson correlation coefficients between the measured mycotoxin concentrations in 2018. A darker blue color indicates a stronger negative correlation, a darker red color indicates a stronger positive correlation. Significant correlations are indicated with asterisks (* $p < 0.05$, *** $p < 0.01$). DON+ = the sum of the concentrations of DON, 3-ADON and 15-ADON.

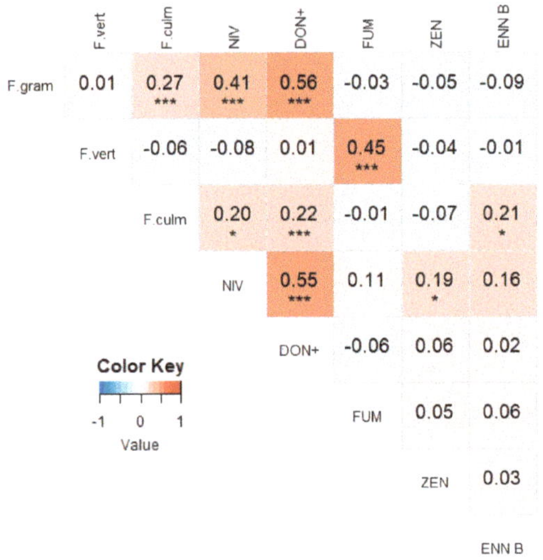

Figure A4. Heat map based on the pairwise Pearson correlation coefficients between the measured mycotoxin concentrations in 2017-2018, with the exclusion of 3 outliers. A darker blue color indicates a stronger negative correlation, a darker red color indicates a stronger positive correlation. Significant correlations are indicated with asterisks (* $p < 0.05$, *** $p < 0.01$). DON+ = the sum of the concentrations of DON, 3-ADON and 15-ADON.

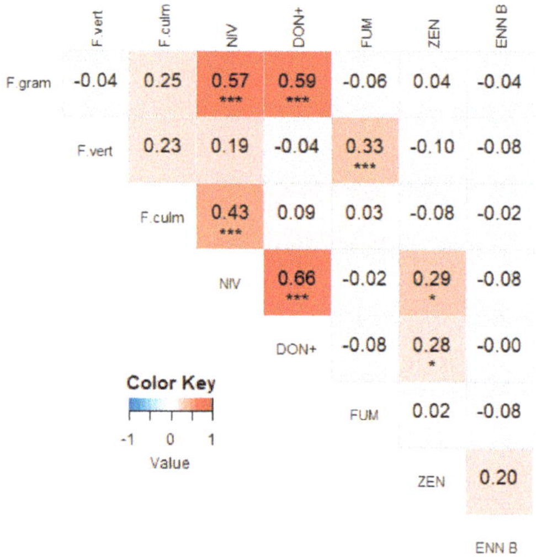

Figure A5. Heat map based on the pairwise Pearson correlation coefficients between measured mycotoxin concentrations and DNA of *F. graminearum*, *F. verticillioides* and *F. culmorum* in 2017. A darker blue color indicates a stronger negative correlation, a darker red color indicates a stronger positive correlation. Significant correlations are indicated with asterisks (* $p < 0.05$, *** $p < 0.01$). DON+ = the sum of the concentrations of DON, 3-ADON and 15-ADON. FUM = the sum of the concentrations of FB1, FB2 and FB3.

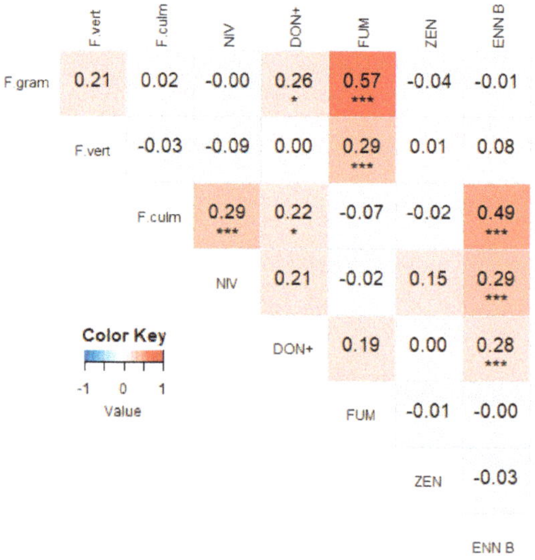

Figure A6. Heat map based on the pairwise Pearson correlation coefficients between measured mycotoxin concentrations and DNA of *F. graminearum*, *F. verticillioides* and *F. culmorum* in 2018. A darker blue color indicates a stronger negative correlation, a darker red color indicates a stronger positive correlation. Significant correlations are indicated with asterisks (* $p < 0.05$, *** $p < 0.01$). DON+ = the sum of the concentrations of DON, 3-ADON and 15-ADON. FUM = the sum of the concentrations of FB1, FB2 and FB3.

References

1. Storm, I.M.L.D.; Sørensen, J.L.; Rasmussen, R.R.; Nielsen, K.F.; Thrane, U. Mycotoxins in silage. *Stewart Postharvest Rev.* **2008**, *4*, 1–12.
2. Tangni, E.K.; Pussemier, L.; van Hove, F. Mycotoxin contaminating maize and grass silages for dairy cattle feeding: Current state and challenges. *J. Anim. Sci. Adv.* **2013**, *3*, 492–511.
3. Rüdelsheim, P.L.J.; Smets, G. *Baseline Information on Agricultural Practices in the EU: Maize (Zea Mays L.)*; Perseus BVBA: Sint-Martens-Latem, Belgium, 2011.
4. Department of Agriculture and Fisheries Voorlopige Arealen Landbouwteelten uit de Verzamelaanvraag 2019. Available online: https://lv.vlaanderen.be/nl/nieuws/voorlopige-arealen-landbouwteelten-uit-de-verzamelaanvraag-2019 (accessed on 20 September 2019).
5. Kiessling, K.H.; Pettersson, H.; Sandholm, K.; Olsen, M. Metabolism of aflatoxin, ochratoxin, zearalenone, and three trichothecenes by intact rumen fluid, rumen protozoa, and rumen bacteria. *Appl. Environ. Microbiol.* **1984**, *47*, 1070–1073.
6. Fink-Gremmels, J. Mycotoxins in cattle feeds and carry-over to dairy milk: A review. *Food Addit. Contam.* **2008**, *25*, 172–180. [CrossRef]
7. Gallo, A.; Giuberti, G.; Frisvad, J.C.; Bertuzzi, T.; Nielsen, K.F. Review on mycotoxin issues in ruminants: Occurrence in forages, effects of mycotoxin ingestion on health status and animal performance and practical strategies to counteract their negative effects. *Toxins* **2015**, *7*, 3057–3111. [CrossRef] [PubMed]
8. Caloni, F.; Spotti, M.; Auerbach, H.; Op Den Camp, H.; Fink Gremmels, J.; Pompa, G. In vitro metabolism of fumonisin B1 by ruminal microflora. *Vet. Res. Commun.* **2000**, *24*, 379–387. [CrossRef] [PubMed]
9. Seeling, K.; Dänicke, S.; Ueberschär, H.; Lebzien, P.; Flachowsky, G. On the effects of *Fusarium* toxin-contaminated wheat and the feed intake level on the metabolism and carry over of zearalenone in dairy cows. *Food Addit. Contam.* **2005**, *22*, 847–855. [CrossRef] [PubMed]
10. Rodrigues, I. A review on the effects of mycotoxins in dairy ruminants. *Anim. Prod. Sci.* **2014**, *54*, 1155–1165. [CrossRef]
11. Mostrom, M.S.; Jacobsen, B.J. Ruminant mycotoxicosis. *Vet. Clin. N. Am. Food Anim. Pract.* **2011**, *27*, 315–344. [CrossRef] [PubMed]
12. Fink-Gremmels, J. The role of mycotoxins in the health and performance of dairy cows. *Vet. J.* **2008**, *176*, 84–92. [CrossRef]
13. Wu, F. Measuring the economic impacts of *Fusarium* toxins in animal feeds. *Anim. Feed Sci. Technol.* **2007**, *137*, 363–374. [CrossRef]
14. Bhat, R.; Rai, R.V.; Karim, A.A. Mycotoxins in food and feed: Present status and future concerns. *Compr. Rev. Food Sci. Food Saf.* **2010**, *9*, 57–81. [CrossRef]
15. Sweeney, M.J.; Dobson, A.D.W. Mycotoxin production by *Aspergillus*, *Fusarium* and *Penicillium* species. *Int. J. Food Microbiol.* **1998**, *43*, 141–158. [CrossRef]
16. Dorn, B.; Forrer, H.R.; Schürch, S.; Vogelgsang, S. *Fusarium* species complex on maize in Switzerland: Occurrence, prevalence, impact and mycotoxins in commercial hybrids under natural infection. *Eur. J. Plant Pathol.* **2009**, *125*, 51–61. [CrossRef]
17. Koehler, B. *Corn Ear Rots in Illinois*; Urbana, Ill., Ed.; Agricultural Experiment Station, University of Illinois: Champaign, IL, USA, 1959.
18. Logrieco, A.; Mulé, G.; Moretti, A.; Bottalico, A. Toxigenic *Fusarium* species and mycotoxins associated with maize ear rot in Europe. *Eur. J. Plant Pathol.* **2002**, *108*, 579–609. [CrossRef]
19. Bottalico, A. *Fusarium* diseases of cereals: Species complex and related mycotoxin profiles, in Europe. *J. Plant Pathol.* **1998**, *80*, 85–103.
20. Oldenburg, E.; Höppner, F.; Ellner, F.; Weinert, J. *Fusarium* diseases of maize associated with mycotoxin contamination of agricultural products intended to be used for food and feed. *Mycotoxin Res.* **2017**, *33*, 167–182. [CrossRef]
21. Scauflaire, J.; Mahieu, O.; Louvieaux, J.; Foucart, G.; Renard, F.; Munaut, F. Biodiversity of *Fusarium* species in ears and stalks of maize plants in Belgium. *Eur. J. Plant Pathol.* **2011**, *131*, 59–66. [CrossRef]
22. Goertz, A.; Zuehlke, S.; Spiteller, M.; Steiner, U.; Dehne, H.W.; Waalwijk, C.; de Vries, I.; Oerke, E.C. *Fusarium* species and mycotoxin profiles on commercial maize hybrids in Germany. *Eur. J. Plant Pathol.* **2010**, *128*, 101–111. [CrossRef]

23. Doohan, F.M.; Brennan, J.; Cooke, B.M. Influence of climatic factors on *Fusarium* species pathogenic to cereals. *Eur. J. Plant Pathol.* **2003**, *109*, 755–768. [CrossRef]
24. Eckard, S.; Wettstein, F.E.; Forrer, H.R.; Vogelgsang, S. Incidence of *Fusarium* species and mycotoxins in silage maize. *Toxins* **2011**, *3*, 949–967. [CrossRef] [PubMed]
25. Picot, A.; Hourcade-Marcolla, D.; Barreau, C.; Pinson-Gadais, L.; Caron, D.; Richard-Forget, F.; Lannou, C. Interactions between *Fusarium verticillioides* and *Fusarium graminearum* in maize ears and consequences for fungal development and mycotoxin accumulation. *Plant Pathol.* **2012**, *61*, 140–151. [CrossRef]
26. Velluti, A.; Marín, S.; Gonzalez, R.; Ramos, A.J.; Sanchis, V. Fumonisin B1, zearalenone and deoxynivalenol production by *Fusarium moniliforme*, *F. proliferatum* and *F. graminearum* in mixed cultures on irradiated maize kernels. *J. Sci. Food Agric.* **2001**, *81*, 88–94. [CrossRef]
27. Reid, L.M.; Nicol, R.W.; Ouellet, T.; Savard, M.; Miller, J.D.; Young, J.C.; Stewart, D.W.; Schaafsma, A.W. Interaction of *Fusarium graminearum* and *F. moniliforme* in maize ears: Disease progress, fungal biomass, and mycotoxin accumulation. *Phytopathology* **1999**, *89*, 1028–1037. [CrossRef]
28. Warren, H.L.; Kommedahl, T. Prevalence and pathogenicity to corn of *Fusarium* species from corn roots, rhizosphere, residues, and soil. *Phytopathology* **1973**, *63*, 1288. [CrossRef]
29. Giorni, P.; Bertuzzi, T.; Battilani, P. Impact of fungi co-occurrence on mycotoxin contamination in maize during the growing season. *Front. Microbiol.* **2019**, *10*, 1265. [CrossRef]
30. Munkvold, G.P. Epidemiology of *Fusarium* diseases and their mycotoxins in maize ears. *Eur. J. Plant Pathol.* **2003**, *109*, 705–713. [CrossRef]
31. Oldenburg, E.; Ellner, F. Distribution of disease symptoms and mycotoxins in maize ears infected by *Fusarium culmorum* and *Fusarium graminearum*. *Mycotoxin Res.* **2015**, *31*, 117–126. [CrossRef]
32. Foley, D.C. Systemic infection of corn by *Fusarium moniliforme*. *Phytopathology* **1962**, *52*, 870–872.
33. Gilbertson, R.L.; Brown, W.M.J.; Ruppel, e.g.,; Capinera, J.L. Association of Corn Stalk Rot *Fusarium* spp. and Western Corn Rootworm Beetles in Colorado. *Phytopathology* **1986**, *76*, 1309–1314. [CrossRef]
34. Munkvold, G.P.; McGee, D.C.; Carlton, W.M. Importance of different pathways for maize kernel infection by *Fusarium moniliforme*. *Phytopathology* **1997**, *87*, 209–217. [CrossRef] [PubMed]
35. Frisvad, J.C.; Thrane, U.; Samson, R.A.; Pitt, J.I. Important mycotoxins and the fungi which produce them. In *Advances in Food Mycology. Advances in Experimental Medicine and Biology*; Hocking, A.D., Pitt, J.I., Samson, R.A., Thrane, U., Eds.; Springer: Boston, MA, USA, 2006; Volume 571, pp. 3–31.
36. Miller, J.D. Fungi and mycotoxins in grain: Implication for stored product research. *J. Stored Prod. Res.* **1995**, *31*, 1–16. [CrossRef]
37. Binder, E.M.; Tan, L.M.; Chin, L.J.; Handl, J.; Richard, J. Worldwide occurrence of mycotoxins in commodities, feeds and feed ingredients. *Anim. Feed Sci. Technol.* **2007**, *137*, 265–282. [CrossRef]
38. Van Asselt, E.D.; Azambuja, W.; Moretti, A.; Kastelein, P.; de Rijk, T.C.; Stratakou, I.; van der Fels-Klerx, H.J. A Dutch field survey on fungal infection and mycotoxin concentrations in maize. *Food Addit. Contam.* **2012**, *29*, 1556–1565. [CrossRef] [PubMed]
39. Kosicki, R.; Błajet-Kosicka, A.; Grajewski, J.; Twaruzek, M. Multiannual mycotoxin survey in feed materials and feedingstuffs. *Anim. Feed Sci. Technol.* **2016**, *215*, 165–180. [CrossRef]
40. Gruber-Dorninger, C.; Jenkins, T.; Schatzmayr, G. Global mycotoxin occurrence in feed: A ten-year survey. *Toxins* **2019**, *11*, 375. [CrossRef]
41. Miller, J.D. Mycotoxins in small grains and maize: Old problems, new challenges. *Food Addit. Contam. Part A* **2008**, *25*, 219–230. [CrossRef]
42. Paterson, R.R.M.; Lima, N. How will climate change affect mycotoxins in food? *Food Res. Int.* **2010**, *43*, 1902–1914. [CrossRef]
43. Battilani, P.; Toscano, P.; Van Der Fels-Klerx, H.J.; Moretti, A.; Camardo Leggieri, M.; Brera, C.; Rortais, A.; Goumperis, T.; Robinson, T. Aflatoxin B1 contamination in maize in Europe increases due to climate change. *Sci. Rep.* **2016**, *6*, 24328. [CrossRef]
44. Basler, R. Diversity of *Fusarium* species isolated from UK forage maize and the population structure of *F. graminearum* from maize and wheat. *PeerJ* **2016**, *4*, e2143. [CrossRef]
45. Bebber, D.P.; Ramotowski, M.A.T.; Gurr, S.J. Crop pests and pathogens move polewards in a warming world. *Nat. Clim. Chang.* **2013**, *3*, 985–988. [CrossRef]
46. Moretti, A.; Pascale, M.; Logrieco, A.F. Mycotoxin risks under a climate change scenario in Europe. *Trends Food Sci. Technol.* **2019**, *84*, 38–40. [CrossRef]

47. Garon, D.; Richard, E.; Sage, L.; Bouchart, V.; Pottier, D.; Lebailly, P. Mycoflora and multimycotoxin detection in corn silage: Experimental study. *J. Agric. Food Chem.* **2006**, *54*, 3479–3484. [CrossRef] [PubMed]
48. Drejer Storm, I.M.L.; Rasmussen, R.R.; Rasmussen, P.H. Occurrence of pre- and post-harvest mycotoxins and other secondary metabolites in Danish maize silage. *Toxins* **2014**, *6*, 2256–2269. [CrossRef]
49. Lepom, P. Occurrence of *Fusarium* species and their mycotoxins in maize: 1. Method of determining zearalenone in maize and maize silage by means of high performance liquid chomatography (HPLC) using fluorescence detection. *Arch. Anim. Nutr.* **1988**, *38*, 799–806.
50. Boudra, H.; Morgavi, D.P. Reduction in *Fusarium* toxin levels in corn silage with low dry matter and storage time. *J. Agric. Food Chem.* **2008**, *56*, 4523–4528. [CrossRef]
51. Richard, E.; Heutte, N.; Sage, L.; Pottier, D.; Bouchart, V.; Lebailly, P.; Garon, D. Toxigenic fungi and mycotoxins in mature corn silage. *Food Chem. Toxicol.* **2007**, *45*, 2420–2425. [CrossRef]
52. Pelhate, J. Maize silage: Incidence of moulds during conservation. *Folia Vet. Lat.* **1977**, *7*, 1–16.
53. Latorre, A.; Dagnac, T.; Lorenzo, B.F.; Llompart, M. Occurrence and stability of masked fumonisins in corn silage samples. *Food Chem.* **2015**, *189*, 38–44. [CrossRef]
54. Zachariasova, M.; Dzuman, Z.; Veprikova, Z.; Hajkova, K.; Jiru, M.; Vaclavikova, M.; Zachariasova, A.; Pospichalova, M.; Florian, M.; Hajslova, J. Occurrence of multiple mycotoxins in European feedingstuffs, assessment of dietary intake by farm animals. *Anim. Feed Sci. Technol.* **2014**, *193*, 124–140. [CrossRef]
55. Keller, L.A.M.; González Pereyra, M.L.; Keller, K.M.; Alonso, V.A.; Oliveira, A.A.; Almeida, T.X.; Barbosa, T.S.; Nunes, L.M.T.; Cavaglieri, L.R.; Rosa, C.A.R. Fungal and mycotoxins contamination in corn silage: Monitoring risk before and after fermentation. *J. Stored Prod. Res.* **2013**, *52*, 42–47. [CrossRef]
56. Mansfield, M.A.; Jones, A.D.; Kuldau, G.A. Contamination of fresh and ensiled maize by multiple *Penicillium* mycotoxins. *Phytopathology* **2008**, *98*, 330–336. [CrossRef] [PubMed]
57. Cheli, F.; Campagnoli, A.; Dell'Orto, V. Fungal populations and mycotoxins in silages: From occurrence to analysis. *Anim. Feed Sci. Technol.* **2013**, *183*, 1–16. [CrossRef]
58. Schatzmayr, G.; Streit, E. Global occurrence of mycotoxins in the food and feed chain: Facts and figures. *World Mycotoxin J.* **2013**, *6*, 213–222. [CrossRef]
59. Kamala, A.; Ortiz, J.; Kimanya, M.; Haesaert, G.; Donoso, S.; Tiisekwa, B.; De Meulenaer, B. Multiple mycotoxin co-occurrence in maize grown in three agro-ecological zones of Tanzania. *Food Control* **2015**, *54*, 208–215. [CrossRef]
60. Kovalsky, P.; Kos, G.; Nährer, K.; Schwab, C.; Jenkins, T.; Schatzmayr, G.; Sulyok, M.; Krska, R. Co-occurrence of regulated, masked and emerging mycotoxins and secondary metabolites in finished feed and maize – An extensive survey. *Toxins* **2016**, *8*, 363. [CrossRef] [PubMed]
61. Dorn, B.; Forrer, H.R.; Jenny, E.; Wettstein, F.E.; Bucheli, T.D.; Vogelgsang, S. *Fusarium* species complex and mycotoxins in grain maize from maize hybrid trials and from grower's fields. *J. Appl. Microbiol.* **2011**, *111*, 693–706. [CrossRef]
62. Schollenberger, M.; Müller, H.M.; Ernst, K.; Sondermann, S.; Liebscher, M.; Schlecker, C.; Wischer, G.; Drochner, W.; Hartung, K.; Piepho, H.P. Occurrence and distribution of 13 trichothecene toxins in naturally contaminated maize plants in Germany. *Toxins* **2012**, *4*, 778–787. [CrossRef]
63. European Commission. Commission regulation (EU) No 574/2011 of 16 June 2011 amending Annex I to directive 2002/32/EC of the European Parliament and of the Council as regards maximum levels for nitrite, melamine, *Ambrosia* spp. and carry-over of certain coccidiostats and histomonostats and consolidating Annexes I and II thereto. *Off. J. Eur. Union* **2011**, *159*, 7–24.
64. European Commission. Commission Regulation (EC) No 1881/2006 of 19 December 2006 setting maximum levels for certain contaminants in foodstuffs. *Off. J. Eur. Union* **2006**, *364*, 5–24.
65. European Commission. Commission recommendation of 27 March 2013 on the presence of T-2 and HT-2 in cereals and cereal products (2013/165/EU). *Off. J. Eur. Union* **2013**, *91*, 12–15.
66. De Boevre, M.; Landschoot, S.; Audenaert, K.; Maene, P.; Di Mavungu, J.D.; Eeckhout, M.; Haesaert, G.; De Saeger, S. Occurrence and within field variability of *Fusarium* mycotoxins and their masked forms in maize crops in Belgium. *World Mycotoxin J.* **2014**, *7*, 91–102. [CrossRef]
67. Streit, E.; Schwab, C.; Sulyok, M.; Naehrer, K.; Krska, R.; Schatzmayr, G. Multi-mycotoxin screening reveals the occurrence of 139 different secondary metabolites in feed and feed ingredients. *Toxins* **2013**, *5*, 504–523. [CrossRef]

68. Speijers, G.J.A.; Speijers, M.H.M. Combined toxic effects of mycotoxins. *Toxicol. Lett.* **2004**, *153*, 91–98. [CrossRef] [PubMed]
69. Grenier, B.; Oswald, I. Mycotoxin co-contamination of food and feed: Meta-analysis of publications describing toxicological interactions. *World Mycotoxin J.* **2011**, *4*, 285–313. [CrossRef]
70. Smith, T.K.; Seddon, I.R. Toxicological synergism between *Fusarium* mycotoxins in feeds. In *Biotechnology in the Feed Industry*; Lyons, T.P., Jacques, K.A., Eds.; Nottingham University Press: Loughborough, UK, 1998; pp. 257–269.
71. Thuvander, A.; Wikman, C.; Gadhasson, I. In vitro exposure of human lymphocytes to trichothecenes: Individual variation in sensitivity and effects of combined exposure on lymphocyte function. *Food Chem. Toxicol.* **1999**, *37*, 639–648. [CrossRef]
72. Fremy, J.M.; Alassane-Kpembi, I.; Oswald, I.P.; Cottrill, B.; Van Egmond, H.P. A review on combined effects of moniliformin and co-occurring *Fusarium* toxins in farm animals. *World Mycotoxin J.* **2019**, *12*, 281–291. [CrossRef]
73. Vanheule, A.; Audenaert, K.; De Boevre, M.; Landschoot, S.; Bekaert, B.; Munaut, F.; Eeckhout, M.; Höfte, M.; De Saeger, S.; Haesaert, G. The compositional mosaic of *Fusarium* species and their mycotoxins in unprocessed cereals, food and feed products in Belgium. *Int. J. Food Microbiol.* **2014**, *181*, 28–36. [CrossRef]
74. Landschoot, S.; Audenaert, K.; Waegeman, W.; Pycke, B.; Bekaert, B.; De Baets, B.; Haesaert, G. Connection between primary Fusarium inoculum on gramineous weeds, crop residues and soil samples and the final population on wheat ears in Flanders, Belgium. *Crop Prot.* **2011**, *30*, 1297–1305. [CrossRef]
75. Monbaliu, S.; Van Poucke, C.; Detavernier, C.T.L.; Dumoultn, F.; Van Velde, M.D.E.; Schoeters, E.; Van Dyck, S.; Averkieva, O.; Van Peteghem, C.; De Saeger, S. Occurrence of mycotoxins in feed as analyzed by a multi-mycotoxin LC-MS/MS method. *J. Agric. Food Chem.* **2010**, *58*, 66–71. [CrossRef]
76. European Commission. Commission Directive 2003/100/EC of 31 October 2003 amending Annex I to Directive 2002/32/EC of the European Parliament an of the Council on undesirable substances in animal feed. *Off. J. Eur. Union* **2003**, *285*, 33–37.
77. Quarta, A.; Mita, G.; Haidukowski, M.; Santino, A.; Mulé, G.; Visconti, A. Assessment of trichothecene chemotypes of *Fusarium culmorum* occurring in Europe. *Food Addit. Contam.* **2005**, *22*, 309–315. [CrossRef] [PubMed]
78. Saghai-Maroof, M.A.; Soliman, K.M.; Jorgensen, R.A.; Allard, R.W. Ribosomal DNA spacer-length polymorphisms in barley: Mendelian inheritance, chromosomal location, and population dynamics. *Proc. Natl. Acad. Sci. USA.* **1984**, *81*, 8014–8018. [CrossRef] [PubMed]
79. Nicolaisen, M.; Suproniene, S.; Nielsen, L.K.; Lazzaro, I.; Spliid, N.H.; Justesen, A.F. Real-time PCR for quantification of eleven individual *Fusarium* species in cereals. *J. Microbiol. Methods* **2009**, *76*, 234–240. [CrossRef] [PubMed]
80. R Core Team. *R: A Language and Environment for Statistical Computing 2017*; R Foundation for Statistical Computing: Vienna, Austria, 2017.
81. Pasquali, M.; Giraud, F.; Brochot, C.; Cocco, E.; Hoffmann, L.; Bohn, T. Genetic *Fusarium* chemotyping as a useful tool for predicting nivalenol contamination in winter wheat. *Int. J. Food Microbiol.* **2010**, *137*, 246–253. [CrossRef]
82. Sampietro, D.A.; Díaz, C.G.; Gonzalez, V.; Vattuone, M.A.; Ploper, L.D.; Catalan, C.A.N.; Ward, T.J. Species diversity and toxigenic potential of *Fusarium graminearum* complex isolates from maize fields in northwest Argentina. *Int. J. Food Microbiol.* **2011**, *145*, 359–364. [CrossRef]
83. Hellin, P.; Dedeurwaerder, G.; Duvivier, M.; Scauflaire, J.; Huybrechts, B.; Callebaut, A.; Munaut, F.; Legrève, A. Relationship between *Fusarium* spp. diversity and mycotoxin contents of mature grains in southern Belgium. *Food Addit. Contam.* **2016**, *33*, 1228–1240. [CrossRef]
84. Ueno, Y. *Trichothecenes: Chemical, Biological and Toxicological Aspects*; Kodansha: Tokyo, Japan, 1983.
85. Wu, W.; Flannery, B.M.; Sugita-Konishi, Y.; Watanabe, M.; Zhang, H.; Pestka, J.J. Comparison of murine anorectic responses to the 8-ketotrichothecenes 3-acetyldeoxynivalenol, 15-acetyldeoxynivalenol, fusarenon X and nivalenol. *Food Chem. Toxicol.* **2012**, *50*, 2056–2061. [CrossRef]
86. Abbas, H.K.; Yoshizawa, T.; Shier, W.T. Cytotoxicity and phytotoxicity of trichothecene mycotoxins produced by *Fusarium* spp. *Toxicon* **2013**, *74*, 68–75. [CrossRef]

87. Minervini, F.; Fornelli, F.; Flynn, K.M. Toxicity and apoptosis induced by the mycotoxins nivalenol, deoxynivalenol and fumonisin B1 in a human erythroleukemia cell line. *Toxicol. Vitr.* **2004**, *18*, 21–28. [CrossRef]
88. Ferreira Lopes, S.; Vacher, G.; Ciarlo, E.; Savova-Bianchi, D.; Roger, T.; Niculita-Hirzel, H. Primary and Immortalized Human Respiratory Cells Display Different Patterns of Cytotoxicity and Cytokine Release upon Exposure to Deoxynivalenol, Nivalenol and Fusarenon-X. *Toxins* **2017**, *9*, 337. [CrossRef] [PubMed]
89. Audenaert, K.; van Broeck, R.; Bekaert, B.; de Witte, F.; Heremans, B.; Messens, K.; Höfte, M.; Haesaert, G. Fusarium head blight (FHB) in Flanders: Population diversity, inter-species associations and DON contamination in commercial winter wheat varieties. *Eur. J. Plant Pathol.* **2009**, *125*, 445–458. [CrossRef]
90. Waalwijk, C.; Kastelein, P.; De Vries, I.; Kerényi, Z.; Van Der Lee, T.; Hesselink, T.; Köhl, J.; Kema, G. Major changes in *Fusarium* spp. in wheat in the Netherlands. *Eur. J. Plant Pathol.* **2003**, *109*, 743–754. [CrossRef]
91. Borutova, R.; Aragon, Y.A.; Nährer, K.; Berthiller, F. Co-occurrence and statistical correlations between mycotoxins in feedstuffs collected in the Asia-Oceania in 2010. *Anim. Feed Sci. Technol.* **2012**, *178*, 190–197. [CrossRef]
92. Ferrigo, D.; Raiola, A.; Causin, R. *Fusarium* toxins in cereals: Occurrence, legislation, factors promoting the appearance and their management. *Molecules* **2016**, *21*, 627. [CrossRef]
93. Zorzete, P.; Castro, R.S.; Pozzi, C.R.; Israel, A.L.M.; Fonseca, H.; Yanaguibashi, G.; Correa, B. Relative populations and toxin production by *Aspergillus flavus* and *Fusarium verticillioides* in artificially inoculated corn at various stages of development under field conditions. *J. Sci. Food Agric.* **2008**, *88*, 48–55. [CrossRef]
94. Cooney, J.M.; Lauren, D.R.; Di Menna, M.E. Impact of competitive fungi on trichothecene production by *Fusarium graminearum*. *J. Agric. Food Chem.* **2001**, *49*, 522–526. [CrossRef]
95. Lamichhane, J.R.; Venturi, V. Synergisms between microbial pathogens in plant disease complexes: A growing trend. *Front. Plant Sci.* **2015**, *6*, 385. [CrossRef]
96. Marín, S.; Sanchis, V.; Ramos, A.J.; Vinas, I.; Magan, N. Environmental factors, in vitro interactions, and niche overlap between *Fusarium moniliforme*, *F. proliferatum*, and *F. graminearum*, *Aspergillus* and *Penicillium* species from maize grain. *Mycol. Res.* **1998**, *102*, 831–837. [CrossRef]
97. Etcheverry, M.; Magnoli, C.; Dalcero, A.; Chulze, S.; Lecumberry, S. Aflatoxin B1, zearalenone and deoxynivalenol production by *Aspergillus parasiticus* and *Fusarium graminearum* in interactive cultures on irradiated corn kernels. *Mycopathologia* **1998**, *142*, 37–42. [CrossRef]
98. Ferrigo, D.; Raiola, A.; Causin, R. Plant stress and mycotoxin accumulation in maize. *Agrochimica* **2014**, *58*, 116–127.
99. Brennan, J.M.; Fagan, B.; Van Maanen, A.; Cooke, B.M.; Doohan, F.M. Studies on *in vitro* growth and pathogenicity of European *Fusarium* fungi. *Eur. J. Plant Pathol.* **2003**, *109*, 577–587. [CrossRef]
100. KMI. Juli 2019: Absolute Warmterecord Gebroken. Available online: https://www.meteo.be/nl/over-het-kmi/contact/gegevens-kmi (accessed on 10 October 2019).
101. Bullerman, L.B.; Schroeder, L.L.; Park, K.Y. Formation and control of mycotoxins in food. *J. Food Prot.* **1984**, *47*, 637–646. [CrossRef] [PubMed]
102. Medina, A.; Rodriguez, A.; Magan, N. Effect of climate change on *Aspergillus flavus* and aflatoxin B 1 production. *Front. Microbiol.* **2014**, *5*, 348. [CrossRef]
103. Miller, J.D. Factors that affect the occurrence of fumonisin. *Environ. Health Perspect.* **2001**, *109*, 321–324.
104. Glenn, A.E. Mycotoxigenic *Fusarium* species in animal feed. *Anim. Feed Sci. Technol.* **2007**, *137*, 213–240. [CrossRef]

microorganisms

MDPI

Article

The Photoreceptor Components FaWC1 and FaWC2 of *Fusarium asiaticum* Cooperatively Regulate Light Responses but Play Independent Roles in Virulence Expression

Ying Tang, Pinkuan Zhu *, Zhengyu Lu, Yao Qu, Li Huang, Ni Zheng, Yiwen Wang, Haozhen Nie, Yina Jiang and Ling Xu *

School of Life Sciences, East China Normal University, Shanghai 200241, China; 15294152681@163.com (Y.T.); 51151300064@stu.ecnu.edu.cn (Z.L.); 51141300067@stu.ecnu.edu.cn (Y.Q.); athenahuangli@163.com (L.H.); 10151910108@stu.ecnu.edu.cn (N.Z.); ywwang@bio.ecnu.edu.cn (Y.W.); hznie@bio.ecnu.edu.cn (H.N.); ynjiang@bio.ecnu.edu.cn (Y.J.)

* Correspondence: pkzhu@bio.ecnu.edu.cn (P.Z.); lxu@bio.ecnu.edu.cn (L.X.); Tel.: +86-021-5434-1012 (L.X.)

Received: 12 February 2020; Accepted: 3 March 2020; Published: 5 March 2020

Abstract: *Fusarium asiaticum* belongs to one of the phylogenetical subgroups of the *F. graminearum* species complex and is epidemically predominant in the East Asia area. The life cycle of *F. asiaticum* is significantly regulated by light. In this study, the fungal blue light receptor white collar complex (WCC), including FaWC1 and FaWC2, were characterized in *F. asiaticum*. The knockout mutants Δ*Fawc1* and Δ*Fawc2* were generated by replacing the target genes via homologous recombination events. The two mutants showed similar defects in light-induced carotenoid biosynthesis, UV-C resistance, sexual fruiting body development, and the expression of the light-responsive marker genes, while in contrast, all these light responses were characteristics in wild-type (WT) and their complementation strains, indicating that FaWC1 and FaWC2 are involved in the light sensing of *F. asiaticum*. Unexpectedly, however, the functions of *Fawc1* and *Fawc2* diverged in regulating virulence, as the Δ*Fawc1* was avirulent to the tested host plant materials, but Δ*Fawc2* was equivalent to WT in virulence. Moreover, functional analysis of FaWC1 by partial disruption revealed that its light–oxygen–voltage (LOV) domain was required for light sensing but dispensable for virulence, and its Zinc-finger domain was required for virulence expression but not for light signal transduction. Collectively, these results suggest that the conserved fungal blue light receptor WCC not only endows *F. asiaticum* with light-sensing ability to achieve adaptation to environment, but it also regulates virulence expression by the individual component FaWC1 in a light-independent manner, and the latter function opens a way for investigating the pathogenicity mechanisms of this important crop disease agent.

Keywords: photobiology; transcription factor; White collar complex; *Fusarium asiaticum*; virulence

1. Introduction

Fusarium head blight (FHB), which is usually caused by the pathogen *Fusarium graminearum*, is known as a global problem devastating small grain cereal crops [1]. Phylogenetic species recognition has revealed that *F. graminearum* sensu lato comprises at least 15 biogeographically structured and phylogenetically distinct species, all of which are known as the *Fusarium graminearum* species complex, or FGSC [2–7]. Among the FGSC, *Fusarium asiaticum* belongs to one sub-lineage and is the predominant FHB agent species in East Asia [8]; it is especially prevalent in the wheat production zones of the Yangtze-Huaihe valley in China [9]. Besides being commonly associated with FHB, *F. asiaticum* has also been found to cause postharvest rot on asparagus spears and produce 3A-DON mycotoxin during host infection [10].

Usually, introgression of the disease resistance genes identified in natural sources into elite cultivars represents a reliable route for plant disease management [11]. However, plant sources for FHB resistance are unfortunately limited, and no fully resistant cultivars are yet available [12]. Controlling FGSC-caused diseases will benefit from an in-depth understanding of how the pathogens infect and spread inside the host. The availability of the *F. graminearum* genome [1] has greatly stimulated the research activity on identification of functional genes as well as pathogenicity factors of this phytopathogen [13–15]. However, the molecular mechanism of development and virulence regulation is less known in *F. asiaticum* than in *F. graminearum*, although there is an ongoing trend in which *F. asiaticum* becomes more aggressive and devastating than *F. graminearum* in the East Asian area [6,14,16].

Light is an important environmental factor that can extensively influence varied aspects of most living organisms on earth [17]. Filamentous fungi can use light as a general signal for regulating development, metabolism, sexual or asexual reproduction, and other life processes to adapt to a specific ecological niche [18–20]. At the molecular level, light is sensed by fungal photoreceptors, leading to activating or suppressing the transcription of photoresponsive genes, which are furthermore considered to result in the accumulation of light-sensing responses.

Light signaling is most extensively studied in the model species *Neurospora crassa*, in which carotenoid biosynthesis and morphological development, including the formation of asexual spores and protoperithecia, are notably regulated by blue light [21–23]. The analysis of blind mutants revealed that the white-collar 1 protein (WC-1), a transcription regulator which contains a light–oxygen–voltage (LOV) domain to bind the flavin chromophore, and the WC-2, a second transcription regulator without a chromophore-binding domain, can form a heterodimer called white-collar complex (WCC) to positively regulate the light-induced genes. Besides *N. crassa*, the molecular components for blue light sensing appeared to be widely conserved in the fungal genomes of Ascomycetes, Mucoromycetes, and Basidiomycetes. Moreover, genes under the control of the WCC can be either light responsive or not light responsive, and WC-1 and WC-2 can also have individual functions besides acting cooperatively as the WCC [24,25].

Many fungal species are causing detrimental diseases to mammals and plants, since the outcomes of all the epidemic diseases on earth can be determined by the triangular interactions among host–pathogen–environment [26]. Whether light signaling in fungi is involved in determining the disease outcomes has attracted considerable research attempts to characterize the photoreceptor functions in pathogenic fungi. However, the WCC regulatory circuit demonstrates functional variation among different species of fungal pathogens, and the significance of fungal light-sensing capacity for virulence expression is concluded on a case-by-case basis [27–30]. Moreover, despite these functional studies of the WCC orthologs in varied phytopathogenic fungi, the following questions remain mysterious: whether sensing light (or the absence of light) by phytopathogenic fungi is essential for pathogenicity, and how to exclude light-independent functions of WCC orthologs when evaluating their contribution in determining fungal pathogenicity.

In this paper, the cloning and characterization of the *Fawc1* and *Fawc2* genes in *F. asiaticum* demonstrate that both *Fawc1* and *Fawc2* are involved in light sensing and regulating pleiotropic fungal development processes, suggesting that FaWC1 and FaWC2 function cooperatively as the WCC to fulfill the photo receptor tasks. However, it is *Fawc1* but not *Fawc2* that is required for virulence. Functional domain analysis of FaWC1 reveals that the LOV and Zinc-finger domains are independently required for light sensing and virulence, respectively. These findings not only expand the knowledge of fungal photobiology but also provide novel insights about the mechanisms for diverged functions of WCC components in determining fungal virulence in a light-independent manner.

2. Materials and Methods

2.1. Fungal Strains and Culture Conditions

The wild-type strain EXAP-08 was previously isolated from postharvest asparagus spears with serious rot symptoms and characterized as *F. asiaticum*, and it is used as the recipient strain for genetic modifications in this study. The wild-type strain and the resulting mutants are listed in Table 1, and all the fungal strains were purified by single spore isolation and were stored as spore suspension in 20% glycerol at −80 °C. The fungal cultures were grown on complete medium (CM) (10 g glucose, 2 g peptone, 1 g yeast extract, 1 g casamino acids, nitrate salts, trace elements, 0.01% of vitamins, 10 g agar and 1 L water, pH 6.5) for mycelial growth, carboxymethyl cellulose (CMC) medium [31] for conidiation, and carrot agar medium [32] for sexual development. For light-responsive phenotypic analyses, plates were incubated for 3–7 days at 25 °C as indicated, either under illumination (20 W m^{-2} white light obtained with fluorescent bulbs) or under dark conditions.

Table 1. *F. asiaticum* strains used in this study. LOV: Light–oxygen–voltage, ZnF: Zinc finger.

Name	Genotype	Reference
EXAP-08	Wild type	[10]
Δ*Fawc1*	Knockout mutant, *Fawc1::Hyg*	This study
Δ*Fawc2*	Knockout mutant, *Fawc2::Hyg*	This study
Δ*Fawc1-C*	*Fawc-1* complemented transformant of Δ*Fawc-1*	This study
Δ*Fawc2-C*	*Fawc-2* complemented transformant of Δ*Fawc-2*	This study
Δ*Fawc1*-$C^{\Delta LOV}$	LOV domain deletion mutant	This study
Δ*Fawc1*-$C^{\Delta ZnF}$	ZnF domain deletion mutant	This study

2.2. Sequence Analysis of Fawc1 and Fawc2 of F. Asiaticum

The amino acid sequences of WC-1and WC-2 of *Neurospora crassa* were used to blast against the *Fusarium graminearum* genome at the Ensembl Fungi database (http://fungi.ensembl.org/index.html). The obtained target sequences were used as references to design primers to amplify the corresponding orthologs in *F. asiaticum* wild-type strain EXAP-08. The amplified products were subcloned to T-vector and subsequently sequenced. The obtained sequences of *Fawc1* and *Fawc2* were further aligned with their orthologs from other fungal species retrieved from publically available databases by DNAMAN version 5.2.2 (LynnonBiosoft Company, Pointe-Claire, QC, Canada).

2.3. Generation of Mutants and Complementation Strains in F. asiaticum

The flanks 5′ and 3′ of *Fawc1* and *Fawc2* were amplified from EXAP-08 genomic DNA, and the *hph* was amplified from plasmid *p22*. Overlap PCR was performed to obtain the 5′-*Fawc1-hph-Fawc1*-3′ gene knockout cassettes by mixing an equimolar ratio of 5′-*Fawc1*, *hph*, and *Fawc1*-3′ as templates, and the 5′-*Fawc2-hph-Fawc2*-3′ cassette was prepared in a similar way. The resulted products were separated by gel electrophoresis and then recovered by a gel extraction kit. The purified gene knockout cassettes (2 μg for each cassette) were subjected to protoplast transformation of the wild-type strain EXAP-08 according to the reported method [33]. To screen for the correct mutants of Δ*Fawc1* and Δ*Fawc2*, the transformants grown on selective media containing 75 μg/mL hygromycin B were purified and subjected to genomic DNA extraction via the standard CTAB protocol. Site specific primers were used to carry out PCR assay to screen for the knockout mutants.

To generate the complementation strains, wild-type *Fawc1* and *Fawc2* connected with their 1.5 kb 5′- and 1 kb 3′-flanking sequences were amplified and subsequently cloned into the flu6 plasmid that contains the geneticin (G418) resistance gene, resulting in *flu6-Fawc1*-com and *flu6-Fawc2*-com expression vectors. Then, the complementation vectors were transformed into the protoplasts of the corresponding gene deletion mutants. CM containing 100 μg/mL of G418 was used to select the successful transformants. To generate mutant strains carrying truncated FaWC1 that lacked either

LOV or ZnF domains, the *Fawc1*$^{\Delta LOV}$ and *Fawc1*$^{\Delta ZnF}$ fragments with deletion of the coding regions for LOV and ZnF domains, respectively, were generated by overlap PCR, and the resulted products were gel-purified and cloned into the flu6 plasmid, leading to the expression vectors *flu6-Fawc1*$^{\Delta LOV}$ and *flu6-Fawc1*$^{\Delta ZnF}$, which were delivered into Δ*Fawc1* mutant to generate the Δ*Fawc1-C*$^{\Delta LOV}$ and Δ*Fawc1-C*$^{\Delta ZnF}$ strains. All the primers used in the mutant generation and diagnosis PCR reactions are listed in supplementary Table S1.

2.4. Extraction of RNA and Quantitative RT-PCR Analysis

For gene expression analysis, samples were prepared as follows: aliquots of 200 μL conidia suspension (10^6/mL) were inoculated on solid CM medium with cellophane overlays and incubated at 23 °C. The mycelium samples were harvested after either 48 h culture in total darkness, or by ending the 48 h culture with light illumination of indicated time duration. Total RNA was extracted using QIAGEN Reagent (Germany), and 1 μg of each RNA sample was used for reverse transcription with the Prime Script™ RT reagent Kit (Perfect Real Time) (TakaRa Biotechnology, Co., Dalian, China). The real-time PCR amplifications were conducted in a CFX96^{TM} Real-Time System (BIO-BAD, Inc., USA) using TakaRa SYBR Premix Ex Taq (TakaRa Biotechnology, Co., Dalian, China). For each sample, the expression of the β-tubulin gene was used as an internal reference. There were three replicates for each sample. The experiment was repeated three times. The gene expression levels were calculated using the $2^{-\Delta\Delta Ct}$ method.

2.5. Growth and Development Phenotyping

Phenotypic differences between the mutant strains and wild type were analyzed by culturing them on CM agar medium in dark and light conditions. Colony morphology was recorded by photographs at four days after inoculation. To assess the self-fertility of each strain, mycelium plugs of 5 mm diameter were excised from the edge of 4-day-old colonies and then inoculated on carrot agar medium. The Petri dishes were put under constant light for four days when the mycelia overgrew the petri dish; then, we added 1 mL of Tween-20 in the petri dish and overwhelmed the aerial mycelia by the back of a spoon. Afterwards, the plates were transferred to near-UV light for homothallic sexual development induction. After two months, the plates were checked for perithecium formation and maturation and photo recorded via a stereomicroscope equipped with a CCD digital camera (SMC168, Moticam Pro 205B Motic, Xiamen, China). For ascospore observation, a very small brick of perithecium was picked off, pressed on a glass slide under a cover glass, and then observed using a microscope (AE30, Motic, Xiamen, China).

2.6. Carotenoid Measurement Assay

Mycelium samples grown in liquid CM medium (23 °C, 150 rpm) for three days were collected for lyophilization. For each strain, 200 mg of lyophilized mycelial biomass was added in 750 μL of hexane and 500 μL of methyl alcohol; then, it was homogenized in a Tissuelyser-24 with stainless-steel beads (Jingxin, Ltd., Shanghai, China). The resulted suspensions were centrifuged at 12,000 rpm for 10 min. The supernatant was used to determine the absorbance value (A) at 445 nm. Carotenoid contents of the samples were quantified via the equation: $X(\text{mg}/100\text{ g}) = \frac{A \times y(\text{mL}) \times 1000}{Af \times g}$, in which "y", "Af", and "g" represent the volume of the extraction buffer, average absorption coefficient of the carotenoid molecular (2500), and the weight of the sample, respectively.

2.7. UV Sensitivity Assay

Five microliters of serial diluted conidial suspensions (10^6, 10^5, 10^4 conidia/mL) from cultures grown in liquid straw media was point-inoculated onto CM agar medium. The plates were exposed to UV-C light (provided by UVB HL-2000 Hybridizer, peaking at 245 nm, 0.2 kJ/m^2) and then allowed to recover in dark or under white light for two days at 23 °C. The UV sensitivity of each strain was determined by comparing the survival of colonial cultures grown in light versus dark after UV exposure.

2.8. Virulence Assay

Fungal conidia were harvested from CMC medium and were suspended in liquid minimum Gamborg B5 medium (or GB5, Coolaber, China, containing 0.6 g·L^{-1} GB5 salts, 10 mM glucose, pH 7.0), and the concentration was adjusted to 1×10^6 conidia/mL. The infection assay was conducted according to the reported method [14] with modifications as follows: the wheat (*Triticum aestivum* cultivars Zhongyuan 98–68) coleoptiles grown for two days were cut with a scissor and inoculated with 5 μL conidial suspension, and then they were kept in a transparent box at 23 °C under a 12 h/12 h light cycle and humid condition. The infected samples were photographed every day, and the lesion areas were calculated using the ImageJ software.

2.9. Statistical Analysis

The data obtained in this study were analyzed with ANOVA followed by Duncan's multiple range tests ($p < 0.05$) for means comparison with the use of SPSS 17.0.

3. Results

3.1. The Orthologs of WC-1 and WC-2 in F. asiaticum

Since *F. asiaticum* belongs to a sub-lineage of the *Fusarium graminearum* species complex, the orthologous genes of *wc1* and *wc2* in *F. asiaticum*, namely *Fawc1* and *Fawc2*, were identified as follows: BLASTP search of the *F. gaminearum* genome in Ensembl Fungi database, using the amino acid sequences of *Neurospora crassa* WC-1 (NCU02356) and WC-2 (NCU00902) proteins [34] as queries, resulted in the corresponding orthologs namely FgWC1 (FGSG_07941) and FgWC2 (FGSG_00710), respectively. The corresponding genes and their flanking sequences were used as references to design specific primers to amplify the open reading frame (ORF) sequences of their orthologs, *Fawc1* and *Fawc2*, in *F. asiaticum* strain EXAP-08 via the *Pfu* DNA polymerase. Sequencing of the cloned PCR products indicated that *Fawc1* (KX905081.1) and *Fawc2* (MT019868) of *F. asiaticum* showed high similarity to *Fgwc1* (99.05%) and *Fgwc2* (99.39%) of *F. graminearum*. The deduced amino acid sequences of FaWC1 and FaWC2 were analyzed via the SMART online tool, and the results indicated that FaWC1 possessed a Zn-finger DNA-binding domain as well as three PAS domains, of which the N-terminal most PAS domain should be the special subclass called the LOV domain (for light, oxygen, and voltage) being responsible for binding the chromophore molecule of flavin adenine dinucleotide (FAD), while the FaWC2 just essentially contained a single PAS domain and a Zn-finger DNA binding domain. In general, FaWC1 and FaWC2 show high similarity with their orthologs in *F. graminearum* and *N. crassa* (Figure 1A). Gene expression analysis showed that both *Fawc1* and *Fawc2* were induced to peak levels by 15 min light exposure, while the long period (12 h) of illumination caused less induction of the transcription levels of these two genes (Figure 1B).

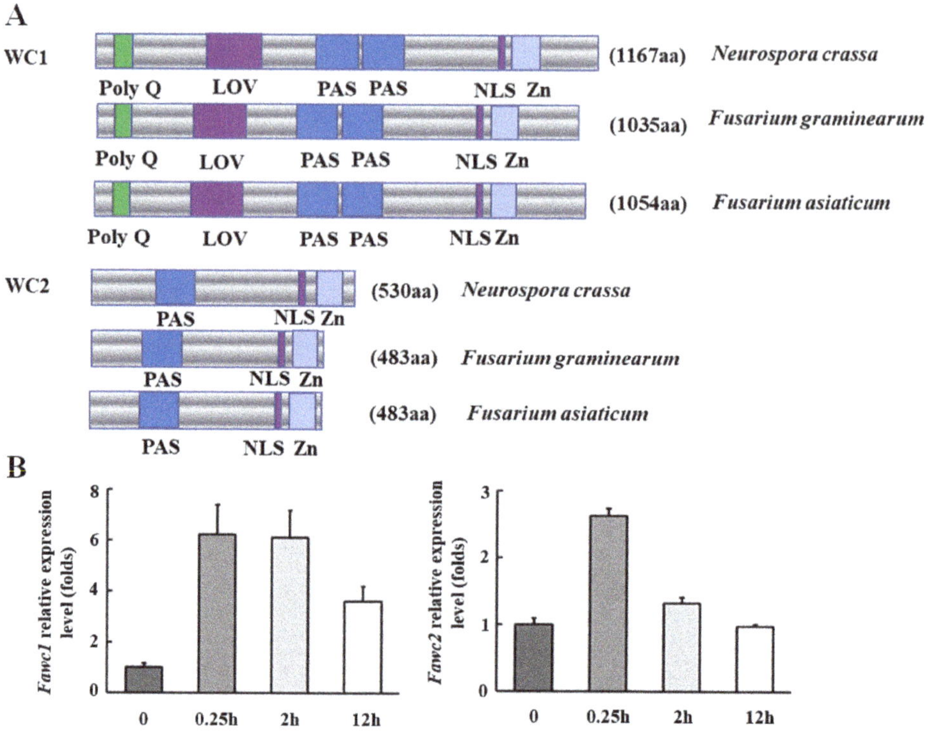

Figure 1. Identification of two photoreceptor genes, *Fawc1* and *Fawc2*, in *F. asiaticum*. (**A**). Schematic demonstration of the domains of WC1 and WC2 orthologs from *N. crassa*, *F. graminearum*, and *F. asiaticum*. Accessions of the amino acid sequences are as follows: WC1 (NCU02356); WC2 (NCU00902); FgWC1 (FGSG_07941); FgWC2 (FGSG_00710); FaWC1 (KX905081.1); FaWC2 (MT019868). Domains of these photoreceptor proteins were analyzed via the SMART online tool (http://smart.embl-heidelberg.de/). PAS: Per-period circadian protein; Arnt: Ah receptor nuclear translocator protein; Sim: Single-minded protein, NLS: Nuclear Location Singal, Zn: Zinc finger binding to DNA consensus sequence. (**B**). Transcript levels of *Fawc1* and *Fawc2* are regulated by light. The horizontal axis indicates light treatment time. The bars present mean values ± SD of three replicate samples.

3.2. Generation and Characterization of the ΔFawc1 and ΔFawc2 Mutants

To reveal the functions of *Fawc1* and *Fawc2*, homologous recombination cassettes used for gene knockout purposes were created as shown in Figure 2. After protoplast transformation, the transformants with hygromycin resistance successfully grew up on the selection medium. To characterize the knockout mutants of Δ*Fawc1* and Δ*Fawc2* in which the hygromycin resistance cassette (*hph*) had correctly replaced each target gene, PCR assays with the specific primer pairs were performed with the genomic DNA, and the correct transformants for each gene were selected for further analysis. To verify that the *Fawc1* or *Fawc2* was completely knocked out, reverse transcription (RT)-PCR was applied, revealing that the transcripts of *Fawc1* and *Fawc2* were present in the EXAP-08 wild-type strain but absent in the mutants of Δ*Fawc1* and Δ*Fawc2*, respectively. To confirm the functions of these two genes, the wild-type *Fawc1* and *Fawc2* connected with their native promoters, and terminators were transformed into Δ*Fawc1* and Δ*Fawc2* to obtain the complementation strains. As FaWC1 possesses both signal input and output domains, the LOV and zinc finger (ZnF) domains, respectively, the truncated versions of FaWC1, lacking either LOV or ZnF domains (Figure 2), were expressed in the Δ*Fawc1* mutant to explore the functions of these domains in light signaling and other life aspects of *F. asiaticum*. All the fungal strains used in this study and their genotypes are listed in Table 1.

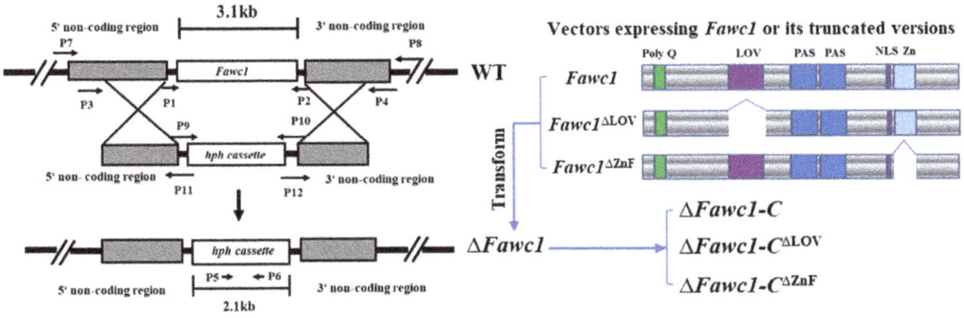

Figure 2. Generation of the transgenic mutants of *F. asiaticum*. Schematic diagrams of homologous recombination occurred between the replacement vector carrying the hygromycin resistance marker (*hph*) and the target gene (*Fawc1*) of *F. asiaticum* strain EXAP-08, resulting in the knockout mutant (Δ*Fawc1*). Meanwhile, the wild-type *Fawc1* or its truncated versions, *Fawc1*$^{\Delta LOV}$ and *Fawc1*$^{\Delta Zn}$, were amplified from genomic DNA of EXAP-08 and cloned into the flu6 plasmid, and then transformed into the Δ*Fawc1* mutant, resulting in the complementation strain Δ*Fawc1-C*, and Δ*Fawc1-C*$^{\Delta LOV}$ and Δ*Fawc1-C*$^{\Delta ZnF}$ mutant strains. The knockout mutant Δ*Fawc2* and its complementation strain Δ*Fawc1-C* were generated via a similar strategy.

3.3. The Marker Responses to Light Signal Are Mediated by WCC and Dependent on LOV but not ZnF Domain of FaWC1 in F. asiaticum

As fungi commonly sense light to indicate the presence of deleterious ultraviolet (UV) radiation, the light-induced survivability through UV damage is recognized as a marker response to light signal in fungi [35,36]. In the present study, the survival of serially diluted spores of all the tested strains was similarly poor if the fungal cultures were kept in the dark after UV-C treatment. However, when white light illumination was applied, the wild type (WT) could survive much better from UV-C irradiation than those cultures in the dark, but no recovery of survival spores was observed with Δ*Fawc1* and Δ*Fawc2*. The complementation strains, Δ*Fawc1-C* and Δ*Fawc2-C*, demonstrated similar photoreactivation levels as the WT, indicating that the defect in the light-induced UV-C resistance in the Δ*Fawc1* and Δ*Fawc2* mutants is indeed caused by the loss of these two genes (Figure 3A). Interestingly, the Δ*Fawc1-C*$^{\Delta LOV}$ showed similar UV-C susceptibility as the Δ*Fawc1* and Δ*Fawc2* mutants, while in contrast, the Δ*Fawc1-C*$^{\Delta ZnF}$ restored UV-C tolerance to the WT level. Gene expression analysis showed that the transcript level of the photolyase gene Faphr1 was significantly induced by light in WT, Δ*Fawc1-C*, Δ*Fawc2-C*, and Δ*Fawc1-C*$^{\Delta ZnF}$ strains; however, no remarkable change of *Faphr1* expression as influenced with light has been observed in the Δ*Fawc1*, Δ*Fawc2*, and Δ*Fawc1-C*$^{\Delta LOV}$ strains (Figure 3B).

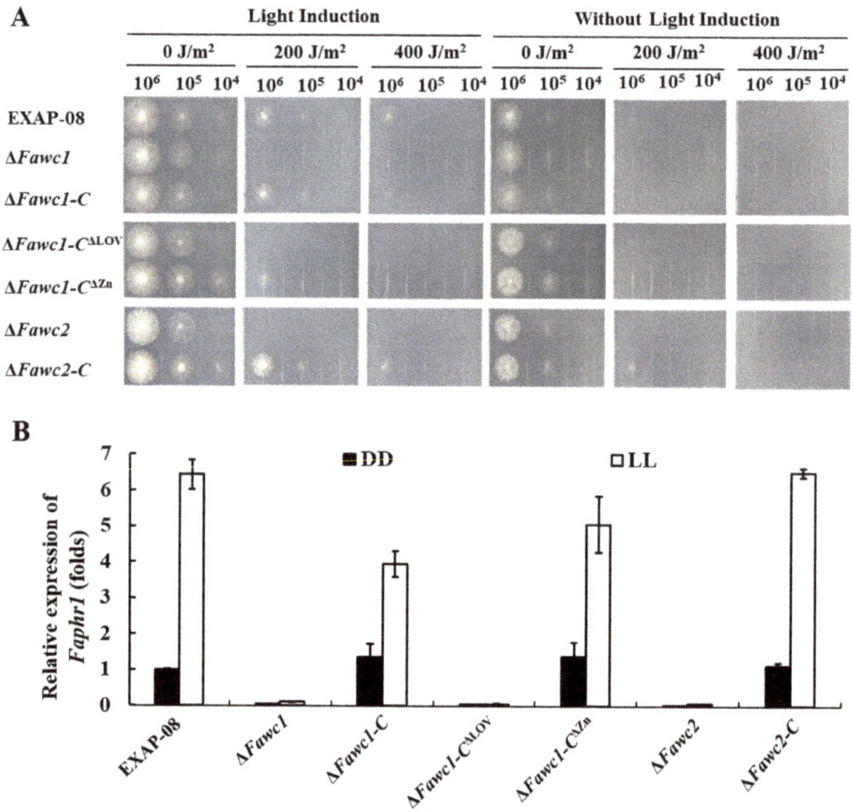

Figure 3. Effect of light on UV-C resistance. (**A**). Serial dilutions of all strains were point-inoculated onto complete agar medium (CM). After the UV irradiation of indicated dosages, the plates were incubated for one day in light (left) or darkness (right). (**B**). The relative expression level of the deduced photolyase gene *Faphr1* in wild-type and mutant strains as influenced by light. DD, samples cultured for 48 h in darkness; LL, samples experienced 47 h culture in darkness followed by one hour of light illumination. The bars present mean values ± SD of three replicate samples.

Another marker response to the light signal in fungi is pigment production [18,37]. The WT strain could produce significantly more orange-colored carotenoid pigment in constant light compared to dark condition after growth in liquid complete medium (CM) for three days (Figure 4A). In contrast, there was no observable orange pigment accumulated by Δ*Fawc1* and Δ*Fawc2* in light and dark conditions (Figure 4A). Similarly, the Δ*Fawc1-C*$^{\Delta LOV}$ mutant, which lacks the LOV domain of FaWC1, showed similar pigmentation phenotypes as the Δ*Fawc1* and Δ*Fawc2* mutants. However, the Δ*Fawc1-C*$^{\Delta ZnF}$ with truncating the ZnF domain of FaWC1 demonstrated enhanced carotenogenesis in response to light, which was similar to the WT strain. Quantitative measurement assay also showed that in darkness, all strains produced basic carotenoid levels, and light treatment caused a significant increment of carotenoid accumulation in WT, Δ*Fawc1-C*, Δ*Fawc2-C*, and Δ*Fawc1-C*$^{\Delta ZnF}$ strains, but light failed to alter the pigmentation behavior in Δ*Fawc1*, Δ*Fawc2-C*, and Δ*Fawc1-C*$^{\Delta LOV}$ strains (Figure 4B). Gene expression analysis showed that the transcript levels of the carotenoid biosynthetic genes *CarRA* and *CarB* [37,38] were up-regulated in light versus dark condition in the WT, ΔFawc1-C, ΔFawc2-C, and Δ*Fawc1-C*$^{\Delta ZnF}$ strains. Contrarily, the expression of *CarRA* and *CarB* could not be induced by light in the Δ*Fawc1*, Δ*Fawc2*, and Δ*Fawc1-C*$^{\Delta LOV}$ strains (Figure 4C,D).

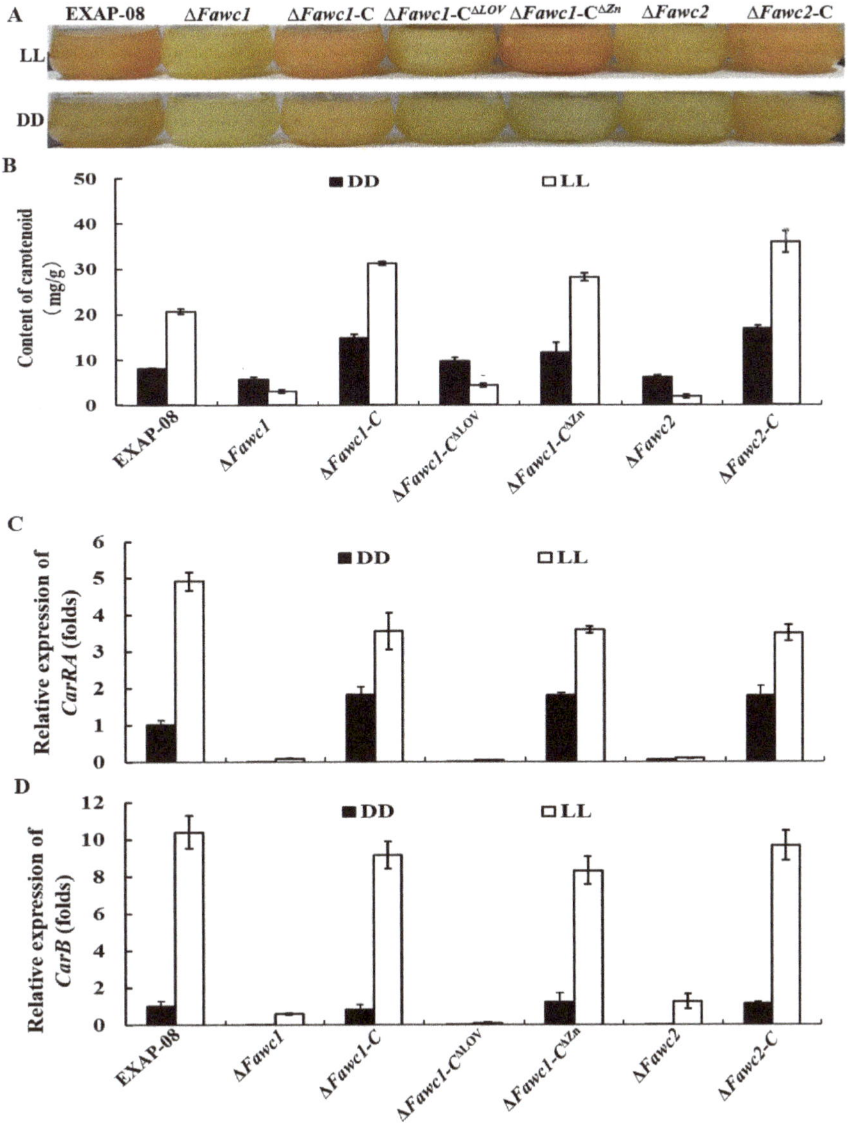

Figure 4. Effect of light on carotenogenesis. (**A**). Carotenoid pigment accumulation of the tested strains cultured in liquid CM for four days under constant light (LL) or darkness (DD). (**B**). Measurement of carotenoid contents in the mycelium of each strain harvested from the liquid shaking culture in (**A**). (**C**) and (**D**). Relative expression levels of deduced carotenoid biosynthesis genes *CarRA* and *CarB* in wild-type and mutant strains under light (LL) or darkness (DD). The bars in **B**, **C**, and **D** present mean values ± SD of three replicate samples.

Collectively, the above data suggest that both FaWC1 and FaWC2 are responsible for light signaling to induce the marker responses, including carotenoid accumulation and UV damage tolerance. Moreover, the LOV and ZnF domains of FaWC1 are required and dispensable, respectively, for mediating the light responses in *F. asiaticum*.

3.4. Perithecia Maturation and Ascospore Development of *F. asiaticum* Are Regulated by WCC Photoreceptor

Sexual reproduction is usually vital for the dissemination of fungal pathogens in their lifecycles, and the near-UV light is known to induce the perithecia maturation and ascospore formation in FGSC [39]. However, two independent studies reported inconsistent effects of the WCC photoreceptor on the sexual development of *F. graminearum*, which was probably due to the difference of the wild-type background strains used in each study [30,40]. The present work with *F. asiaticum* showed that the WT strain was able to form mature perithecia, in which the sexual ascospores could be found apparently. In contrast, the Δ*Fawc1* and Δ*Fawc2* mutants could produce a comparable amount of perithecia as WT, but these mutants' perithecia failed to develop into the black-pigmented mature stage, being deficient in ascospore formation (Figure 5). Re-introducing the wild-type *Fawc1* and *Fawc2* into the corresponding gene deletion mutants had fully recovered their perithecia maturation and ascospore formation abilities, suggesting that these WCC photo receptor components, FaWC1 and FaWC2, are indeed required for sexual reproduction development in this fungus. Additionally, the Δ*Fawc1-C*$^{\Delta ZnF}$ and Δ*Fawc1-C*$^{\Delta LOV}$ demonstrated phenotypes in sexual development as WT and Δ*Fawc1*, respectively, indicating that it should be the LOV domain, but not the ZnF domain, that is required for mediating the light signal to regulate the sexual reproduction of *F. asiaticum*.

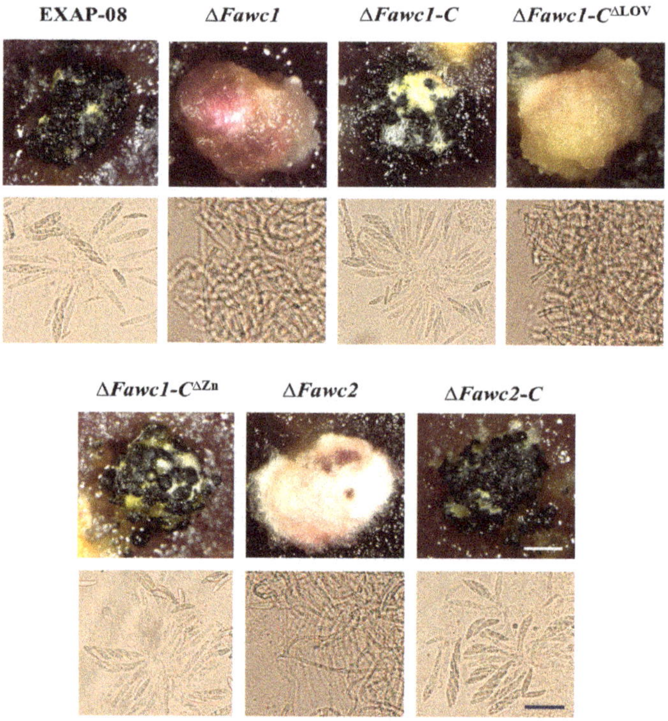

Figure 5. Regulation of *Fawc1* and *Fawc2* on the sexual reproduction development processes of *F. asiaticum*. The perithecium formed by each strain on carrot agar medium was observed via stereomicroscope. Via picking up the perithecia and pressing them on glass slides for microscopic analysis, mature ascospores were observed from the perithecia of EXAP-08, Δ*Fawc1-C*, Δ*Fawc1-C*$^{\Delta ZnF}$, and Δ*Fawc2-C* strains, while in contrast, only mycelium biomass could be found in the immature perithecium of Δ*Fawc1*, Δ*Fawc2*, and Δ*Fawc1-C*$^{\Delta LOV}$ mutant strains.

3.5. FaWC1 and FaWC2 Play Different Roles in Regulating Virulence Expression

In the infection assay with wheat coleoptiles, inoculation with the WT caused apparent brown rot symptom in the host plant materials (Figure 6). In contrast, the Δ*Fawc1* mutant showed more than 80% reduction in pathogenicity in comparison with WT. Meanwhile, the complementation strain Δ*Fawc1-C* had a recovered pathogenicity level similar to that of the WT strain, suggesting that FaWC1 is involved in regulating the pathogenicity of *F. asiaticum*. However, the mutant with the deletion of *Fawc2* caused equivalent disease severity to the WT. These data suggested that FaWC1 and FaWC2 played independently different roles in regulating the pathogenicity of this fungus.

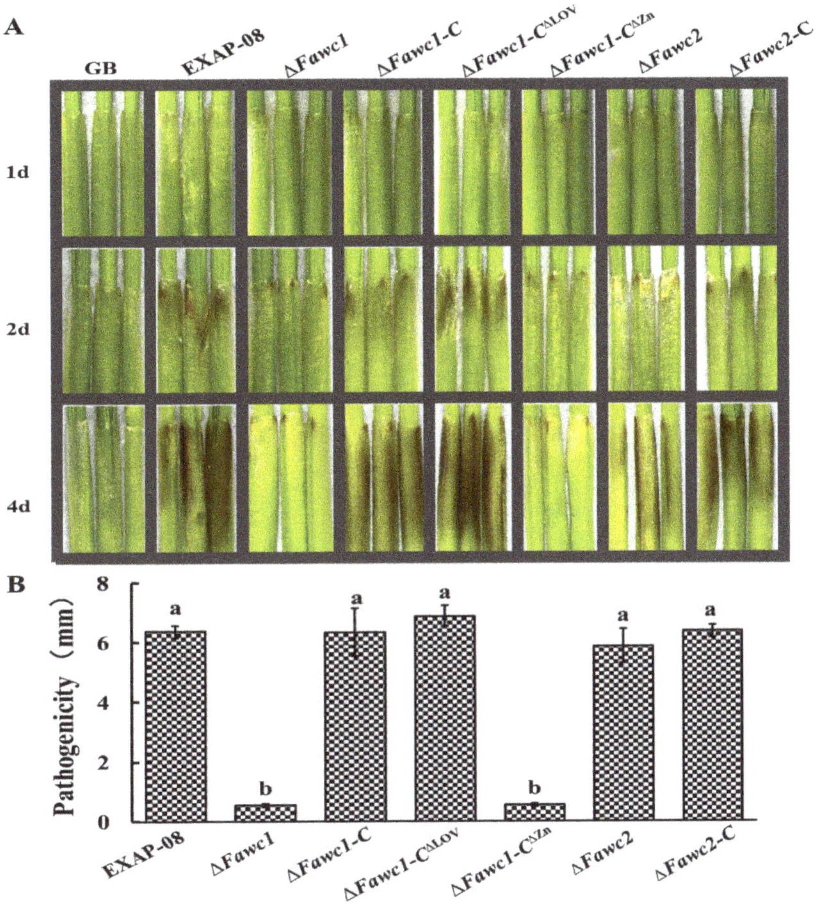

Figure 6. Virulence assay on wheat confirms that *Fawc1* regulate the virulence of *F. asiaticum* in a light-independent manner. (**A**). Wheat coleoptiles were inoculated with 5 μL conidial suspensions and were kept humid inside a plastic box. Fungal strains for test include the wild-type EXAP-08, Δ*Fawc1*, Δ*Fawc1-C*, Δ*Fawc1-C*$^{\Delta LOV}$, Δ*Fawc1-C*$^{\Delta ZnF}$, Δ*Fawc2*, and Δ*Fawc2-C*. Photographs were taken at four days post inoculation. (**B**). Statistical analysis of lesion sizes caused by each fungal genotype. Different letters represent a significant difference at $p < 0.05$. The bars present mean values ± SD ($n = 20$).

In order to assess whether and how light signaling via the WCC pathway could be involved in fungal pathogenicity expression, the mutant strains lacking the LOV and ZnF domains of FaWC1 were further analyzed in the host infection assay. Unexpectedly, the Δ*Fawc1-C*$^{\Delta LOV}$ mutant showed similar pathogenicity as the WT and complementation strains. While in contrast, the pathogenicity of the Δ*Fawc1*-C$^{\Delta ZnF}$ mutant was similar to the Δ*Fawc1* mutant, being significantly reduced compared to WT (Figure 6). Consequently, it can be concluded that the FaWC1 LOV domain, which is required for sensing light signals, exerts no influence on pathogenicity; on the other hand, the ZnF domain of FaWC1 is involved in regulating pathogenicity, although this domain is dispensable for mediating light signals in *F. asiaticum*.

4. Discussion

Light is a strong environment cue orchestrating numerous biological processes in various organisms. Understanding the photobiology mechanisms in fungi can be useful in designing feasible strategies to decrease the detrimental effects of fungi while enhancing those qualities that are beneficial. Most known fungal responses to light are mediated by the well-conserved fungal blue light receptor white collar complex (WCC) [41,42]. In this paper, we functionally investigated the WCC orthologs, *Fawc1* and *Fawc2*, in the plant pathogenic species *F. asiaticum*.

The FaWC1 and FaWC2 characterized in *F. asiaticum* strain EXAP-08 showed high similarity to their counterparts in *F. graminearum* sensu stricto, although they are distinct from each other in geographic, metabolic, and pathogenic phenotypes [9,16,43,44]. The FaWC1 contains a LOV domain for binding the flavin chromophore, a nuclear localization domain (NLS), a Zinc finger domain (ZnF) for DNA binding and transcription factor functions, and a PAS domain for protein–protein interaction. The FaWC2 also contains a PAS domain, NLS domain, and ZnF domain. All these protein domain characteristics of FaWC1 and FaWC2 are highly similar to WC-1 and WC-2 of *N. crassa*, respectively. Light generally increases cellular metabolism and at the same time causes significant oxidative stress to the organism, and to deal with this stress, protective pigments, e.g., carotenoids, are made [45]. Stimulated carotenogenesis as influenced with light treatment is a common photoresponsive phenotype among filamentous fungi [25], including several *Fusarium* species. Experiments in this study showed a clear dependence on WCC for light-induced carotenoid biosynthesis in *F. asiaticum*.

Another common reason for fungi to sense light should be to survive through the environmental stress posed by sunlight, especially the radiation of ultraviolet (UV), which can cause DNA damage or induce the accumulation of toxic reactive oxygen species (ROS) [45]. The ability to sense less harmful wavelengths of visible light can help the organisms to anticipate the emergence of, and promote tolerance to, the harmful UV. This premise has been shown among diverse fungal species through functional study with the orthologs of blue light receptor WCC [46]. It is the photoinduction of DNA repair enzymes, photolyases, and UV endonucleases, by the conserved blue light signaling pathway (WCC) that should be mainly responsible for the light-induced UV resistance in varied fungi [40,47–50]. In the present study, this reliable discipline has been confirmed in *F. asiaticum* in which the loss of *Fawc1* or *Fawc2* resulted in a deficiency of light-induced expression of photolyase gene (*Faphr1*) and consequently hypersensitivity to UV. Taken together, the two marker responses to light signals, including the expression of pigment synthesis and DNA repair enzymes to protect the fungal cells from the damage caused by UV during sunlight exposure, are indeed mediated cooperatively by FaWC1 and FaWC2 in *F. asiaticum*.

Fungal development can also be regulated by light signals. According to varied reports, light signals should have specific roles in certain species including *Aspergillus nidulans* [51], *Cercospora zeae-maydis* [27], *Botrytis cinerea* [29], and *Alternaria alternate* [52]. A key factor required in the regulation of fungal reproductive development events is the velvet complex composed of VeA, VelB, and LaeA [53], and the interactions between photoreceptors and velvet proteins have been demonstrated in *A. nidulans* [54]. As for the FGSC, the development of perithecia and production of ascospores are recognized to be important for the success of these pathogens in their disease cycle [55]. As found in the *F. graminearum* strain Z-3639, WCC should be required for the normal maturity of perithecia during sexual development [40]. However, another group later reported that WCC could negatively regulate sexual development in another strain of *F. graminearum*; the Z-3643 [30]. In the present study, the perithecial maturation and ascospore formation of *F. asiaticum* are dependent on the presence of the WCC. Consequently, the role of the conserved light receptors in regulating fungal reproductive development could be varied case by case.

Besides fungal development and metabolism, the involvement of light receptors in fungal pathogenesis is increasingly attracting research attention [20]. The orthologs of WCC have been recognized in several pathogenic species. However, functional differences have also been reported among them [27–30,40,52]. For example, in *Magnaporthe oryzae*, constant light suppresses disease development, which is mediated via MGWC-1 [28]. In the maize leaf pathogen *C. zeae-maydis*, the WC-1 ortholog Crp-1 is required for stomata tropism and for appressorium and lesion formation, implying that Crp-1 positively contributed to its virulence [27]. While in the necrotrophic pathogen *B. cinerea*, the WCC and its direct target BcLTF1, a light-responsive transcription factor homologous to the SUB1 in *N. crassa*, are required to cope with the oxidative stress that is caused by either excessive exposure to light or arising during host invasion, and thus required for achieving full virulence under excessive light [29].The present study with *F. asiaticum* shows that the deletion of *Fawc1* resulted in decreased virulence. This is in contrast with the reports of *F. graminearum*, in which the WCC seems to be not involved in regulating virulence [30,40]. Even in those fungal species in which WCC are involved in determining virulence expression, the ways are quite different from each other; for example, WCC orthologs are involved in affecting the infection activities of *B. cinerea*, *C. zeae-maydis*, and *M. oryzae* in a light-dependent manner [27–29]. While in contrast, WCC are required for the full virulence of *C. neoformans* and *F. oxysporum* in mammals in a light-independent way [20,47,49]. Since the WCC proteins possess domains for both signal input (LOV domain) and output (Zn-finger domain), it occurs in several fungi that the WCC orthologs have both light-dependent and independent regulatory roles; for example, in *Trichoderma reesei*, the WCC photoreceptors, BLR1 and BLR2, play a role in the alteration of carbohydrate metabolic functions and the transport of compounds with distinct, both positive and negative targets in darkness [25].

To address the issue that whether the light signal and light-perceiving ability by fungi can affect pathogenicity, we have further revealed that the LOV domain of FaWC1 is required, while the ZnF domain is dispensable for inducing carotenoid synthesis and UV tolerance. These findings imply that the WCC-mediated light signaling in *F. asiaticum* does rely on the LOV domain for perceiving light signals, while the ZnF domain of FaWC1 seems to be not involved in regulating the downstream light responses. This phenomenon can be interpreted according to what has been found in several other fungi that indicates that the WC-2 ZnF domain is indeed responsible for the light signal output to regulate the expression of downstream genes [56]. More importantly, the ZnF domain of FaWC1 should have its downstream targets, which are not for light signaling but most likely for pathogenicity regulation. These findings have raised two open questions for future study: (1) to identify the light-independent targets of FaWC1 that are involved in pathogenicity expression; and (2) to resolve the upstream mechanisms that can orchestrate the photoreceptor to well balance between sensing light and regulating light-independent pathogenicity behaviors.

5. Conclusions

In the present study with *F. asiaticum*, the conserved WCC photoreceptor orthologs, namely FaWC1 and FaWC2, have been characterized. Both FaWC1 and FaWC2 are required for perceiving light signals to regulate UV resistance, secondary metabolism, and sexual reproduction. However, FaWC1 and FaWC2 performed diverged roles in virulence expression. Based on the genetics data, we conclude that the WCC orthologs in *F. asiaticum* are responsible for light signaling to adapt to the environmental niche; meanwhile, moonlighting functions of the individual component FaWC1 are required for regulating pathogenicity in a light-independent manner.

Supplementary Materials: The following are available online at http://www.mdpi.com/2076-2607/8/3/365/s1, Table S1: All the primers used in the mutant generating and diagnosis PCR reactions.

Author Contributions: Conceptualization, P.Z. and L.X.; methodology, P.Z., Y.J. and Y.T.; software, Y.T.; validation, P.Z., Y.T. and L.X.; formal analysis, P.Z.; investigation, Y.T., P.Z., Z.L., Y.Q., L.H., N.Z. and Y.W.; resources, P.Z. and L.X.; data curation, Y.T. and P.Z.; writing—original draft preparation, Y.T. and P.Z.; writing—review and editing, Y.T., P.Z., H.N., Y.J. and L.X.; visualization, Y.T. and P.Z.; supervision, P.Z. and Y.J. and L.X.; project administration, P.Z. and L.X.; funding acquisition, P.Z. and L.X. All authors have read and agreed to the published version of the manuscript.

Funding: This research was financially supported by the National Key Research and Development Program of China (2016YFD0400105), National Natural Science Foundation of China (31972121), Shanghai Municipal Science and Technology Commission (18391901400), and Shanghai Municipal Agricultural and Rural Committee (2019-02-08-00-02-F01146).

Acknowledgments: We thank Weihua Tang (Institute of Plant Physiology and Ecology, Chinese Academy of Sciences, Shanghai, China) for kindly providing plasmid pflu6.

Conflicts of Interest: The authors declare no conflict of interest.

References

1. Cuomo, C.A.; Güldener, U.; Xu, J.R.; Trail, F.; Turgeon, B.G.; Pietro, A.D.; Walton, J.D.; Ma, L.J.; Baker, S.E.; Rep, M.; et al. The *Fusarium graminearum* genome reveals a link between localized polymorphism and pathogen specialization. *Science* **2007**, *317*, 1400–1402. [CrossRef] [PubMed]
2. O'Donnell, K.; Kistler, H.C.; Tacke, B.K.; Casper, H.H. Gene genealogies reveal global phylogeographic structure and reproductive isolation among lineages of *Fusarium graminearum*, the fungus causing wheat scab. *PNAS* **2000**, *97*, 7905–7910. [CrossRef]
3. O'Donnell, K.; Ward, T.J.; Geiser, D.M.; Kistler, H.C.; Aoki, T. Genealogical concordance between the mating type locus and seven other nuclear genes supports formal recognition of nine phylogenetically distinct species within the *Fusarium graminearum* clade. *Fungal Genet. Biol.* **2004**, *41*, 600–623. [CrossRef] [PubMed]
4. O'Donnell, K.; Ward, T.J.; Aberra, D.; Kistler, H.C.; Aoki, T.; Orwig, N.; Kimura, M.; Bjørnstad, Å.; Klemsdal, S.S. Multilocus genotyping and molecular phylogenetics resolve a novel head blight pathogen within the *Fusarium graminearum* species complex from Ethiopia. *Fungal Genet. Biol.* **2008**, *45*, 1514–1522. [CrossRef]
5. Starkey, D.E.; Ward, T.J.; Aoki, T.; Gale, L.R.; Kistler, H.C.; Geiser, D.M.; Suga, H.; Tóth, B.; Varga, J.; O'Donnell, K. Global molecular surveillance reveals novel *Fusarium* head blight species and trichothecene toxin diversity. *Fungal Genet. Biol.* **2007**, *44*, 1191–1204. [CrossRef]
6. Yli-Mattila, T.; Gagkaeva, T.; Ward, T.J.; Aoki, T.; Kistler, H.C.; O'Donnell, K. A novel Asian clade within the *Fusarium graminearum* species complex includes a newly discovered cereal head blight pathogen from the Russian Far East. *Mycologia* **2009**, *101*, 841–852. [CrossRef]
7. Sarver, B.A.J.; Ward, T.J.; Gale, L.R.; Broz, K.; Kistler, H.C.; Aoki, T.; Nicholson, P.; Carter, J.; O'Donnell, K. Novel *Fusarium* head blight pathogens from Nepal and Louisiana revealed by multilocus genealogical concordance. *Fungal Genet. Biol.* **2011**, *48*, 1096–1107. [CrossRef]
8. Zhang, H.; Van der Lee, T.; Waalwijk, C.; Chen, W.; Xu, J.; Xu, J.; Zhang, Y.; Feng, J. Population analysis of the *Fusarium graminearum* species complex from wheat in China show a shift to more aggressive isolates. *PLoS ONE* **2012**, *7*, e31722. [CrossRef]
9. Zhang, X.; Ma, H.; Zhou, Y.; Xing, J.; Chen, J.; Yu, G.; Sun, X.; Wang, L. Identification and genetic division of *Fusarium graminearum* and *Fusarium asiaticum* by species-specific SCAR markers. *J. Phytopathol.* **2014**, *162*, 81–88. [CrossRef]
10. Zhu, P.; Wu, L.; Liu, L.; Huang, L.; Wang, Y.; Tang, W.; Wu, L. *Fusarium asiaticum*: An Emerging Pathogen Jeopardizing Postharvest Asparagus Spears. *J. Phytopathol.* **2013**, *161*, 696–703. [CrossRef]
11. Steiner, B.; Kurz, H.; Lemmens, M.; Buerstmayr, H. Differential gene expression of related wheat lines with contrasting levels of head blight resistance after *Fusarium graminearum* inoculation. *Theor. Appl. Genet.* **2009**, *118*, 753–764. [CrossRef] [PubMed]
12. Rawat, N.; Pumphrey, M.O.; Liu, S.X.; Zhang, X.F.; Tiwari1, V.K.; Ando, K.; Trick, H.N.; Bockus, W.W.; Akhunov, E.; Anderson, J.A.; et al. Wheat *Fhb1* encodes a chimeric lectin with agglutinin domains and a pore-forming toxin-like domain conferring resistance to Fusarium head blight. *Nat. Genetics.* **2016**, *48*, 1576–1580. [CrossRef] [PubMed]

13. Ma, L.J.; Van der Does, H.C.; Borkovich, B.A.; Coleman, J.J.; Daboussi, M.J.; Pietro, A.D.; Dufresne, M.; Freitag, M.; Grabherr, M.; Henrissat, B.; et al. Comparative genomics reveals mobile pathogenicity chromosomes in *Fusarium*. *Nature* **2010**, *464*, 367–373. [CrossRef] [PubMed]
14. Zhang, X.W.; Jia, L.J.; Zhang, Y.; Jiang, G.; Li, X.; Zhang, D.; Tang, W.H. In planta stage-specific fungal gene profiling elucidates the molecular strategies of *Fusarium graminearum* growing inside wheat coleoptiles. *Plant Cell.* **2012**, *24*, 5159–5176. [CrossRef] [PubMed]
15. Taylor, R.D.; Saparno, A.; Blackwell, B.; Anoop, V.; Gleddie, S.; Tinker, N.A.; Harriset, L.J. Proteomic analyses of *Fusarium graminearum* grown under mycotoxin-inducing conditions. *Proteomics* **2008**, *8*, 2256–2265. [CrossRef] [PubMed]
16. Zhang, H.; Zhang, Z.; Van der Lee, T.; Chen, W.Q.; Xu, J.; Xu, J.S.; Yang, L.; Yu, D.; Waalwijk, C.; Feng, J. Population genetic analyses of *Fusarium asiaticum* populations from barley suggest a recent shift favoring 3ADON producers in southern China. *Phytopathology* **2010**, *100*, 328–336. [CrossRef]
17. Bahn, Y.S.; Xue, C.; Idnurm, A.; Rutherford, J.C.; Heitman, J.; Cardenas, M.E. Sensing the environment: Lessons from fungi. *Nat. Rev. Microbiol.* **2007**, *5*, 57–69. [CrossRef]
18. Tisch, D.; Schmoll, M. Light regulation of metabolic pathways in fungi. *Appl. Microbiol. Biotechnol.* **2010**, *85*, 1259–1277. [CrossRef]
19. Heintzen, C.; Loros, J.J.; Dunlap, J.C. The PAS protein VIVID defines a clock-associated feedback loop that represses light input, modulates gating, and regulates clock resetting. *Cell* **2001**, *104*, 453–464. [CrossRef]
20. Idnurm, A.; Crosson, S. The photobiology of microbial pathogenesis. *PLoS Pathog.* **2009**, *5*, e1000470. [CrossRef]
21. Ballario, P.; Vittorioso, P.; Magrelli, A.; Talora, C.; Cabibbo, A.; Macino, G. White collar-1, a central regulator of blue light responses in *Neurospora*, is a zinc finger protein. *EMBO J.* **1996**, *15*, 1650–1657. [CrossRef] [PubMed]
22. Harding, R.W.; Melles, S. Genetic Analysis of Phototropism of *Neurospora crassa* Perithecial Beaks Using White Collar and Albino Mutants. *Plant Physiol.* **1983**, *72*, 996–1000. [CrossRef] [PubMed]
23. Lauter, F.R.; Russo, V.E. Blue light induction of conidiation-specific genes in *Neurospora crassa*. *Nucleic Acids. Res.* **1991**, *19*, 6883–6886. [CrossRef] [PubMed]
24. Schmoll, M.; Tian, C.; Sun, J.; Tisch, D.; Glass, N.L. Unravelling the molecular basis for light modulated cellulase gene expression - the role of photoreceptors in *Neurospora crassa*. *BMC Genomics.* **2012**, *13*, 127. [CrossRef]
25. Tisch, D.; Schmoll, M. Targets of light signalling in *Trichoderma reesei*. *BMC Genomics.* **2013**, *14*, 657. [CrossRef]
26. Scholthof, K.B.G. The disease triangle: Pathogens, the environment and society. *Nat. Rev. Microbiol.* **2007**, *5*, 152–156. [CrossRef]
27. Kim, H.; Ridenour, J.B.; Dunkle, L.D.; Bluhm, B.H. Regulation of stomatal tropism and infection by light in *Cercospora zeae-maydis*: Evidence for coordinated host/pathogen responses to photoperiod? *PLoS Pathog.* **2011**, *7*, e1002113. [CrossRef]
28. Kim, S.; Singh, P.; Park, K.; Park, S.; Friedman, A.; Zheng, T.; Lee, Y.H.; Lee, K. Genetic and molecular characterization of a blue light photoreceptor MGWC-1 in *Magnaporth oryzae*. *Fungal Genet. Biol.* **2011**, *48*, 400–407. [CrossRef]
29. Canessa, P.; Schumacher, J.; Hevia, M.A.; Tudzynski, P.; Larronodo, L.F. Assessing the effects of light on differentiation and virulence of the plant pathogen *Botrytis cinerea*: Characterization of the White Collar Complex. *PLoS ONE* **2013**, *8*, e84223. [CrossRef]
30. Kim, H.; Kim, H.K.; Lee, S.; Yun, S.H. The white collar complex is involved in sexual development of *Fusarium graminearum*. *PLoS ONE* **2015**, *10*, e0120293. [CrossRef]
31. Xu, Y.B.; Li, H.P.; Zhang, J.B.; Song, B.; Chen, F.F.; Duan, X.J.; Xu, H.Q.; Liao, Y.C. Disruption of the chitin synthase gene CHS1 from *Fusarium asiaticum* results in an altered structure of cell walls and reduced virulence. *Fungal Genet. Biol.* **2010**, *47*, 205–215. [CrossRef] [PubMed]
32. Leslie, J.F.; Summerell, B.A. *Fusarium* laboratory workshops-A recent history. *Mycotoxin Res.* **2006**, *22*, 73–74. [CrossRef] [PubMed]

33. Desmond, O.J.; Manners, J.M.; Stephens, A.E.; Maclean, D.J.; Schenk, P.M.; Gardiner, D.M.; Munn, A.L.; Kazan, K. The *Fusarium* mycotoxin deoxynivalenol elicits hydrogen peroxide production, programmed cell death and defence responses in wheat. *Mol. Plant Pathol.* **2008**, *9*, 435–445. [CrossRef] [PubMed]
34. Crosthwaite, S.K.; Dunlap, J.C.; Loros, J.J. *Neurospora wc-1* and *wc-2*: Transcription, photoresponses, and the origins of circadian rhythmicity. *Science* **1997**, *276*, 763–769. [CrossRef]
35. Thoma, F. Light and dark in chromatin repair: Repair of UV- induced DNA lesions by photolyase and nucleotide excision repair. *EMBO J.* **1991**, *18*, 6585–6598. [CrossRef]
36. Corrochano, L.M. Light in the Fungal World: From Photoreception to Gene Transcription and Beyond. *Annu. Rev. Genet.* **2019**, *53*, 1–22. [CrossRef]
37. Avalos, J.; Pardo-Medina, J.; Parra-Rivero, O.; Ruger-Herreros, M.; Rodríguez-Ortiz, R.; Hornero-Méndez, D.; Limón, M.C. Carotenoid Biosynthesis in *Fusarium*. *J. Fungi.* **2017**, *3*, 39. [CrossRef]
38. Prado, M.M.; Prado-Cabrero, A.; Fernández-Martín, R.; Avalos, J. A gene of the opsin family in the carotenoid gene cluster of *Fusarium fujikuroi*. *Curr. Genet.* **2004**, *46*, 47–58. [CrossRef]
39. Tschanz, A.T.; Horst, R.K.; Nelson, P.E. The Effect of Environment on Sexual Reproduction of *Gibberella Zeae*. *Mycologia.* **1976**, *68*, 327–340. [CrossRef]
40. Kim, H.; Son, H.; Lee, Y.W. Effects of light on secondary metabolism and fungal development of *Fusarium graminearum*. *J. Appl. Microbiol.* **2014**, *116*, 380–389. [CrossRef]
41. Chen, C.H.; Ringelberg, C.S.; Gross, R.H.; Dunlap, J.C.; Loros, J.J. Genome-wide analysis of light-inducible responses reveals hierarchical light signalling in *Neurospora*. *EMBO J.* **2009**, *28*, 1029–1042. [CrossRef] [PubMed]
42. Wu, C.; Yang, F.; Smith, K.M.; Peterson, M.; Dekhang, M.; Zhang, Y.; Zucker, J.; Bredeweg, E.L.; Mallappa, C.; Zhou, X.; et al. Genome-wide characterization of light-regulated genes in *Neurospora crassa*. *Genes* **2014**, *4*, 1731–1745. [CrossRef] [PubMed]
43. Van der Lee, T.; Zhang, H.; van Diepeningenc, A.; Waalwijk, C. Biogeography of *Fusarium graminearum* species complex and chemotypes: A review. *Food Addit. Contam.* **2015**, *32*, 453–460. [CrossRef] [PubMed]
44. Backhouse, D. Global distribution of *Fusarium graminearum*, *F. asiaticum* and *F. boothii* from wheat in relation to climate. *Eur. J. Plant Pathol.* **2014**, *139*, 161–173. [CrossRef]
45. Fuller, K.K.; Loros, J.J.; Dunlap, J.C. Fungal photobiology: Visible light as a signal for stress, space and time. *Curr. Genet.* **2015**, *61*, 275–288. [CrossRef]
46. Yu, Z.; Fischer, R. Light sensing and responses in fungi. *Nat. Rev. Microbiol.* **2019**, *17*, 25–36. [CrossRef]
47. Zhu, P.; Idnurm, A. The contribution of the White Collar complex to *Cryptococcus neoformans* virulence is independent of its light-sensing capabilities. *Fungal Genet. Biol.* **2018**, *121*, 56–64. [CrossRef]
48. Berrocal-Tito, G.M.; Esquivel-Naranjo, E.U.; Horwitz, B.A.; Herrera-Estrella, A. *Trichoderma atroviride* PHR1, a fungal photolyase responsible for DNA repair, autoregulates its own photoinduction. *Eukaryot Cell.* **2007**, *6*, 1682–1692. [CrossRef]
49. Ruiz-Roldán, M.C.; Garre, V.; Guarro, J.; Mariné, M.; Roncero, M.I.G. Role of the white collar 1 photoreceptor in carotenogenesis, UV resistance, hydrophobicity, and virulence of *Fusarium oxysporum*. *Eukaryot Cell.* **2008**, *7*, 1227–1230. [CrossRef]
50. Verma, S.; Idnurm, A. The Uve1 endonuclease is regulated by the white collar complex to protect *cryptococcus neoformans* from UV damage. *PLoS Genet.* **2013**, *9*, e1003769. [CrossRef]
51. Bayram, Ö.; Braus, G.H.; Fischer, R.; Rodriguez-Romero, J. Spotlight on *Aspergillus nidulans* photosensory systems. *Fungal Genet. Biol.* **2010**, *47*, 900–908. [CrossRef] [PubMed]
52. Pruβ, S.; Fetzner, R.; Seither, K.; Herr, A.; Pfeiffer, E.; Metzler, M.; Lawrence, C.B.; Fischer, R. Role of the *Alternaria alternata* blue-light receptor *LreA* (white-collar 1) in spore formation and secondary metabolism. *Appl. Environ. Microbiol.* **2014**, *80*, 2582–2591. [CrossRef]
53. Bayram, Ö.; Krappmann, S.; Ni, M.; Bok, J.W.; Helmstaedt, K.; Valerius, O.; Braus-Stromeyer, S.; Kwon, N.J.; Keller, N.P.; Yu, J.H.; et al. VelB/VeA/LaeA Complex Coordinates Light Signal with Fungal Development and Secondary Metabolism. *Science* **2008**, *320*, 1504–1506. [CrossRef]
54. Purschwitz, J.; Müller, S.; Kastner, C.; Schöser, M.; Haas, H.; Espeso, E.A.; Atoui, A.; Calvo, A.M.; Fischer, R. Functional and physical interaction of blue- and red-light sensors in *Aspergillus nidulans*. *Curr. Biol.* **2008**, *18*, 255–259. [CrossRef] [PubMed]

55. Trail, F.; Xu, H.; Loranger, R.; Gadoury, D. Physiological and environmental aspects of ascospore discharge in *Gibberella zeae* (anamorph *Fusarium graminearum*). *Mycologia.* **2002**, *94*, 181–189. [CrossRef] [PubMed]
56. Wang, B.; Zhou, X.; Loros, J.J.; Dunlap, J.C. Alternative Use of DNA Binding Domains by the *Neurospora* White Collar Complex Dictates Circadian Regulation and Light Responses. *Mol. Cell Biol.* **2015**, *36*, 781–793. [CrossRef]

microorganisms

MDPI

Review

Pathogenicity and Virulence Factors of *Fusarium graminearum* Including Factors Discovered Using Next Generation Sequencing Technologies and Proteomics

Molemi E. Rauwane [1], Udoka V. Ogugua [1], Chimdi M. Kalu [1], Lesiba K. Ledwaba [1,2], Adugna A. Woldesemayat [3] and Khayalethu Ntushelo [1,*]

1 Department of Agriculture and Animal Health, Science Campus, University of South Africa, Corner Christiaan de Wet Road and Pioneer Avenue, Private Bag X6, Florida 1710, South Africa; rauwame@unisa.ac.za (M.E.R.); oguguatalks@gmail.com (U.V.O.); kaluchimdimang@gmail.com (C.M.K.); LedwabaL@arc.agric.za (L.K.L.)

2 Agricultural Research Council-Vegetable and Ornamental Plants (ARC-VOP), Plant Breeding Division, Roodeplaat, Private Bag X293, Pretoria 0001, South Africa

3 Department of Biotechnology, Bioinformatics, Research Unit, College of Biological and Chemical Engineering, Addis Ababa Science and Technology University, Building Block 73, Room 106, Addis Ababa 1000, Ethiopia; adugnaabdi@gmail.com

* Correspondence: ntushk@unisa.ac.za

Received: 15 October 2019; Accepted: 29 November 2019; Published: 22 February 2020

Abstract: *Fusarium graminearum* is a devasting mycotoxin-producing pathogen of grain crops. *F. graminearum* has been extensively studied to understand its pathogenicity and virulence factors. These studies gained momentum with the advent of next-generation sequencing (NGS) technologies and proteomics. NGS and proteomics have enabled the discovery of a multitude of pathogenicity and virulence factors of *F. graminearum*. This current review aimed to trace progress made in discovering *F. graminearum* pathogenicity and virulence factors in general, as well as pathogenicity and virulence factors discovered using NGS, and to some extent, using proteomics. We present more than 100 discovered pathogenicity or virulence factors and conclude that although a multitude of pathogenicity and virulence factors have already been discovered, more work needs to be done to take advantage of NGS and its companion applications of proteomics.

Keywords: pathogenicity; virulence; *Fusarium graminearum*; next-generation sequencing; proteomics

1. Introduction

Plant pathogens have developed sophisticated penetration, infection, and colonization strategies to suppress plant defense mechanisms of susceptible hosts and cause disease. The display of minor symptoms by the plant is of least concern but excessive tissue damage and crop loss result in serious economic losses with adverse implications for society, especially in poor countries. To gain insights into how these plant pathogens develop the sophisticated strategies of attack, platforms such as next-generation sequencing (NGS), and its companion applications like proteomics, are currently being exploited. Pertinent scientific inquiry of pathogenicity and virulence processes of pathogens, comprehensive and integrated investigations remain imperative and must take advantage of such high-tech platforms as NGS to generate multitudes of useful datasets. Among microbes associated with the plant are fungi which encounter plants, penetrate and colonize plant tissue to either cause disease or live within the colonized plant in a symbiotic relationship with the host. Fungi which cause diseases are called fungal pathogens and their adverse effects can range from tiny spots on the plant surface to significant symptom expression and tissue collapse. In causing diseases, fungi employ various strategies. They can penetrate the host using penetration structures and develop a large network of hyphae within the intercellular spaces. Through the hyphae, these fungi draw nutrients from the living plant cells. This mode of attack is called biotrophy. Alternatively, fungal pathogens kill the host tissue and derive nutrients from the dead tissue. This mode of attack is called necrotrophy. In the hemi-biotrophic mode of attack, the fungus successively adopts a necrotrophic lifestyle after biotrophy. The invasion of the plant by the pathogen is usually carried out through the production of factors used for plant tissue manipulation to gain physical access into the tissues as well as draw nutrients. These factors enable the plant to cause disease as well as advance the infection within the interior of the tissue, ultimately worsening the disease condition. The ability of a fungus to cause disease is termed pathogenicity, and the ability to worsen the disease is called virulence. The terms pathogenicity factors and virulence factors are loosely used to refer to any substance a pathogen uses to parasitize the plant. We will also use these concepts jointly and will not attempt to classify the various disease-causing factors as either pathogenicity or virulence factors. As a necrotrophic fungus, the wheat pathogen *Fusarium graminearum* Schwabe (teleomorph *Gibberellae zeae*) produces, among other things, cell wall-degrading enzymes (CWDEs) and toxins. Plant tissue exposed to the CWDEs and the toxins loses firmness, macerates and cell contents leak. As the cell contents provide nourishment to the fungus, it grows and advances into the inner tissues. The disease spreads and eventually, depending on other factors, the affected part dies. With its ability to produce toxins, *F. graminearum*, a homothallic (self-fertile) ascomycetous fungus, of the Order: Sordariomycetes and Family: Nectriaceae causes Fusarium head blight (FHB), one of the most economically important diseases of wheat, barley, rice and other grain crops worldwide [1–5]. Also known as scab, this devastating disease was first described in 1884 in England and since then, its prominence has increased worldwide with outbreaks reported in temperate and semitropical areas globally [1,6–9]. Under favorable conditions, FHB can advance from initial infection to the destruction of the entire crop within a few weeks [1]. A combination of factors, global warming, increase in wheat production under irrigation and the concomitant increase in no-till practices is the likely cause of the recent resurgence of FHB. *F. graminearum* is favored by high temperatures and irrigation splash disseminates its propagules. Wheat spikes infected by *F. graminearum* have a bleached appearance and grain infected with this fungus is shriveled with pale grey color and an occasional pinkish discoloration. Over and above these undesirable qualities, ingestion of significant amounts of mycotoxin-contaminated grain may cause vomiting, headache, and dizziness in humans. Animals may lose weight and suffer anorexia. Serious effects of ingestion of large amounts of mycotoxins include leukoencephalomalacia in horses, pulmonary edema in swine, and kidney and liver cancers in mice (mentioned in Proctor et al. [10]), and a plethora of records link mycotoxin consumption to cancer in humans. However, some of these sicknesses are linked to mycotoxins produced by other *Fusarium* species, and not *F. graminearum.* The primary mycotoxin produced by *F. graminearum* is the trichothecene deoxynivalenol (DON) (see Figure 1 for the classification of DON) and other toxins produced include zearalenone, nivalenol (NIV), 4-acetylnivalenol (4-ANIV), and DON derivatives 3- and 15-acetyldeoxynivalenol (3-ADON and 15-ADON) [2,11–14]. The serious effects of crop infection by *F. graminearum* require in-depth studies on the pathogen, primarily its pathogenicity and virulence factors which can be comprehensively studied using, among other techniques, whole-genome sequencing, transcriptomics using the convenient, reliable, and large data-generating tools of NGS and proteomics. Recently, a multitude of data has been generated for *F. graminearum* either to understand its genome organization or its gene composition as well as genes involved in plant attack. Although more work is still required, the pace at which this information is generated is too high to allow sufficient time for synthesis, organization, and communication. Meanwhile, with the dynamics of climate change, the relations between the pathogen and the plant require active and rapid utilization of the data which is generated for use to ensure plant health and ultimately good human livelihood. Given this background, it remains important to reflect on the work which has been undertaken in *F. graminearum* research and collate it as a useful resource for various *F. graminearum* workers. Furthermore, the work to understand the genome of *F. graminearum* and its pathogenicity and virulence factors requires acceleration taking advantage of NGS technologies and its supplementing proteomics applications. This is the purpose of this current review, which aimed to trace progress made in discovering pathogenicity and virulence factors, in general, as well as using NGS, and to some extent, proteomics. We present more than 100 *F. graminearum* factors which either directly perform pathogenicity and virulence functions or are indirectly linked to pathogenicity and virulence. We conclude that a multitude of pathogenicity and virulence factors have been discovered, however, more work needs to be done taking advantage of NGS and its companion applications of proteomics. This review article is organized as follows: We provide a narrative background of pathogenicity and virulence, information on *F. graminearum* and information on NGS. We then discuss in detail the various pathogenicity and virulence factors of *F. graminearum* and elaborate on the pathogenicity and virulence factors discovered using NGS (see Figure 2 for the thought process and article organization) and, to some extent, those discovered using proteomics. We collated information from various recent scientific publications to uniquely reflect the pathogenicity and virulence factors. We conclude that a multitude of pathogenicity and virulence factors have been discovered, however, more work needs to be done taking advantage of NGS and its companion applications of proteomics.

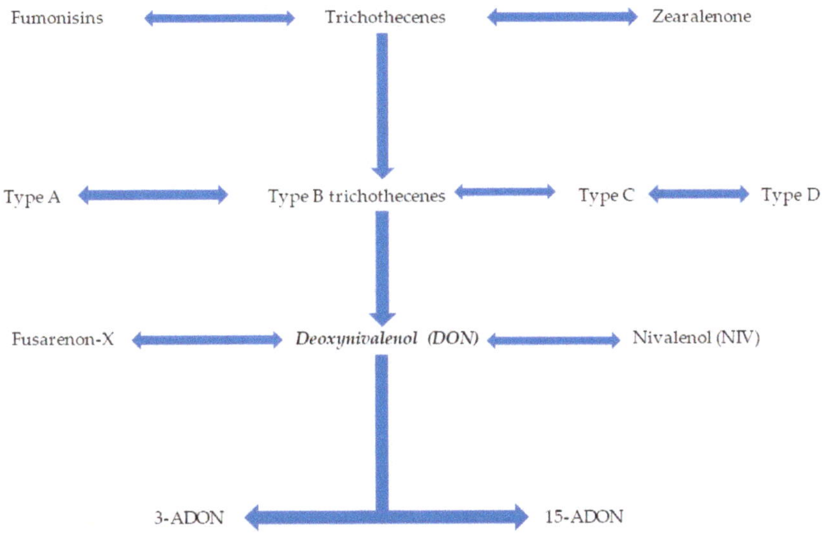

Figure 1. The classification of the major toxin produced by *Fusarium graminearum*, DON. The major toxins produced by *Fusarium* species are fumonisins, trichothecenes, and zearalenone. DON is a type B trichothecene which has derivatives 3-ADON and 15-ADON.

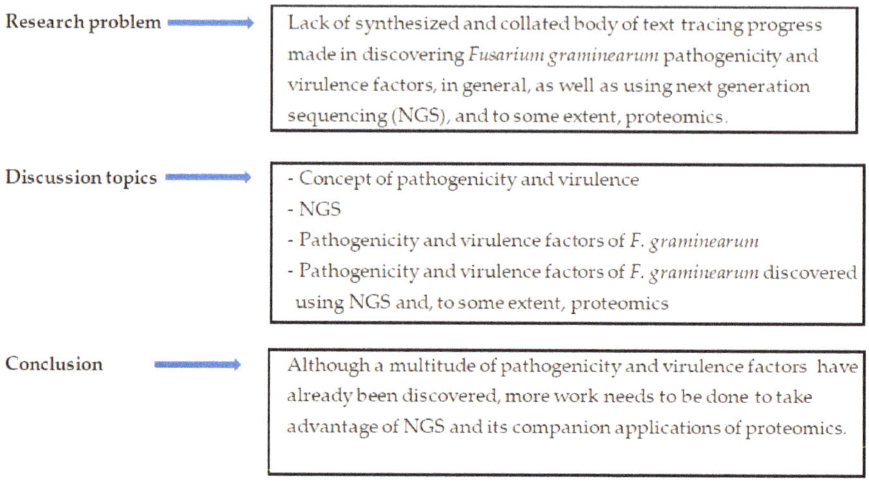

Figure 2. An illustration of concept development and the organization of the article.

2. Next-Generation Sequencing, Its Relevance in Studying Plant Pathogenic Fungi and *Fusarium graminearum*

Nucleic acid sequencing has evolved from determining a few nucleotides of a nucleic acid fragment to its current form of generating large datasets of millions of reads which cover significant genomes of organisms. The generation of these large genomic and RNA datasets has been exploited by researchers to unravel various complexities of genome organization and function in many organisms. Nucleic acid sequencing is the establishment of the order of nucleotides of a given DNA or RNA fragment [15,16]. Earlier efforts to sequence nucleic acids, such as Sanger sequencing, would determine a stretch of a few nucleotides. These earlier sequencing efforts were slow and costly and the need for cheaper and faster sequencing methods grew. This demand drove the development of NGS. NGS, also termed high-throughput or massively parallel sequencing, is a genre of technologies that allow for thousands to billions of DNA fragments to be sequenced simultaneously and independently [17,18]. Although different platforms are commercially available, the NGS workflow features are generally similar, with four major steps: (i) DNA library preparation, (ii) clonal amplification, (iii) massive parallel sequencing, and (iv) data analysis [19,20]. In just the past five years, a number of instruments have been invented. The NGS platform Illumina NextSeq was invented in 2014, followed by the Ion Torrent S5/S5XL 540 in 2015. These were followed by Pacific Biosciences PacBio Sequel in 2016 [21]. These are among the instruments which have been recently invented and utilized to generate large datasets for organisms. One of the first inventions was the 454 GS FLX (454 Life Sciences, Branford, CT, USA). It utilizes an amplicon-based technology to create libraries from genetic regions selected with pairs of multiple primers [18,22,23]. On the contrary, Ion Torrent utilizes the pH variation sequencing method [18,24], whereas the modern platforms, such as Illumina MiSeq and Gene Reader System, utilize the fluorescence emission sequencing method [24,25]. NGS technologies have impressively accelerated research and are becoming employed routinely to seek solutions in different areas of research, as well as provide insights into pathogenicity and virulence of various plant pathogens [26–34]. Although NGS is more affordable than first-generation sequencing (considering time and money), it is still not within the reach of many laboratories in undeveloped countries [16]. However, sequencing services are offered by many sequencing companies at relatively low prices, and therefore, the need to buy the sequencers is less compelling for many. Despite the many challenges associated with NGS, such as inaccurate sequencing of homopolymer regions on certain NGS platforms which can lead to sequencing errors [16], as well as the lack of expertise in downstream processes such as data analysis, there is still success in the generation of sequences. NGS has been exploited to uncover genome organization and sequence for various plant pathogenic fungi. Plant pathogenic fungi were identified as important in some of the ambitious earlier projects in the area of characterization of fungi, such as the Fungal Genome Initiative [35,36], the Fungal Genomics Program [37,38] and its extension, the 1000 Fungal Genomes (1KFG) Project [39]. By extension, the plant pathogenic fungus *F. graminearum* has benefited tremendously and still does so from the various sequencing initiatives. The genome of *F. graminearum* has been sequenced completely and published for the benefit of *Fusarium* workers worldwide. This is in addition to the numerous gene sequences available in various databases. The complete genome and sequence resources have provided useful information on the biology, pathogenicity, and virulence of *F. graminearum*. A multitude of genes coding for pathogenicity and virulence factors of *F. graminearum* has been uncovered and the valuable information analyzed by researchers to deduce the much-needed insight into this devastating plant pathogen. We trace the important pathogenicity and virulence factors discovered and focus our attention on those which were discovered using NGS, and to some extent those which were discovered using proteomics. Within the context of this review article, pathogenicity factors and virulence factors are any substances produced by the fungus to gain access into the plant, manipulate it for its benefit, thus causing disease, as well as advance pathogenicity, thus causing the severity of the disease. Genes encoding these pathogenicity and virulence factors are themselves called pathogenicity and virulence factors. The terms pathogenicity and virulence are used jointly in this article. There does not seem to be clarity on the distinction between pathogenicity factors and virulence factors, and therefore, many publications seem to utilize these terms loosely. There is great justification to generate genome sequence information on *F. graminearum*. In order to offset the serious pathogenic problems caused by *F. graminearum* and to develop insights into the virulence and antagonistic defense mechanisms in host plants, it appears imperative to undertake the identification of the pathogenicity and virulence factors which make up the attack arsenal of *F. graminearum*. For this reason, the MIPS Fusarium graminearum Genome Database (FGDB) with an updated set with an estimated 14,000 genes and downstream analysis in a live gene validation process was established to provide a comprehensive genomic and molecular analysis [40]. Because *Fusarium* species, including *F. graminearum*, are among the most important phytopathogenic and toxigenic fungi, it is essential to understand the molecular underpinnings of their pathogenicity and virulence. Based on the comparative genomic analysis conducted on three phenotypically diverse species that include *F. graminearum*, it was revealed that among others, this particular homothallic fungal species, *F. graminearum*, has shown a relatively narrow host range that includes important crops like wheat (mentioned in Ma et al. [41]). Before discussing pathogenicity and virulence factors of *F. graminearum*, we look at pathogenicity and virulence factors of plant pathogens generally.

3. Pathogenicity and Virulence Factors of Plant Pathogens

Pathogenicity is the ability of an organism to cause disease, virulence, on the other hand, refers to the magnitude of the disease caused. There are three modes of attack by fungi. One mode requires 'cooperation' from the plant and living plant tissues. The fungus forms structures for penetration and a network of hyphae intertwined within the colonized plant tissues. Nutrients in the plants are extracted by this mass of hyphal structures formed in the intercellular spaces of the plant. This mode is called biotrophy (Figure 3). Other fungi release a barrage of degradative enzymes and toxins to lyse plant tissues and in so doing feed on the leaking cell contents and degraded tissue. This is called necrotrophy (Figure 4). Hemi-biotrophy is a hybrid mode between biotrophy and necrotrophy. Biotrophs produce proteins which circumvent the defense responses of the plant. These secreted proteins may have a counteracting effect on the plant's defense proteins. This protein interaction in a match-match situation plays out as gene-for-gene interaction. This was found to be the case with flax rust of wheat caused by *Melampsora lini* and wheat rusts caused by the Puccinias. This gene-for-gene interaction is evidence of co-evolution between the plant and its pathogen. The mechanism of attack is less systematic in necrotrophs in which toxin production can be unspecific. Necrotrophs flood the plant with CWDEs and toxins which "chew up" the tissue. The hard physical defense structures of the plant, such as pectin, lignin, and glucans, crumble and the cells collapse. Pathogenicity and virulence-associated with plant pathogens are made possible by the secretion of pathogenicity and virulence factors which are molecules that help the pathogen colonize the plant [42,43]. For the purpose of this review, genes which are involved in the production of pathogenicity and virulence factors produced are themselves called pathogenicity and virulence factors.

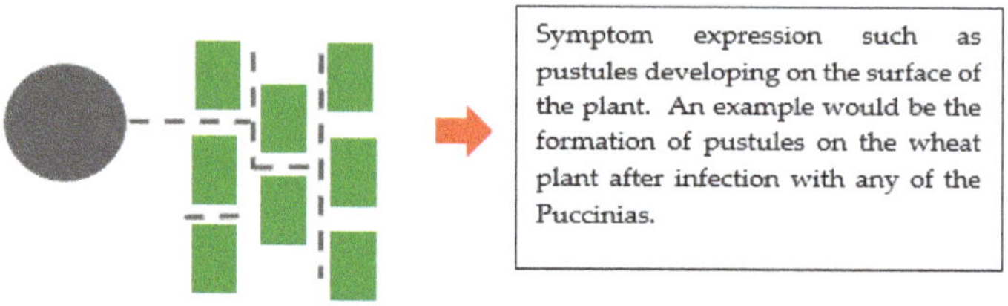

Figure 3. An illustration of biotrophy between a plant pathogenic fungus and a plant. The grey circle represents a fungal spore. Dotted lines represent the germ tube which has developed to form a network of hyphae, and the green rectangles represent plant cells. Finally, symptoms are expressed.

Figure 4. An illustration of necrotrophy between a plant pathogenic fungus and a plant. The grey circle represents a fungal spore, the orange arrows represent secreted pathogenicity and virulence factors. The green rectangles represent plant cells, the plant cells become deformed, ripped, and cell contents leak. Finally, symptoms are expressed.

Decades of research have disclosed that plant pathogens have used a notable assortment of proteins and toxins as virulence factors [44]. Several categories of pathogenicity genes, namely (i) gene signals, (ii) genes generating or detoxifying toxins, (iii) metabolic enzymes, (iv) genes involved in the generation of particular infection structures, and (v) CWDEs, can be identified (mentioned in Duyvesteijn et al. [45]). Vadlapudi and Naidu [46] have identified several fungal processes and molecules which contribute to fungal pathogenicity or virulence which have the ability to harm the plants, and these include cell-wall-degrading proteins and toxins. Reception and transduction of signals in the early phase of plant infection play a vital role in triggering development and morphogenesis mechanisms involved in the penetration of the plant host. Several plant pathogens produce toxins that can harm the tissues of plants. Toxins may be either non-specific or host selective. The link between host selective toxin (HST) secretion and pathogenicity in plant pathogens and between sensitivity to toxins and susceptibility to disease in plants provides convincing proof that HSTs can be accountable for host-selective infection and disease growth [47]. Non-specific toxins harm phylogenetically unrelated plant cells, while HSTs are both pathogenicity determinants on various hosts and their genetic varieties. *Loculoascomycetes* produce the most recognized fungal HSTs. Trichothecenes are also compounds that are toxic to plants, and they are believed to be pathogenicity and/or virulence factors. The protein region of the oomycete *Phytophthora* family of the *phytotoxin*-like scr74 gene, the C-terminal of *Phytophthora* RXLR effector paralogues and a single amino acid polymorphism in the *Phytophthora* EPIC1 effector were related to the capacity to specialize in a fresh host [48–50]. Stukenbrock and McDonald [51] also reported polymorphism data showing the spread of two codons in the host-specific necrotrophic effector ToxA produced by *Pyrenophora tritici-repentis* and *Stagonospora nodorum*. Various fungi are models for studying plant-directed toxins. These include *Alternaria alternata*, *Sclerotinia sclerotiorum*,

Botrytis cineria, and *Fusarium*. Certain plant species can also be used as models to study the effect of toxins on plants as it was demonstrated with trichothecenes on the unicellular plant *Chlamydomonas reinhardtii* [52]. The production of CWDEs (e.g., cellulases, hemicellulases, xylanases, and pectinases) is one of the processes used by fungi to invade plants. CWDEs degrade the complex polymers, cellulose, hemicellulose, xylan, and pectin to gain access to the contents of the plant cell. The enzymatic action of the CWDEs crumbles the cell wall, its integrity is weakened and eventually it collapses. Coupled with toxin production, the production of CWDEs is part of the infection process of *F. graminearum* when it attacks wheat spikelets. These compounds have been reported to help the infection spread to other tissues [53]. Several pathogenicity factors identified in *Fusarium* are also part of preserved complexes or pathways, such as mitogen-activated protein kinases. However, specific genes of *Fusarium* involved in host-pathogen interactions such as HST, elicitors, or Avr genes are mainly undefined, apart from the effectors secreted in xylem. For the purposes of this review article, attention is focused on pathogenicity and virulence factors which are produced by *F. graminearum* when it infects its various hosts which are primarily grain crops.

3.1. Pathogenicity and Virulence Factors of Fusarium graminearum

F. graminearum causes FHB, a devastating disease of wheat and other crops which include maize (*Zea mays* L.) and barley (*Hordeum vulgare* L.) [1,54] (mentioned in Desjardins et al. [54]). The rate at which this fungus attacks and infects its host plant is dependent on different mechanisms involving the secretion of extracellular enzymes and mycotoxins. Wanjiru and colleagues [53] show that the pathogenicity of *F. graminearum* is dependent on the extracellular enzymes secreted. The fungus enters the host via the epidermal cell walls with the help of infection-hyphae, leading to the reduction of the host plant's cellulose, pectin, and xylan. Upon gaining access into the plant, *F. graminearum* produces various depolymerase enzymes such as CWDEs that macerate plant tissue. Jenczmionka and Schäfer [55] showed that the production of CWDEs such as pectinases, amylases, cellulases, xylanases, proteases, and lipases by *F. graminearum* enhances its penetration and proliferation in the attacked plant. The production of CWDEs is systematic, regulated, and programmed by the pathogen during attack. To prove regulation and co-ordination, Jenczmionka and Schäfer [55] showed that the *F. graminearum* Gpmk1 MAP kinase regulates the ability of the fungus to induce extracellular endoglucanase, xylanolytic as well as proteolytic activities.

With the disruption of the cell wall, *F. graminearum* uses the polysaccharides in the plant cell wall as a source of nutrients when advancing infection into the deeper underlying tissue [56]. These CWDEs have also been reported to elicit primary immune responses in plants [57–61]. According to Kikot and colleagues [62], various studies have shown a link between the presence of pectic enzymes, disease symptoms, and virulence, indicating CWDEs as decisive factors in the process of phytopathogenic fungi. Plant defense requires the pathogen to engage in a more sophisticated attack mode to overcome this defense. The production of CWDEs alone does not suffice, and therefore, other pathogenicity and virulence factors must come into play. Over and above CWDEs, *F. graminearum* attacks host plants through the release of mycotoxins which include primarily the trichothecene DON, also known as vomitoxin [63]. Sella and colleagues [64] reported that *F. graminearum* secretes lipase, DON, OS-2 (a stress-activated kinase) and Gpmk1 MAP kinases as essential pathogenicity and virulent factors necessary for the development of full disease symptoms in soybean seedlings, and OS-2 is involved in overcoming the resistance possessed by the soybean phytoalexin. Voigt and colleagues [65] observed that the lipase secreted by *F. graminearum* encoded by the FGL1 gene is a virulent factor that enhances the fungal pathogenicity against wheat and maize. Through functional annotation by gene deletion, Zhang and colleagues [66] also discovered the involvement of *FgNoxR* in conidiation, germination, sexual development and pathogenicity of *F. graminearum*. In another study, Jia and colleagues [67], through gene deletion and mutant analysis, identified a putative secondary metabolite biosynthesis gene cluster *fg3_54* responsible for cell-to-cell penetration. The invasiveness of the *fg3_54*-deleted *F. graminearum* strains is restored by fusaoctaxin A (an octapeptide), leading to the conclusion that fusaoctaxin A is a virulence factor necessary for cell-to-cell invasion of wheat by *F. graminearum*. Gardiner and colleagues [68] point out that DON is associated with fungal virulence in wheat and barley and plays a crucial role in the spread of FHB. Moreover, DON elicits various defense responses in wheat which include hydrogen peroxide production, programmed cell death [69]. Trichothecenes are secondary metabolites secreted by many fungi belonging to the genera *Fusarium*, *Myrothecium*, *Stachybotrys*, and *Trichoderma*. Ueno [70] classifies trichothecenes into four types based on the unique characteristics of their chemical structures. *Fusarium* spp. secrete mostly type A and type B trichothecenes and not type C and type D [71]. These two types are characterized by the presence of a carbonyl group attached to C-8 of the sequiterpenoid backbone of trichothecenes. The presence of additional 7, 8 epoxides differentiates type C from the other types. Type D contains a macrocyclic ring that connects C-4 and C-15 of the sequiterpenoid backbone. Not a matter of discussion in the current review is the formation of toxin glycosides, also called masked mycotoxins, which are formed by cereal crops which are infected with mycotoxins. Toxin glycosides may be converted to toxins during digestion by the consumers of grain. Of great concern is the fact that toxin glycosides may escape detection by routine laboratory procedures (taken from McCormick et al. [72]. *F. graminearum* secretes type B trichothecenes DON, 15-ADON and 3-ADON or NIV. Lanoseth and Elen [73] reported the presence of DON, 3-ADON and NIV in infected oats. In *F. graminearum*-infected barley, wheat and corn, DON is the most common toxin produced by the fungus [74]. Ueno [70] indicates that the mechanism of action of trichothecenes involves binding the toxin to the ribosomes and inhibiting the synthesis of proteins.

A wide variety of proteins plays essential roles in the synthesis of trichothecenes, and most are encoded by genes within a trichothecene biosynthetic cluster (*Tri* cluster) [75,76]. The *Tri* cluster encodes 13 proteins that are responsible for the secretion of toxins in *F. graminearum*. Jenczmionka and colleagues [77] reported the role of MAP kinase Gpmk1/MAP1 in the production of *F. graminearum* conidia, which implies that the removal of the enzyme could result in a reduction in conidia production and eventually disease spread and pathogenicity. Urban and colleagues [78] observed that in their study the mutants of Δgpmk1/map1 did not cause any disease to wheat, although they were able to produce DON in the infected plant. It was further discovered that Gpmk1 was involved in the regulation of CWDEs (endo-1,4-b-glucanase and other proteolytic, xylanolytic, and lipolytic enzymes) expression [55]. Ochiai and colleagues [79] reported the activities of a protein kinase cascade in trichothecene biosynthesis. According to the authors, the protein kinases are actively involved in histidine kinase signal transduction. Alteration of the genes leads to the reduction in the expression of *Tri6*, *Tri4* and NIV production in rice cultures. Perithecia, ascospore, conidia formation and spore germination are impacted by the activities of the serine/threonine-protein kinase gene (*GzSNF1*) [80]. *GzSNF1* is also involved in the expression of *F. graminearum* genes which encode for endo-1,4-ß-xylanase 1 precursor (GzXYL1), an endo-1,4-ß-xylanase 2 precursor (GzXYL2) and an extracellular ß-xylosidase (GzXLP) which are involved in depolymerization and plant cell wall degradation. Lee and colleagues [80] also report a reduction in the virulence of Δgzsnf1 mutants on barley, proving the importance of the GzSNF1 gene in pathogenicity and virulence. The docking and fusion of secretory vesicles are required for vegetative and sexual growth in fungi. In the process of docking and fusion, v-SNARE proteins are fused with cargo proteins into vesicles. SNARE proteins are soluble N-ethylmaleimide-sensitive fusion protein (NSF) attachment protein receptors.

Hong [81] reports that the fusion of the apposing membranes which belong to the transport intermediate together with the target compartment is sped up by the interaction between the membrane-integrated t-SNARE protein and v-SNARE protein. In *F. graminearum*, syntaxin-like t-SNARE proteins are encoded by *GzSYN1* and *GzSYN2* and the deletion of these genes results in decreased virulence of the fungus on barley [82]. Syntaxin a is a multidomain protein with a globular amino terminal domain, a SNARE domain, and a carboxyl terminal transmembrane domain. In Δgzsyn1 mutants, perithecia were found to be unevenly distributed despite retaining reproductive ability with a decrease in the radial hyphal growth in culture. The Δgzsyn2 mutants are sterile females that could act like males in outcrosses with Δmat1-2 strains, resulting in the normal formation of perithecia. Radial hyphal growth of Δgzsyn2 mutants looks like the wild-type strains although the mycelia of both Δgzsyn1 and Δgzsyn2 mutants are generally thinner compared to those of wild-type strains [82].

Secreted lipase in *F. graminearum* is encoded by Fgl1 genes and the alteration of these genes in the fungus can reduce the virulence of the fungus in wheat with symptoms not spreading beyond the spikelets adjacent to inoculated spikelets [65]. Voigt and colleagues [65] further reported that in their study the maize plant infected with Δfgl1 mutants showed slight symptoms and the disease severity estimation was lower in the cobs infected with the mutant compared to the wild-type. The ability of fungi to colonize plants and initiate disease formation is dependent on the transmembrane transport proteins. These protein transporters can also exhibit virulence activities via the provision of protection from poisonous secondary metabolites secreted by host plants or even the fungus itself [83]. Many studies have indicated the roles of multiple ATP-binding cassette transporters or major facilitator superfamily (MFS) proteins in virulence activities of the fungi such as *Botrytis cinerea, Cochliobolus carbonum, Cercospora kikuchii,* Fusaria, *Magnaporthe grisea, Mycosphaerella graminicola* and *Penicillium digitatum* [83–92]. The same applies to *F. graminearum*, ATP-Binding Cassette Transporter Gene *FgABCC9* was found to be involved in the fungus's pathogenicity towards wheat [93]. *F. graminearum* attacks its host plant through the secretion of mycotoxins characterized as virulence factors. These secreted virulence factors and the level of secretion depend on the presence of the relevant genes that facilitate their secretion. Presently, trichothecene mycotoxins have been identified as the major virulence factors associated with *F. graminearum* [94,95]. The trichothecene gene cluster has diverse allelic genes with some of the genes found outside of the main cluster [95]. The allelic variations in *fgTri8* are responsible for producing the DON derivatives [96]. Furthermore, allelic differences in *fgTri1* are linked to the formation of the alternate trichothecene NX-2, whereas *fgTri13* and *fgTri7* are responsible for the trichothecene NIV [97,98]. The genes associated with the trichothecene production are not the only genes discovered to be responsible for the pathogenicity and virulence of *F. graminearum*. *FGSG_04694*, a gene-encoding polyketide synthase (PKS2) responsible for mycelial growth and fungi virulence, was found in *F. graminearum* [99]. Gaffoor and colleagues [100] confirm the function of PKS2 as they observed decreased mycelial growth and virulence in a mutant PKS2 found in *F. graminearum*. PKS2 is an accessory gene with irregular patterns of conservation that is also found in *F. graminearum* responsible for secondary metabolism and virulence [101,102]. The genes within these accessory gene clusters include terpene synthase-encoding genes *FGSG_08181*, *FGSG_08182*, which encode a putative transcription factor and three putative cytochrome P450 genes *FGSG_17088*, *FGSG_08183* and *FGSG_08187* [101,102]. Harris and colleagues [103] observed the expression of these accessory genes in cereals when infected by the fungus, which implies that they are involved in the virulence of the fungus.

ZIF1 that is conserved in filamentous ascomycetes encodes a bZIP (basic leucine zipper) transcription factor. The absence of *FgZIF1* affects the virulence and reproduction ability of *F. graminearum*. Expression of *FgZIF1* in mozif1 of *M. oryzae* mutants shows clearly that this transcription factor is functionally conserved in *F. graminearum* and *M. oryzae* [104]. Topoisomerase I encoded by *TOP1* is another enzyme involved in the pathogenicity and conidia formation of fungi. The removal of *TOP1* in *F. graminearum* was found to reduce disease expression in infected spikelets and abolition of spore formation [105]. *FGSG_10057* encodes a Zn(II)2Cys6-type transcription factor in *F. graminearum* that enhances radial growth and virulence activity of the fungus in wheat [106]. Most of these factors that enhance virulence are proteins and these proteins need to be transported to the areas in the host plant where disease symptoms are expressed. Most recently, it was found that *F. graminearum* *FgCWM1* encodes a cell wall mannoprotein which plays a role in pathogenicity. *FgCWM1* mutants exhibit reduced pathogenicity in wheat [107]. Another recent discovery is that of a SNARE gene *FgSec22* which was found to be required for vegetative growth, pathogenesis and DON biosynthesis in *F. graminearum* [108]. Similarly, an *F. graminearum* mitochondrial gene mitochondrial gene *FgEch1* was found to be important for conidiation, DON production, and plant infection [109]. Another recently discovered pathogenicity and virulence factor is FgPEX4, which is important for development, cell wall integrity and pathogenicity [110]. Elucidation and complete delineation of the pathogenicity and virulence factors of *F. graminearum* must still be done by increasing efforts in whole-genome sequencing of strains from various parts of the world and by conducting in planta gene expression studies to assess genes which are differently expressed during the infection of a host plant by *F. graminearum*. Nowadays these can be conveniently done using NGS. The pathogenicity and virulence factors of *F. graminearum* can be conveniently summed under CWDEs, toxins, toxin biosynthesis genes, and other pathogenicity and virulence determinants. In this section, an elaborate explanation of the different pathogenicity and virulence factors was provided. Below we provide a convenient list of these pathogenicity and virulence factors under different topics, namely, CWDEs, toxins, toxin biosynthesis genes and other pathogenicity and virulence determinants (Table 1). The pathogenicity and virulence factors are therefore summarized under their respective headings. In trying to categorize pathogenicity and virulence factors there may be some ambiguities which may lead to unclear categorizations.

Table 1. List of pathogenicity and virulence factors of *Fusarium graminearum*.

Cell Wall Degrading Enzymes			
	Category/Type/Classification	**Function**	**At Least One Reference Where Mentioned**
Depolymerase enzyme	Degradative enzyme	Catalyses depolymerization reactions	[53]
Pectinase	Degradative enzyme	Breaks down pectin	[55]
Cellulase	Degradative enzyme	Breaks down cellulose	[55]
Extracellular endoglucanase	Degradative enzyme	Breaks down glucan	[55]
Endo-1,4-b-glucanase	Degradative enzyme	Breaks down glucan	[55]
Proteolytic enzyme	Degradative enzyme	Breaks down proteins	[55]
Xylanolytic enzyme	Degradative enzyme	Breaks down xylan	[55]
Lipolytic enzyme	Degradative enzyme	Breaks down lipids	[55]
Toxins			
Trichothecene NX-2	Type A trichothecene (toxin)	Toxicity	[97]
Deoxynivalenol (DON)	Type B trichothecene (toxin)	Toxicity	[2]
3-acetyldeoxynivalenol (3-ADON)	Type B trichothecene (toxin)	Toxicity	[11]
15-ADON	Type B trichothecene (toxin)	Toxicity	[11]
Nivalenol (NIV)	Type B trichothecene (toxin)	Toxicity	[11]
Fusaoctaxin A	Toxin	Responsible for cell-to-cell invasion of wheat by *F. graminearum*	[67]
Toxin biosynthesis genes			
Tri cluster genomic region	Trichothecene biosynthesis genes	Involved in the synthesis of trichothecene mycotoxins	[75]
Other pathogenicity and virulence determinants			
Gpmk1 MAP kinase	MAP kinase	Involved in mating, conidiation, and pathogenicity	[77]
OS-2	Stress activated kinase	Involved in conferring resistance to a soybean phytoalexin	[64]
FGL1	Lipase gene	Enhances the fungal pathogenicity against wheat and maize	[65]
FgNoxR	A regulatory subunit of NADPH oxidase	Involved in conidiation, sexual development and pathogenicity of *F. graminearum*	[66]

Table 1. *Cont.*

Cell Wall Degrading Enzymes			
	Category/Type/Classification	**Function**	**At Least One Reference Where Mentioned**
fg3_54	Putative secondary metabolite biosynthesis gene cluster	Responsible for cell-to-cell invasiveness	[67]
Protein kinase	Kinase cascade in trichothecene biosynthesis	Involved in trichothecene production	[79]
Histidine kinase	Kinase	Involved in trichothecene production	[79]
GzSNF1	Serine/threonine-protein kinase	Responsible sexual, asexual development, and virulence	[80]
GzXYL1	Endo-1,4-ß-xylanase 1 precursor gene	Involved in plant cell wall degradation	[80]
GzXYL2	Endo-1,4-ß-xylanase 2 precursor gene	Involved in plant cell wall degradation	[80]
GzXLP	Extracellular ß-xylosidase gene	Involved in depolymerization and plant cell wall degradation	[80]
v-SNARE protein	Vesicle related proteins SNARE	Interacts with t-SNARE to catalyze the fusion of the apposing membranes of the transport intermediate and the target compartment)	[82]
t-SNARE protein	Target membrane-related SNARE	Interacts with v-SNARE to catalyze the fusion of the apposing membranes of the transport intermediate and the target compartment	[82]
Syntaxin-like t-SNARE protein	Syntaxin-like membrane-integrated protein	Required for vegetative growth, sexual reproduction, and virulence in *Gibberella zeae.* Proteins are also encoded by *GzSYN1* and *GzSYN2* in *F. graminearum* that enhanced virulence of the fungus on barley	[82]
GzSYN1	Syntaxin-like SNARE gene	Enhances perithecia and radial hyphal growth	[82]
GzSYN2	Syntaxin-like SNARE gene	Enhances perithecia and radial hyphal growth	[82]
Multiple ATP-binding cassette transporter	Major facilitator superfamily of membrane transporter	Involved in virulence	[83]
FgABCC9	ATP-binding cassette transporter gene	involved in the fungal pathogenicity towards wheat	[93]

Table 1. *Cont.*

Cell Wall Degrading Enzymes			
	Category/Type/Classification	**Function**	**At Least One Reference Where Mentioned**
FGSG_04694	Gene-encoding polyketide synthase PKS2	Responsible for mycelial growth and fungi virulence	[99]
PKS2	Polyketide synthase gene	Responsible for secondary metabolism and virulence	[101]
FGSG_08181	Terpene synthase-encoding gene	Involved in the virulence of the fungus	[99]
FGSG_08182	Terpene synthase-encoding gene	Involved in the virulence of the fungus	[103]
FGSG_17088	Putative cytochrome P450 gene	Responsible for the expression of disease in fungus-infected cereals	[99]
FGSG_08183	Putative cytochrome P450 gene	Responsible for the expression of disease in fungus-infected cereals	[99]
FGSG_08187	Putative cytochrome P450 gene	Responsible for the expression of disease in fungus-infected cereals	[99]
ZIF1	Encodes bZIP transcription factor	Involved in virulence and reproduction ability of *F. graminearum*	[104]
bZIP transcription factor	Transcription factor	Enhances the virulence of *F. graminearum* in infected plants	[104]
TOP1 I	Enzyme	Involved in sporulation and pathogenicity	[105]
FGSG_10057	Conserved hypothetical protein	Involved in growth and virulence	[106]
Zn(II)2Cys6-type transcription factor	Transcription factor	Regulates fungal reproduction and pathogenicity	[106]

3.2. Comparative Genomics and Molecular Basis of Pathogenicity and Virulence in Fusarium graminearum

In order to offset the serious pathogenic problems caused by *F. graminearum* and to develop insights into the virulence and antagonistic defense mechanisms in host plants, it appears imperative to undertake the identification of fungal pathogenicity and virulence factors which make up the arsenal of this fungal plant pathogen. Presently this is conveniently done using the high-throughput genome sequencing technologies which generate large datasets to reveal genome organization and the various genes present in *F. graminearum*. The information generated from genome sequencing and RNA sequencing projects can be utilized to understand the mechanisms employed by *F. graminearum* to gain entry into plant tissue as well as advance with the plant. Various commendable projects have been launched in the previous years and have advanced our understanding of pathogenicity and virulence of *F. graminearum*. Among these significant projects was the MIPS *Fusarium graminearum* Genome Database (FGDB) with an estimated 14,000 genes and downstream analysis in a live gene validation process was established to provide a comprehensive genomic and molecular analysis [40]. Because *Fusarium* species are among the most important phytopathogenic and toxigenic fungi, it is essential to understand the molecular underpinnings of their pathogenicity. In addition, the production of mycotoxins by these fungi put animals and humans who consume the crop product at risk. Given this alarming situation, aggressive research in *F. graminearum* is imperative and it must take advantage of the presence of convenient tools to generate a multitude of data to elucidate the various pathogenicity and virulence processes of *F. graminearum*. Based on the comparative genomic analysis conducted on three phenotypically diverse species that include *F. graminearum*, it was revealed that among others this particular homothallic fungal species, *F. graminearum*, has shown a relatively narrow host range which includes important cereals [41]. It is therefore important to note that this pathogen is particularly notorious on wheat, barley, rice, oats by causing head blight or 'scab' and on maize causing mainly stalk and ear rot disease [2]. However, genomic analysis shows that this fungus may also infect other plant species without causing disease symptoms. Further genome analysis revealed that 67 gene clusters with significant enrichment of predicted secondary metabolites and with functional enzymes were shown to be expressed among which 30% with gene overexpression were likely in virulence [102]. While the exchange of genes between the core and supernumerary genomes bestows significant opportunities for adaptation and evolution on the organism, it appears reminiscent in *F. graminearum*, to the compartmentalization of genetic material where non-conserved regions are found at various places on the four core chromosomes [111]. Studies from comparative genomics indicate that these mobile pathogenicity chromosomes exist in most *Fusarium* species with lineage-specific genomic regions [41], nevertheless, the molecular foundation of pathogenicity in *F. graminearum* was shown to be closely associated with the MAP1 gene which is also responsible for the development of perithecia in the same fungal species [78]. Furthermore, 29 *F. graminearum* genes are rapidly evolving, in planta-induced and encode secreted proteins, strongly pointing toward effector function [112], implicating genomic footprints that can be used in predicting gene sets likely to be involved in host–pathogen interactions. In association with this, as forward and reverse genetics have improved our understanding of molecular mechanisms involved in pathogenesis, it was revealed that mitogen-activated protein kinase and cyclic AMP-protein kinase A cascades both regulate virulence in *Fusarium* species and it has been postulated that cell wall integrity might be necessary for invasive growth and/or resistance to plant defense compounds [113]. These snippets which have been discovered to give clues on the pathogenicity and virulence of *F. graminearum* necessitated a deeper and comprehensive interrogation of the genome of this fungal pathogen to uncover all pathogenicity and virulence genes. NGS was conveniently used to unravel various pathogenicity and virulence factors of *F. graminearum* to the benefit of *F. graminearum* researchers worldwide.

3.3. Pathogenicity and Virulence Factors of *Fusarium graminearum* Discovered Using NGS Technologies

Understanding the molecular mechanisms involved in fungal pathogens in plants has been accelerated over the past decade. Notably, NGS has contributed immensely towards the generation of vast datasets of genomes and transcriptomes. The availability of fungal genome sequences from a majority of plant fungal pathogens has contributed to these discoveries [28,29,114,115]. Genomics, transcriptomics, proteomics, and metabolomics approaches were introduced, allowing possible identification of genes, proteins, and metabolites of fungi in various artificial cultures and during infection of plants under different experimental conditions [116,117]. Fungal pathogens cause diseases in plants, resulting in tissue damage and disease due to pathogenicity and virulence factors which assist fungal pathogen survival and persistence [118]. The pathogens affect their host by adapting to their environment and secreting/producing pathogenicity-related toxins, pectic enzymes, and hormone-like compounds. These products can have devastating effects on the quality and yield of crops in the field and can also cause postharvest diseases [119]. The mechanisms involved in fungal pathogenesis in plants are therefore being studied broadly, to protect plants against diseases of economic importance. *F. graminearum* is one of the plant pathogens that affect grain cereal crops globally, causing different diseases in different crops [80,120,121]. The pathogen produces metabolites that are toxic or non-toxic, which enables it to manipulate the plant to acquire nutrition. Inhibition of pathogenicity and virulence factors enhanced by the pathogenic fungus results in the development of diseases in plants. Pathogenicity factors involved in plant-pathogen interactions have been investigated extensively in different plants, and genes, proteins, and metabolites have been identified [81,97,122–125], including those involved in response to *F. graminearum* infections [55,104,113,120,126,127]. Pathogenicity and virulence genes that have been largely identified belong to the trichothecene biosynthesis gene cluster, as described by Proctor and colleagues [95]. For *F. graminearum*, it is largely known that pathogenicity and virulence follow a path of germ tube emergence from conidia, production of cell wall-degrading enzymes and the production of trichothecene mycotoxins [53,94,95,128].

However, complete delineation of the infection process of *F. graminearum* requires sequencing and analysis of the entire fungal genome, conveniently and preferably using NGS. From the various efforts to study pathogenicity and virulence factors of *F. graminearum* using NGS, a few studies are worth noting. The first is the study of King and colleagues [115] (sequencing was done using the Illumina HiSeq 2000 sequencing platform) which provided a complete genome sequence from a combined genome analysis from various sources which had *F. graminearum* genome sequence information. Through the modification the gene model set, the FGRRES_17235_M gene was identified to be of particular interest because it is a virulence factor, it encodes for a cysteine-rich secretory protein, allergen V5/Tpx-1-related with CAP and signal peptide domains, with a previous link with plant pathogenesis proteins of the PR-1 family [129] and had been identified in the highly virulent *F. graminearum* strain CS3005 (gene ID: FG05_09548). Another gene 15917_M was identified as an endo-1,4-beta-xylanase enzyme, which hydrolyzes (1- > 4)-beta-D-xylosidic linkages in xylans, of the cell walls. The second study by Wang and colleagues [130] identified eight genes responsible for *F. graminearum*-wheat interactions. Three of the genes had already been identified in various studies [112,131]. Their gene annotation revealed largely polymer degrading function i.e., xylanase, catalase, protease and lipase. The third study was by Cuomo and colleagues [114], who identified a variety of pathogenicity and virulence factors belonging to the gene classes cutinases, pectate lyases, pectin lyases and other genes encoding secreted proteins. The fourth study reports, among the findings, the presence of 616 potential effector genes, 126 of which are expressed in a host-specific manner. This same study by Laurent and colleagues [132] which utilized the Illumina HiSeq 2000 identified 252 variants within the genic sequences and the intergenic sequences of *Tri* genes. *Tri* genes are involved in the production of type B trichothecenes. Given the mammoth task of elucidating pathogenicity and virulence factors of *F. graminearum*, increasing efforts in whole-genome sequencing of strains from various parts of the world is necessary. These efforts must be coupled with in planta gene expression studies to assess genes which are differently expressed during the infection of a host plant by *F. graminearum*. Traditionally, common techniques for gene expression studies included northern blotting, real-time PCR and microarrays. Nowadays most studies which could be done using northern blotting, real-time PCR and microarrays can be conducted conveniently using NGS. However, these relatively old techniques paved the way for NGS.

3.3.1. Notable Studies Which Paved Way for NGS

Before NGS technologies were commonplace, Northern blotting, real-time PCR, and microarrays were used to discover pathogenicity and virulence factors of fungi including *F. graminearum*. These studies paved the way for studies based on NGS. Some of the studies (reviewed in this section, none of which were performed in the past two years) utilized proteomic approaches and metabolomics. *F. graminearum* genome analysis reports that the pathogen consists of 1250 genes that encode secreted protein effectors [133]. These genes associated with fungal pathogenesis in vitro and in planta have been identified, and a large number are activated during infection. The *F. graminearum* classes of genes overlap with other plant-microbe interaction studies. These genes include the trichothecene gene cluster [95], bZIP transcription factor [104], syntaxin-like t-SNARE proteins [81], PKS [134], lipase [65], EBR1 [127], FgATG15 [135], among others. These genes play different roles in fungal pathogenesis. Among these types of genes, the *Tri* gene cluster has been characterized in the *F. graminearum* species complex, with the type B cluster being the most studied, due to their ability to cause diseases in both animals and humans [136]. In addition, proteins such as the five TRI proteins TRI1, FG00071; TRI3, FG03534; TRI4, FG03535; TRI14, FG03543; TRI101, FG07896 [137]; kinases [138] and FG00028, metallopeptidase MEP1; FG00060, KP4 killer toxin; FG00150, NADP-dependent oxidoreductase (COG2130); FG00192, peptidase S8 (pfam00082); FG00237, O-acyltransferase (pfam02458), among others [139], were found to be involved in *F. graminearum* pathogenesis [124,140,141]. The study by Dhokane et al. [124] demonstrates the interface between metabolomics and NGS. The role of the genes found to be involved in *F. graminearum* pathogenesis [124,140,141] has been identified and characterized using high throughput sequencing approaches, confirming their roles in fungal pathogenesis. NGS studies can employ either studying the genome or gene expression by means of transcriptomics. Both have been instrumental in discovering pathogenicity and virulence factors of *F. graminearum*.

3.3.2. *Fusarium graminearum* Pathogenesis-Related Genes Discovered Using RNA-Seq Transcriptomics

Many technologies have been employed for the detection, identification, and quantification of mycotoxins secreted by *F. graminearum* infections in grain cereals. A few studies, including those by Pasquali and Migheli [136], report the most important fungal mycotoxins belonging to the type B trichothecenes that are produced by the *Fusarium* spp. Identification of differentially expressed genes regulated by mycotoxins using transcriptomics is one of the approaches to identify and catalog pathogenicity and virulence factors in response to *F. graminearum* in grain cereals. Using comparative transcriptomic analysis, Walkowiak and colleagues [142] identified 1500 differentially expressed genes of two *F. graminearum* strains with 3-ADON and 15-ADON trichothecene toxin chemotypes. Furthermore, a whole-genome sequencing and comparative genomics study investigated four *Fusarium* strains and reported few pathogenicity and virulence genes [99]. These included the g8968 gene, which was predicted to contain the *Tri5* domain. The *Tri5* is a terpene synthase gene that catalyzes the first step of trichothecene biosynthesis in *F. graminearum* [95]. Furthermore, *Tri8* was also identified in this study, and was reported to have exhibited a high frequency of SNPs and indels. The importance of *Tri5* in pathogenicity and virulence was also supported in non-NGS studies. Boddu and colleagues [143] report *Tri5* encoding a DON enzyme and revealed that loss-of-function *F. graminearum tri5* mutants were unable to produce DON in wheat and barley. Similar results of the importance of *Tri5* were also observed in a study by Jonkers and colleagues [144] whereby the Wor1-like Protein Fgp1 regulated pathogenicity, toxin synthesis and reproduction in *F. graminearum*. The study predicted that the loss of mycotoxin accumulation alone may be enough to explain the associated loss of pathogenicity to wheat.

Using transcriptomic analyses, differentially expressed genes (DEGs) were identified in infected spikelets and rachis wheat samples following *F. graminearum* infections [145]. From the list of the DEGs identified, a few trichothecene biosynthesis genes of the *F. graminearum Tri* cluster were mostly upregulated in the pathogen when infecting the resistant near isogeneic lines (NILs). Interestingly, another transcriptomic study was conducted between three host plants infected with *F. graminearum* strain to identify DEGs during colonization [103]. The study discovered that some genes were only expressed in a specific host, and there was also a difference in the genes' functional categories identified in each host. In summary, the pathogenicity and virulence factors (listed in Section 3.3) of *F. graminearum* discovered using NGS technologies are provided in Table 2 below:

Table 2. List of pathogenicity and virulence factors of *Fusarium graminearum*, including genes discovered using comparative genomics methods.

Cell Wall Degrading Enzymes			
	Category/Type/Classification	**Function**	**At Least One Reference Where Mentioned**
15917_M	Endo-1,4-beta-xylanase enzyme	Hydrolyses (1- > 4)-beta-D-xylosidic linkages in xylans of the cell walls	[115]
Xylanase	Degradative enzyme	Bring about the disintegration of xylan and cell wall penetration	[55]
Protease	Degradative enzyme	Responsible for the breakdown of protein	[55]
Lipase	Degradative enzyme	Responsible for the breakdown of lipids	[55]
Cutinases	Degradative enzyme	Plays polymer degrading function	[114]
Pectate lyases	Degradative enzyme	Plays polymer degrading function	[114]
Pectin lyases	Degradative enzyme	Plays polymer degrading function	[114]
β-amylase protein	Degradative enzyme	Involved in *F. graminearum* pathogenesis	[138]
Metallopeptidase	Degradative enzyme	Involved in *F. graminearum* pathogenesis	[139]
Peptidase	Degradative enzyme	Involved in *F. graminearum* pathogenesis	[139]
Toxins			
Type B trichothecenes	Trichothecene mycotoxin	Toxicity	[2]
KP4 killer toxin	Toxic polypeptide	Toxicity	[146]
Genes for toxin biosynthesis			
TRI1 gene	*Tri* cluster gene	Involved in *F. graminearum* pathogenesis	[137]
FGRRES_17235_M	Virulence-related gene	Encodes cysteine-rich secretory protein, allergen V5/Tpx-1-related with CAP and PR-1 family	[115]
15917_M	Endo-1,4-beta-xylanase enzyme	Hydrolyses (1- > 4)-beta-D-xylosidic linkages in xylans of the cell walls	[115]
g8968 gene (predicted to contain the *Tri5* domain)	Pathogenicity and virulence gene	Predicted to contain the *Tri5* domain	[99]
Tri5	*Tri* cluster gene	Involved in trichothecene biosynthesis	[95]
Tri8	*Tri* cluster gene	Involved *Fusarium* trichothecene phytotoxicity	[96]

Table 2. *Cont.*

Cell Wall Degrading Enzymes			
	Category/Type/Classification	**Function**	**At Least One Reference Where Mentioned**
Pathogenicity and virulence proteins			
TRI3	Trichothecene biosynthesis protein	Involved in *F. graminearum* pathogenesis	[137]
TRI4	Trichothecene biosynthesis protein	Involved in *F. graminearum* pathogenesis	[137]
TRI101	Trichothecene biosynthesis protein	Involved in *F. graminearum* pathogenesis	[137]
Other pathogenicity and virulence determinants			
Hormone-like compounds	Compounds with hormone-like properties	Enhances the adaptation of the fungi to the host plant environment	[119]
PR-1 family proteins	Pathogenicity related protein	Involved in pathogenicity	[129]
Basic leucine zipper (bZIP) transcription factor	Transcription factor	Enhances the virulence and reproduction ability of *F. graminearum* in infected plants	[104]
Syntaxin-like t-SNARE proteins	Syntaxin-like membrane-integrated proteins	Required for vegetative growth, sexual reproduction, and virulence in *G. zeae*. Proteins are also encoded by *GzSYN1* and *GzSYN2* in *F. graminearum* that enhanced virulence of the fungus on barley	[82]
Polyketide synthase (PKS) gene	Polyketide synthase gene	Responsible for secondary metabolism and virulence	[101]
Enhanced branching 1 (EBR1)	Zn(2)Cys(6) transcription factor	Involved in *F. graminearum* pathogenesis	[127]
NADP-dependent oxidoreductase	Oxidoreductase enzyme	Involved in *F. graminearum* pathogenesis	[139]
O-acyltransferase	Transferase enzyme	Involved in *F. graminearum* pathogenesis	[139]
Wor1-like Protein Fgp1	Regulatory protein	Regulates pathogenicity, toxin synthesis, and reproduction in *F. graminearum*	[144]

3.4. *Fusarium Graminearum* Pathogenesis Proteins Discovered Using Proteomics Approaches

Proteins are macromolecular machines which undertake various biochemical functions either building blocks, transporters, enzymes, and other functions. Proteins functions are coordinated, they are intertwined with other constituents of organisms like genes, RNA, and metabolites.

Proteomics is a large-scale study of sets of proteins produced by organisms. The set of total proteins produced by organisms is termed the proteome. The proteome varies across cells, and to some extent, it is defined by the underlying transcriptome. Traditionally, proteins were studied using low-throughput methods which focus on a relatively small set of proteins and provide qualitative data on structure, function, and interaction which other cell constituents. The small windows of knowledge opened by these traditional techniques denied biochemists of a broader bigger of the entire proteome of a cell. From the traditional methods of studying proteins emerged proteomics, a large-scale study of the proteome which is able to provide a snapshot of total proteins in an organism. As opposed to gel-based and antibody-based methods of studying proteins, mass spectrometry (MS) has been utilized to produce large datasets on the proteome. The basic workflow followed in proteomics is the extraction from the tissue of total proteins, followed by trypsin digestion, separation chromatography of short peptides from the digestion, and mass analysis by MS. From then onwards is the identification of proteins in the studied sample and the generation of the protein list. Similar to NGS, proteomic studies have been accelerated by the invention of various instruments which perform both the separation of the digested peptides, mass analysis, and other downstream applications. The instruments had to meet a number of qualities which include high throughput and high confidence in the identification of peptides, notably Orbitrap and time-of-flight mass analyzers [147–150]. The common trend in improving the performance of the MSs was to create hybrid systems. The hybrid systems make use of different ion analyzers or separators to enhance the capability, quality and usefulness of the results obtained. The advent of a triple quadrupole instrument enhanced MS capability over a single quadrupole. With the triple quadrupole, data on m/z values are combined with data on molecule fragmentation patterns to improve accuracy. The fragmentation pattern data is made possible by the presence of the second quadrupole which acts as a collision cell.

Fungi produce proteins for pathogenicity and virulence. Usually, some of these proteins are secreted into the intercellular spaces of plant and may either degrade the cell wall or act as effectors to perform various other pathogenicity and virulence functions. The secreted proteins are part of what is called the secretome. For comprehensive studies of pathogenicity and virulence proteins, studying the whole-organism proteome becomes necessary and it is made possible using high-throughput proteomics instruments. Yang and colleagues [138] pointed out that the invention of "omics" and bioinformatics tools has enhanced the proteome analysis of phytopathogenic fungi and their host interactions. Paper and colleagues [139] identified 120 fungal proteins of *F. graminearum* which include CWDEs from infected wheat heads through vacuum filtration. Among the identified proteins, about 56% controlled putative secretion signals. Transcriptomics data can be complemented with other omics approaches such as proteomics. The functions of proteins expressed at a given time can be identified and understood using proteomics approaches [138,151]. Although high throughput sequencing technologies have been available for over a decade, identifying differentially expressed proteins involved in fungal pathogenicity in cereals has not been widely investigated. Several proteins have been identified and characterized in *F. graminearum* and associated infections. However, they focused on the secretome and the impact of DON [137,139,152]. When the expression of *F. graminearum* proteins was investigated in response to in vitro stimulation of biosynthesis of the mycotoxin, trichothecene, 130 *F. graminearum* proteins that showed changes in expression were reported [137]. Many of the proteins identified were involved in fungal virulence. Moreover, investigation of a secretome of *F. graminearum* annotated secreted pathogenesis proteins related to the KP4 killer toxin and gEgh 16 proteins, among others, which were associated with pathogenicity [146].

Recently, a study by Lu and Edwards [153] reported about 190 small secreted cysteine-rich proteins (SSCPs) found in the genome of *F. graminearum* using genome-wide analysis. From the list of the SSCPs reported, five belonging to the cysteine-rich secretory proteins, Antigen 5 and pathogenesis-related 1 proteins were established. These SSCPs were observed to contain homologies to proteins that have established crystal structures. The authors also maintained that previous studies had not reported these SSCP associations with pathogenicity or virulence in plants. Moreover, in planta expression patterns showed upregulation of nine proteins associated with pathogenesis. These proteins contain conserved domains of Ecp-2-like panels 1 and 4, CFEM-like panel 3, Kp4-like panels 10, 14 and 9, PR-1-like panel 11, hydrophobin-like panel 12 and glycol_61 family panel 13, which are linked to fungal pathogenicity. This is in line with a study by Paper and colleagues [137], who identified 229 in vitro and 120 in planta proteins secreted by *F. graminearum* during infection of a wheat head using a high-throughput MS/MS comparative study. The study reported that 49 in planta proteins were not present in vitro, indicating that fungal lysis occurred during pathogenesis. Rampitsch and colleagues [154], on the other hand, reported 29 proteins whose relative abundance was affected in their secretome following infection by *F. graminearum* using a comparative secretome analysis. These proteins included metabolic enzymes, proteins of unknown function and pathogenesis-related proteins. Other studies involved in the identification of *F. graminearum* pathogenesis-related proteins in vitro and in planta also include the PR-3 and PR-5 proteins [155]. Various forms of proteomics are significant in studying plant-pathogen relations and other factors which include elicitors. An example of this is phosphoproteomics which has to some extent been studied in necrotrophic pathogens like *B. cinerea* [156–158], *Septoria tritici* [159]. These studies need to be extended to *F. graminearum* to deepen the understanding of its pathogenicity and virulence.

4. Summary and Conclusions

The infection mechanism of the plant pathogenic fungus *F. graminearum* is complex and intricate. It involves the production of a germ tube from conidia, the production of CWDEs, and eventually, the production of mycotoxins, chiefly type B trichothecene mycotoxins, DON, and NIV. Within the context of this review article, we regarded the CWDEs and the mycotoxins as pathogenicity and virulence factors which enable *F. graminearum* to gain entry into the plant and advance within the interior of the infected tissue. Our focus was on pathogenicity and virulence factors so far discovered, and we also devoted attention to those discovered using NGS and, to a limited extent, proteomics. We conclude that a multitude of pathogenicity and virulence factors have been discovered, however, more work needs to be done taking advantage of NGS and its companion applications of proteomics. Discovery of more pathogenicity and virulence and factors may facilitate the newer methods of control of *F. graminearum* infection of wheat and DON accumulation, for instance, as it has been shown by Machado et al. [160] using RNAi. Progress on the use of RNAi may depend greatly on the discovery of more pathogenicity and virulence factors of *F. graminearum*.

Author Contributions: M.E.R. wrote parts on the use of next generation sequencing in discovering *F. graminearum* pathogenicity and virulence factors. U.V.O. wrote about pathogenicity and virulence factors. C.M.K. wrote about pathogenicity and virulence factors of *F. graminearum*. L.K.L. wrote about next generation sequencing in general. A.A.W. wrote about the significance of NGS in studying fungal pathogenicity and K.N. wrote the remaining parts including the abstract, the introduction and the conclusions, and organized the manuscript to be a coherent article. All the authors participated in the formulation of the concept. All authors have read and agreed to the published version of the manuscript.

Funding: This research received no external funding.

Conflicts of Interest: The authors declare no conflict of interest.

References

1. McMullen, M.P.; Jones, R.; Gallenberg, D. Scab of wheat and barley: Are-emerging disease of devastating impact. *Plant Dis.* **1997**, *81*, 1340–1348. [CrossRef] [PubMed]
2. Goswami, R.S.; Kistler, H.C. Heading for disaster: *Fusarium graminearum* on cereal crops. *Mol. Plant Pathol.* **2004**, *5*, 515–525. [CrossRef] [PubMed]

3. Lee, J.; Chang, I.; Kim, H.; Yun, S.; Leslie, J.F.; Lee, Y. Genetic diversity and fitness of *Fusarium graminearum* populations from rice in Korea. *Appl. Environ. Microbiol.* **2009**, *84*, 3285–3295. [CrossRef] [PubMed]
4. Dean, R.; van Kan, J.A.L.; Pretorius, Z.A.; Hammond-Kosack, K.E.; Di Pietro, A.; Spanu, P.D.; Rudd, J.J.; Dickman, M.; Kahmann, R.; Ellis, J.; et al. The top-10 fungal pathogens in molecular plant pathology. *Mol. Plant Pathol.* **2012**, *13*, 414–430. [CrossRef] [PubMed]
5. Kazan, K.; Gardiner, D.M. Transcriptomics of cereal-*Fusarium graminearum* interactions: What we have learned so far. *Mol. Plant Pathol.* **2018**, *19*, 764–778. [CrossRef]
6. Stack, R.W. Return of an old problem: *Fusarium* head blight of small grains. *Plant Health Prog.* **2000**, *1*, 19. [CrossRef]
7. Muriuki, J.G. Deoxynivalenol and Nivalenol in Pathogenesis of *Fusarium* Head Blight in Wheat. Ph.D. Thesis, University of Minnesota, Minneapolis, MN, USA, 2001.
8. Schmale, D.G.; Bergstrom, G.C. Fusarium Head Blight in Wheat. The Plant Health Instructor. Available online: http://www.apsnet.org/edcenter/intropp/lessons/fungi/ascomycetes/Pages/Fusarium.aspx (accessed on 1 October 2019).
9. Stack, R.W. History of *Fusarium* head blight with emphasis on North America. In *Fusarium Head Blight of Wheat and Barley*; Leonard, K.J., Bushnell, W.R., Eds.; APS Press: St. Paul, MN, USA, 2003; pp. 1–34.
10. Proctor, R.H.; Desjardins, A.E.; Plattner, R.D.; Hohn, T.M. A polyketide synthase gene required for biosynthesis of fumonisin mycotoxins in *Gibberella fujikuroi* mating population A. *Fungal Genet. Biol.* **1999**, *27*, 100–112. [CrossRef]
11. Tralamazza, S.M.; Bemvenuti, R.H.; Zorzete, P.; de Souza Garcia, F.; Corrêa, B. Fungal diversity and natural occurrence of deoxynivalenol and zearalenone in freshly harvested wheat grains from Brazil. *Food Chem.* **2016**, *196*, 445–450. [CrossRef]
12. Lee, T.; Lee, S.H.; Shin, J.Y.; Kim, H.K.; Yun, S.H.; Kim, H.Y.; Lee, S.; Ryu, J.G. Comparison of trichothecene biosynthetic gene expression between *Fusarium graminearum* and *Fusarium asiaticum*. *Plant Pathol. J.* **2014**, *30*, 33–42. [CrossRef]
13. Desjardins, A.E.; Hohn, T.M.; McCormick, S.P. Trichothecene biosynthesis in *Fusarium* species: Chemistry, genetics, and significance. *Microbiol. Mol. Biol. Rev.* **1993**, *157*, 595–604. [CrossRef]
14. Desjardins, A.E.; Hohn, T.M. Mycotoxins in plant pathogenesis. *Mol. Plant Microbe Interact.* **1997**, *10*, 147–152. [CrossRef]
15. Sanger, F.; Nicklen, S.; Coulson, A.R. DNA sequencing with chain-terminating inhibitors. *Proc. Natl. Acad. Sci. USA* **1977**, *74*, 5463–5467. [CrossRef] [PubMed]
16. Grada, A.; Weinbrecht, K. Next-generation sequencing: Methodology and application. *J. Investig. Dematol.* **2013**, *133*, 11–14. [CrossRef] [PubMed]
17. Bentley, D.R.; Balasubramanian, S.; Swerdlow, H.P.; Smith, G.P.; Milton, J.; Brown, C.G.; Hall, K.P.; Evers, D.J.; Barnes, C.L.; Bignell, H.R.; et al. Accurate whole human genome sequencing using reversible terminator chemistry. *Nature* **2008**, *456*, 53–59. [CrossRef] [PubMed]
18. Rothberg, J.M.; Hinz, W.; Rearick, T.M.; Schultz, J.; Mileski, W.; Davey, M.; Leamon, J.H.; Johnson, K.; Milgrew, M.J.; Edwards, M.; et al. An integrated semiconductor device enabling non-optical genome sequencing. *Nature* **2011**, *47*, 348–352. [CrossRef] [PubMed]
19. Vigliar, E.; Malapelle, U.; de Luca, C.; Bellevicine, C.; Troncone, G. Challenges and opportunities of next-generation sequencing: A cytopathologist's perspective. *Cytopathology* **2015**, *26*, 271–283. [CrossRef]
20. Li, J.; Batcha, A.M.; Grüning, B.; Mansmann, U.R. An NGS workflow blueprint for DNA sequencing data and its application in individualized molecular oncology. *Cancer Inform.* **2016**, *14*, 87–107. [CrossRef]
21. Kchouk, M.; Gibrat, J.F.; Elloumi, M. Generations of sequencing technologies: From first to next generation. *Biol. Med.* **2017**, *9*, 1–8. [CrossRef]
22. Margulies, M.; Egholm, M.; Altman, W.E.; Attiya, S.; Bader, J.S.; Bemben, L.A.; Berka, J.; Braverman, M.S.; Chen, Y.J.; Chen, Z.; et al. Genome sequencing in microfabricated high-density picolitre reactors. *Nature* **2005**, *437*, 376–380. [CrossRef]
23. Rothberg, J.M.; Leamon, J.H. The development and impact of 454 sequencing. *Nat. Biotechnol.* **2008**, *26*, 1117–1124. [CrossRef]
24. Malapelle, U.; Pisapia, P.; Passiglia, F.; Dendooven, A.; Pauwels, P.; Salgado, R.; Rolfo, C. Next–generation sequencing in clinical practice. In *Oncogenomics*; Dammacco, F., Silvestris, F., Eds.; Academic Press: Cambridge, MA, USA, 2019; Volume 4, pp. 57–64.

25. Koitzsch, U.; Heydt, C.; Attig, H.; Immerschitt, I.; Merkelbach-Bruse, S.; Fammartino, A.; Büttner, R.H.; Kong, Y.; Odenthal, M. Use of the GeneReader NGS System in a clinical pathology laboratory: A comparative study. *J. Clin. Pathol.* **2017**, *70*, 725–728. [CrossRef] [PubMed]
26. Salanoubat, M.; Genin, S.; Artiguenave, F.; Gousy, J.; Mangenot, S.; Arlat, M.; Billault, A.; Brottier, P.; Camus, J.C.; Cattolico, L.; et al. Genome sequence of the plant pathogen *Ralstonia solanacearum*. *Nature* **2002**, *415*, 497–502. [CrossRef] [PubMed]
27. Buell, C.R.; Joardar, V.; Lindeberg, M.; Selengut, J.; Paulsen, I.T.; Gwinn, M.L.; Dodson, R.J.; Deboy, R.T.; Durkin, A.S.; Kolonay, J.F.; et al. The complete genome sequence of the *Arabidopsis* and tomato pathogen *Pseudomonas syringae* pv. *tomato DC3000*. *Proc. Natl. Acad. Sci. USA* **2003**, *100*, 10181–10186. [CrossRef] [PubMed]
28. Galagan, J.E.; Calvo, S.E.; Borkovich, K.A.; Selker, E.U.; Read, N.D.; Jaffe, D.; FitzHugh, W.; Ma, L.J.; Smirnov, S.; Purcell, S.; et al. The genome sequence of the filamentous fungus *Neurospora crassa*. *Nature* **2003**, *422*, 859–868. [CrossRef] [PubMed]
29. Dean, R.A.; Tabolt, N.J.; Ebbole, D.J.; Farman, M.L.; Mitchell, T.K.; Orbach, M.J.; Thon, M.; Kulkarni, R.; Xu, J.R.; Pan, H.Q.; et al. The genome sequence of the rice blast fungus *Magnaporthe grisea*. *Nature* **2005**, *434*, 980–986. [CrossRef] [PubMed]
30. Wu, Q.; Ding, S.W.; Zhang, Y.; Zhu, S. Identification of viruses and viroids by Next-Generation Sequencing and homology-dependent and homology-independent algorithms. *Annu. Rev. Phytopathol.* **2015**, *53*, 425–444. [CrossRef]
31. Aylward, J.; Steenkamp, E.T.; Dreyer, L.L.; Roets, F.; Wingfield, B.D.; Wingfield, M.J. A plant pathology perspective of fungal genome sequencing. *IMA Fungus* **2017**, *8*, 1–15. [CrossRef]
32. Jones, S.; Baizan-Edge, A.; MacFarlane, S.; Torrance, L. Viral diagnostics in plants using Next generation sequencing: Computational analysis in practice. *Front. Plant Sci.* **2017**, *8*, 1770. [CrossRef]
33. Pecman, A.; Kutnjak, D.; Gutiérrez-Aguirre, I.; Adams, I.; Fox, A.; Boonham, N.; Ravnikar, M. Next generation sequencing for detection and discovery of plant viruses and viroids: Comparison of two approaches. *Front. Microbiol.* **2017**, *8*, 1998. [CrossRef]
34. Knief, C. Analysis of plant microbe interactions in the era of next generation sequencing technologies. *Front. Plant Sci.* **2014**, *5*, 216. [CrossRef]
35. Birren, B.; Fink, G.; Lander, E. *Fungal Genome Initiative: White Paper Developed by the Fungal Research Community*; Whitehead Institute Center for Genome Research: Cambridge, MA, USA, 2002.
36. The Fungal Genome Initiative Steering Committee. *A White Paper for Fungal Comparative Genomics*; Whitehead Institute: Cambridge, MA, USA, 2003.
37. Grigoriev, I.V.; Cullen, D.; Goodwin, S.B.; Hibbett, D.; Jeffries, T.W.; Kubicek, C.P.; Kuske, C.; Magnuson, J.K.; Martin, F.; Spatafora, J.W.; et al. Fuelling the future with fungal genomics. *Mycology* **2011**, *2*, 192–209.
38. Martin, F.; Cullen, D.; Hibbett, D.; Pisabarro, A.; Spatafora, J.W.; Bakerm, S.E.; Grigoriev, I.V. Sequencing the tree of life. *New Phytol.* **2011**, *190*, 818–821. [CrossRef] [PubMed]
39. Spatafora, J. 1000 fungal genomes to be sequenced. *IMA Fungus* **2011**, *2*, 41–45. [CrossRef]
40. Güldener, U.; Mannhaupt, G.; Münsterkötter, M.; Haase, D.; Oesterheld, M.; Stümpflen, V.; Mewes, H.W.; Adam, G. FGDB: A comprehensive fungal genome resource on the plant pathogen *Fusarium graminearum*. *Nucleic Acids Res.* **2006**, *34*, 456–458. [CrossRef]
41. Ma, L.J.; van der Does, H.C.; Borkovich, K.A.; Coleman, J.J.; Daboussi, M.J.; Di Pietro, A.; Dufresne, M.; Freitag, M.; Grabherr, M.; Henrissat, B.; et al. Comparative genomics reveals mobile pathogenicity chromosomes in *Fusarium*. *Nature* **2010**, *464*, 367–373. [CrossRef]
42. Yoder, O.C. Toxins in pathogenesis. *Ann. Rev. Phytopathol.* **1980**, *18*, 103–129. [CrossRef]
43. Sharma, A.K.; Dhasamana, N.; Dubey, N.; Kumar, N.; Gangwal, A.; Gupta, M.; Singh, Y. Bacterial virulence factors: Secreted for survival. *Indian J. Microbiol.* **2017**, *57*, 1–10. [CrossRef]
44. Speth, E.B.; Lee, Y.N.; He, S.Y. Pathogen virulence factors as molecular probes of basic plant cellular functions. *Curr. Opin. Plant Biol.* **2007**, *10*, 580–586. [CrossRef]
45. Duyvesteijn, R.G.; van Wijk, R.; Boer, Y.; Rep, M.; Cornelissen, B.J.; Haring, M.A. Frp1 is a *Fusarium oxysporum* F-box protein required for pathogenicity on tomato. *Mol. Microbiol.* **2005**, *57*, 1051–1063. [CrossRef]
46. Vadlapudi, V.; Naidu, K.C. Fungal pathogenicity of plants: Molecular approach. *Eur. J. Exp. Biol.* **2011**, *1*, 38–42.

47. Tsuge, T.; Harimoto, Y.; Akimitsu, K.; Ohtani, K.; Kodama, M.; Akagi, Y.; Egusa, M.; Yamamoto, M.; Otani, H. Host-selective toxins produced by the plant pathogenic fungus *Alternaria alternata*. *FEMS Microbiol. Rev.* **2013**, *37*, 44–66. [CrossRef] [PubMed]
48. Liu, Z.; Bos, J.I.; Armstrong, M.; Whisson, S.C.; da Cunha, L.; Torto-Alalibo, T.; Win, J.; Avrova, A.O.; Wright, F.; Birch, P.R.J.; et al. Patterns of diversifying selection in the phytotoxin-like scr74 gene family of *Phytophthora infestans*. *Mol. Biol. Evol.* **2005**, *22*, 659–672. [CrossRef] [PubMed]
49. Win, J.; Morgan, W.; Bos, J.; Krasileva, K.V.; Cano, L.M.; Chaparro-Garcia, A.; Ammar, R.; Staskawicz, B.J.; Kamoun, S. Adaptive evolution has targeted the C-terminal domain of the RXLR effectors of plant pathogenic oomycetes. *Plant Cell* **2007**, *19*, 2349–2369. [CrossRef] [PubMed]
50. Dong, S.; Stam, R.; Cano, L.M.; Song, J.; Sklenar, J.; Yoshida, K.; Bozkurt, T.O.; Oliva, R.; Liu, Z.; Tian, M.Y.; et al. Effector specialization in a lineage of the Irish potato famine pathogen. *Science* **2014**, *343*, 552–555. [CrossRef]
51. Stukenbrock, E.H.; McDonald, B.A. Geographical variation and positive diversifying selection in the host-specific toxin SnToxA. *Mol. Plant Pathol.* **2007**, *8*, 321–332. [CrossRef]
52. Alexander, N.J.; McCormick, S.P.; Ziegenhorn, S.L. Phytotoxicity of selected trichothecenes using *Chlamydomonas reinhardtii* as a model system. *Nat. Toxins* **1999**, *7*, 265–269. [CrossRef]
53. Wanjiru, W.M.; Zhensheng, K.; Buchenauer, H. Importance of cell wall degrading enzymes produced by *Fusarium graminearum* during infection of wheat heads. *Eur. J. Plant Pathol.* **2002**, *108*, 803–810. [CrossRef]
54. Desjardins, A.E.; Proctor, H.R.; Bairoch, A.; McCormick, S.P.; Shaner, G.; Buechley, G.; Hohne, T.M. Reduced virulence of trichothecene-nonproducing mutants of *Gibberella zeae* in wheat field tests. *Mol. Plant Microbe Interact.* **1996**, *9*, 775–781. [CrossRef]
55. Jenczmionka, N.J.; Schäfer, W. The *Gpmk1 MAP* kinase of *Fusarium graminearum* regulates the induction of specific secreted enzymes. *Curr. Genet.* **2005**, *47*, 29–36. [CrossRef]
56. Fernandez, J.; Marroquin-Guzman, M.; Wilson, R.A. Mechanisms of nutrient acquisition and utilization during fungal infections of leaves. *Annu. Rev. Phytopathol.* **2014**, *52*, 155–174. [CrossRef]
57. Ryan, C.A.; Farmer, E.E. Oligosaccharide signals in plants: A current assessment. *Annu. Rev. Plant Physiol. Plant Mol. Biol.* **1991**, *42*, 651–674. [CrossRef]
58. Felix, G.; Regenass, M.; Boller, T. Specific perception of subnanomolar concentrations of chitin fragments by tomato cells: Induction of extracellular alkalinization, changes in protein phosphorylation and establishment of a refractory state. *Plant J.* **1993**, *4*, 307–316. [CrossRef]
59. Zhang, B.; Ramonell, K.; Somerville, S.; Stacey, G. Characterization of early, chitin-induced gene expression in *Arabidopsis*. *Mol. Plant Microbe Interact.* **2002**, *15*, 963–970. [CrossRef] [PubMed]
60. Altenbach, D.; Robatzek, S. Pattern recognition receptors: From the cell surface to intracellular dynamics. *Mol. Plant Microbe Interact.* **2007**, *20*, 1031–1039. [CrossRef] [PubMed]
61. De Wit, P.J. How plants recognize pathogens and defend themselves. *Cell Mol. Life Sci.* **2007**, *64*, 2726–2732. [CrossRef] [PubMed]
62. Kikot, G.E.; Hours, R.A.; Alconada, T.M. Contribution of cell wall degrading enzymes to pathogenesis of *Fusarium graminearum*: A review. *J. Basic Microbiol.* **2009**, *49*, 231–241. [CrossRef] [PubMed]
63. Yazar, S.; Omurtag, G.Z. Fumonisins, trichothecenes and zearalenone in cereals. *Int. J. Mol. Sci.* **2008**, *9*, 2062–2090. [CrossRef]
64. Sella, L.; Gazzetti, K.; Castiglioni, C.; Schäfer, W.; Favaron, F. *Fusarium graminearum* possesses virulence factors common to *Fusarium* Head Blight of wheat and seedling rot of soybean but differing in their impact on disease severity. *Phytopathology* **2014**, *104*, 1201–1207. [CrossRef]
65. Voigt, C.A.; Schäfer, W.; Salomon, S. A secreted lipase of *Fusarium graminearum* is a virulence factor required for infection of cereals. *Plant J.* **2005**, *42*, 364–375. [CrossRef]
66. Zhang, C.; Lin, Y.; Wang, J.; Wang, Y.; Chen, M.; Norvienyeku, J.; Li, G.; Yu, W.; Wang, Z. FgNoxR, a regulatory subunit of NADPH oxidases, is required for female fertility and pathogenicity in *Fusarium graminearum*. *FEMS Microbiol. Lett.* **2016**, *363*. [CrossRef]
67. Jia, L.J.; Tang, H.Y.; Wang, W.Q.; Yuan, T.L.; Wei, W.Q.; Pang, B.; Gong, X.M.; Wang, S.F.; Li, Y.J.; Zhang, D.; et al. A linear nonribosomal octapeptide from *Fusarium graminearum* facilitates cell-to-cell invasion of wheat. *Nat. Commun.* **2019**, *10*, 922. [CrossRef] [PubMed]

68. Gardiner, S.A.; Boddu, J.; Berthiller, F.; Hametner, C.; Stupar, R.M.; Adam, G.; Muehlbauer, G.J. Transcriptome analysis of the barley-deoxynivalenol interaction: Evidence for a role of glutathione in deoxynivalenol detoxification. *Mol. Plant Microbe Interact.* **2010**, *23*, 962–976. [CrossRef] [PubMed]
69. Desmond, O.J.; Manners, J.M.; Stephens, A.E.; Maclean, D.J.; Schenk, P.M.; Gardiner, D.M.; Munn, A.L.; Kazan, K. The Fusarium mycotoxin deoxynivalenol elicits hydrogen peroxide production, programmed cell death and defence responses in wheat. *Mol. Plant Pathol.* **2008**, *9*, 435–445. [CrossRef] [PubMed]
70. Ueno, Y. Toxicological features of T-2 toxin and related trichothecenes. *Fundam. Appl. Toxicol.* **1984**, *4*, 124–132. [CrossRef]
71. Kimura, M.; Tokai, T.; Takahashi-Ando, N.; Ohsato, S.; Fujimura, M. Molecular and genetic studies of *Fusarium trichothecene* biosynthesis: Pathways, genes, and evolution. *Biosci. Biotechnol. Biochem.* **2007**, *71*, 2105–2123. [CrossRef]
72. McCormick, S.P.; Kato, T.; Maragos, C.M.; Busman, M.; Lattanzio, V.M.; Galaverna, G.; Dall-Asta, C.; Crich, D.; Price, N.P.; Kurtzman, C.P. Anomericity of T-2 toxin-glucoside: Masked mycotoxin in cereal crops. *J. Agric. Food Chem.* **2015**, *63*, 731–738. [CrossRef]
73. Lanoseth, W.; Elen, O. Differences between barley, oats and wheat in the occurrence of deoxynivalenol and other trichothecenes in Norwegian grain. *J. Phytopathol.* **1996**, *144*, 113–118. [CrossRef]
74. Rotter, B.A.; Prelusky, D.B.; Pestka, J.J. Toxicology of deoxynivalenol (vomitoxin). *J. Toxicol. Environ. Health* **1996**, *48*, 1–34. [CrossRef]
75. Kimura, M.; Tokai, T.; O'Donnell, K.; Ward, T.J.; Fujimura, M.; Hamamoto, H.; Shibata, T.; Yamaguchi, I. The trichothecene biosynthesis gene cluster of *Fusarium graminearum* F15 contains a limited number of essential pathway genes and expressed non-essential genes. *FEBS Lett.* **2003**, *539*, 105–110. [CrossRef]
76. Wong, P.; Walter, M.; Lee, W.; Mannhaupt, G.; Münsterkötter, M.; Mewes, H.W.; Adam, G.; Güldener, U. FGDB: Revisiting the genome annotation of the plant pathogen *Fusarium graminearum*. *Nucleic Acids Res.* **2011**, *39*, 637–639. [CrossRef]
77. Jenczmionka, N.J.; Maier, F.J.; Lösch, A.P.; Schäfer, W. Mating, conidiation and pathogenicity of *Fusarium graminearum*, the main causal agent of the head-blight disease of wheat, are regulated by the MAP kinase Gpmk1. *Curr. Genet.* **2003**, *43*, 87–95. [CrossRef] [PubMed]
78. Urban, M.; Mott, E.; Farley, T.; Hammond-Kosack, K. The *Fusarium graminearum* MAP1 gene is essential for pathogenicity and development of perithecia. *Mol. Plant Pathol.* **2003**, *4*, 347–359. [CrossRef] [PubMed]
79. Ochiai, N.; Tokai, T.; Nishiuchi, T.; Takahashi-Ando, N.; Fujimura, M.; Kimura, M. Involvement of the osmosensor histidine kinase and osmotic stress-activated protein kinases in the regulation of secondary metabolism in *Fusarium graminearum*. *Biochem. Biophys. Res. Commun.* **2007**, *363*, 639–644. [CrossRef] [PubMed]
80. Lee, S.H.; Lee, J.; Lee, S.; Park, E.H.; Kim, K.W.; Kim, M.D.; Yun, S.H.; Lee, Y.W. GzSNF1 is required for normal sexual and asexual development in the ascomycete *Gibberella zeae*. *Eukaryot. Cell* **2009**, *8*, 116–127. [CrossRef]
81. Hong, W. SNAREs and traffic. *Biochim. Biophys. Acta BBA Mol. Cell Res.* **2005**, *1744*, 120–144. [CrossRef]
82. Hong, S.Y.; So, J.; Lee, J.; Min, K.; Son, H.; Park, C.; Yun, S.H.; Lee, Y.W. Functional analyses of two syntaxin-like SNARE genes, GzSYN1 and GzSYN2, in the ascomycete *Gibberella zeae*. *Fungal Genet. Biol.* **2010**, *47*, 364–372. [CrossRef]
83. Stergiopoulos, I.; Zwiers, L.H.; De Waard, M.A. Secretion of natural and synthetic toxic compounds from filamentous fungi by membrane transporters of the ATP-binding cassette and major facilitator superfamily. *Eur. J. Plant Pathol.* **2002**, *108*, 719–734. [CrossRef]
84. Pitkin, J.W.; Panaccione, D.G.; Walton, J.D. A putative cyclic peptide efflux pump encoded by the ToxA gene of the plant-pathogenic fungus *Cochliobolus carbonum*. *Microbiology* **1996**, *142*, 1557–1565. [CrossRef]
85. Alexander, N.J.; McCormick, S.P.; Hohn, T.M. Tri12, a trichothecene efflux pump from *Fusarium sporotrichioides*: Gene isolation and expression in yeast. *Mol. Gen. Genet.* **1999**, *261*, 977–984. [CrossRef]
86. Callahan, T.M.; Rose, M.S.; Meade, M.J.; Ehrenshaft, M.; Upchurch, R.G. CFP, the putative cercosporin transporter of *Cercospora kikuchii*, is required for wild-type cercosporin production, resistance, and virulence on soybean. *Mol. Plant Microbe Interact.* **1999**, *12*, 901–910. [CrossRef]
87. Urban, M.; Bhargava, T.; Hamer, J.E. An ATP-driven efflux pump is a novel pathogenicity factor in rice blast disease. *EMBO J.* **1999**, *18*, 512–521. [CrossRef] [PubMed]

88. Zwiers, L.H.; De Waard, M.A. Characterization of the ABC transporter genes MgAtr1 and MgAtr2 from the wheat pathogen *Mycosphaerella graminicola*. *Fungal Genet. Biol.* **2000**, *30*, 115–125. [CrossRef]
89. Nakaune, R.; Hamamoto, H.; Imada, J.; Akutsu, K.; Hibi, T. A novel ABC transporter gene, PMR5, is involved in multidrug resistance in the phytopathogenic fungus *Penicillium digitatum*. *Mol. Genet. Genom.* **2002**, *267*, 179–185. [CrossRef] [PubMed]
90. Fleissner, A.; Sopalla, C.; Weltring, K.M. An ATP-binding cassette multidrug-resistance transporter is necessary for tolerance of *Gibberella pulicaris* to phytoalexins and virulence on potato tubers. *Mol. Plant Microbe Interact.* **2002**, *15*, 102–108. [CrossRef]
91. Schoonbeek, H.J.; van Nistelrooy, J.G.M.; de Waard, M.A. Functional analysis of ABC transporter genes from *Botrytis cinerea* identifies BcatrB as a transporter of eugenol. *Eur. J. Plant Pathol.* **2003**, *109*, 1003–1011. [CrossRef]
92. Skov, J.; Lemmens, M.; Giese, H. Role of a *Fusarium culmorum* ABC transporter (*FcABC1*) during infection of wheat and barley. *Physiol. Mol. Plant Pathol.* **2004**, *64*, 245–254. [CrossRef]
93. Qi, P.F.; Zhang, Y.Z.; Liu, C.H.; Zhu, J.; Chen, Q.; Guo, Z.R.; Wang, Y.; Xu, B.J.; Zheng, T.; Jiang, Y.F.; et al. *Fusarium graminearum* ATP-binding cassette transporter gene FgABCC9 is required for its transportation of salicylic acid, fungicide resistance, mycelial growth and pathogenicity towards wheat. *Int. J. Mol. Sci.* **2018**, *19*, 2351. [CrossRef] [PubMed]
94. Jansen, C.; von Wettstein, D.; Schäfer, W.; Kogel, K.H.; Felk, A.; Maier, F.J. Infection patterns in barley and wheat spikes with wildtype and trichodiene synthase gene disrupted *Fusarium graminearum* expressing a green fluorescence protein marker. *Proc. Natl. Acad. Sci. USA* **2005**, *102*, 16892–16897. [CrossRef]
95. Proctor, R.H.; McCormick, S.P.; Alexander, N.J.; Desjardins, A.E. Evidence that a secondary metabolic biosynthetic gene cluster has grown by gene relocation during evolution of the filamentous fungus *Fusarium*. *Mol. Microbiol.* **2009**, *74*, 1128–1142. [CrossRef]
96. McCormick, S.P.; Alexander, N.J. *Fusarium Tri8* encodes a trichothecene C-3 esterase. *Appl. Environ. Microbiol.* **2002**, *68*, 2959–2964. [CrossRef]
97. Varga, E.; Wiesenberger, G.; Hametner, C.; Ward, T.J.; Dong, Y.; Schöfbeck, D.; McCormick, S.; Broz, K.; Stückler, R.; Schuhmacher, R.; et al. New tricks of an old enemy: Isolates of *Fusarium graminearum* produce a type A trichothecene mycotoxin. *Environ. Microbiol.* **2015**, *17*, 2588–2600. [CrossRef] [PubMed]
98. Lee, T.; Han, Y.K.; Kim, K.H.; Yun, S.H.; Lee, Y.W. *Tri13* and *Tri7* determine deoxynivalenol-and nivalenol-producing chemotypes of *Gibberella zeae*. *Appl. Environ. Microbiol.* **2002**, *68*, 2148–2154. [CrossRef] [PubMed]
99. Walkowiak, S.; Rowland, O.; Rodrigue, N.; Subramaniam, R. Whole genome sequencing and comparative genomics of closely related Fusarium Head Blight fungi: *Fusarium graminearum*, *F. meridionale* and *F. asiaticum*. *BMC Genom.* **2016**, *17*, 1014. [CrossRef] [PubMed]
100. Gaffoor, I.; Brown, D.W.; Trail, F. Functional analysis of the polyketide synthase genes in the filamentous fungus *Gibberella zeae* (anamorph *Fusarium graminearum*). *Eukaryot. Cell* **2005**, *4*, 1926–1933. [CrossRef] [PubMed]
101. Hansen, F.T.; Gardiner, D.M.; Lysøe, E.; Fuertes, P.R.; Tudzynski, B.; Wiemann, P.; Sondergaard, T.E.; Giese, H.; Brodersen, D.E.; Sørensen, J.L. An update to polyketide synthase and non-ribosomal synthetase genes and nomenclature in *Fusarium*. *Fungal Genet. Biol.* **2015**, *75*, 20–29. [CrossRef]
102. Sieber, C.M.K.; Lee, W.; Wong, P.; Münsterkötter, M.; Mewes, H.W.; Schmeitzl, C.; Varga, E.; Berthiller, F.; Adam, G.; Güldener, U. The *Fusarium graminearum* genome reveals more secondary metabolite gene clusters and hints of horizontal gene transfer. *PLoS ONE* **2014**, *9*, e0110311. [CrossRef]
103. Harris, L.J.; Balcerzak, M.; Johnston, A.; Schneiderman, D.; Ouellet, T. Host-preferential *Fusarium graminearum* gene expression during infection of wheat, barley, and maize. *Fungal Biol.* **2016**, *120*, 111–123. [CrossRef]
104. Wang, Y.; Liu, W.; Hou, Z.; Wang, C.; Zhou, X.; Jonkers, W.; Ding, S.; Kistler, H.C.; Xu, J.R. A novel transcriptional factor important for pathogenesis and ascosporogenesis in *Fusarium graminearum*. *Mol. Plant Microbe Interact.* **2011**, *24*, 118–128. [CrossRef]
105. Baldwin, T.K.; Urban, M.; Brown, N.; Hammond-Kosack, K.E. A role for topoisomerase in *Fusarium graminearum* and *F. culmorum* pathogenesis and sporulation. *Mol. Plant Microbe Interact.* **2010**, *23*, 566–577. [CrossRef]

106. Dufresne, M.; van der Lee, T.; Ben M'Barek, S.; Xu, X.D.; Zhang, X.; Liu, T.G.; Waalwijk, C.; Zhang, W.; Kema, G.H.; Daboussi, M.J. Transposon-tagging identifies novel pathogenicity genes in *Fusarium graminearum*. *Fungal Genet. Biol.* **2008**, *45*, 1552–1561. [CrossRef]
107. Zhang, Y.Z.; Chen, Q.; Liu, C.H.; Lei, L.; Li, Y.; Zhao, K.; Wei, M.Q.; Guo, Z.R.; Wang, Y.; Xu, B.J.; et al. *Fusarium graminearum* FgCWM1 Encodes a Cell Wall Mannoprotein Conferring Sensitivity to Salicylic Acid and Virulence to Wheat. *Toxins* **2019**, *11*, 628. [CrossRef] [PubMed]
108. Adnan, M.; Fang, W.; Sun, P.; Zheng, Y.; Abubakar, Y.S.; Zhang, J.; Lou, Y.; Zheng, W.; Lu, G.D. R-SNARE FgSec22 is essential for growth, pathogenicity and DON production of *Fusarium graminearum*. *Curr. Genet.* **2019**, 1–15. [CrossRef] [PubMed]
109. Tang, L.; Yu, X.; Zhang, L.; Zhang, L.; Chen, L.; Zou, S.; Liang, Y.; Yu, J.; Dong, H. Mitochondrial FgEch1 is responsible for conidiation and full virulence in *Fusarium graminearum*. *Curr. Genet.* **2019**, 1–11. [CrossRef] [PubMed]
110. Zhang, L.; Wang, L.; Liang, Y.; Yu, J. FgPEX4 is involved in development, pathogenicity, and cell wall integrity in *Fusarium graminearum*. *Curr. Genet.* **2019**, *65*, 747–758. [CrossRef]
111. Waalwijk, C.; Vanheule, A.; Audenaert, K.; Zhang, H.; Warris, S.; van de Geest, H.; van der Lee, T. *Fusarium* in the age of genomics. *Trop. Plant Pathol.* **2017**, *42*, 184–189. [CrossRef]
112. Sperschneider, J.; Gardiner, D.M.; Thatcher, L.F.; Lyons, R.; Singh, K.B.; Manners, J.M.; Taylor, J.M. Genome-wide analysis in three *Fusarium* pathogens identifies rapidly evolving chromosomes and genes associated with pathogenicity. *Genome Biol. Evol.* **2015**, *7*, 1613–1627. [CrossRef]
113. Michielse, C.B.; van Wijk, R.; Reijnen, L.; Cornelissen, B.J.; Rep, M. Insight into the molecular requirements for pathogenicity of *Fusarium oxysporum* f. sp. lycopersici through large-scale insertional mutagenesis. *Genome Biol.* **2009**, *10*. [CrossRef]
114. Cuomo, C.A.; Guldener, U.; Xu, J.R.; Trail, F.; Turgeon, B.G.; Di Pietro, A.; Walton, J.D.; Ma, L.J.; Baker, S.E.; Rep, M.; et al. *Fusarium graminearum* genome reveals a link between localized polymorphism and pathogen specialization. *Science* **2007**, *317*, 1400–1402. [CrossRef]
115. King, R.; Urban, M.; Hammond-Kosack, M.C.; Hassani-Pak, K.; Hammond-Kosack, K.E. The completed genome sequence of the pathogenic ascomycete fungus *Fusarium graminearum*. *BMC Genom.* **2015**, *16*, 544–564. [CrossRef]
116. Chen, F.; Zhang, J.; Song, X.; Yang, J.; Li, H.; Tang, H.; Liao, Y.C. Combined metabolomic and quantitative real-time PCR analyses reveal systems metabolic changes of *Fusarium graminearum* induced by *Tri5* gene deletion. *J. Proteome Res.* **2011**, *10*, 2273–2285. [CrossRef]
117. Fall, L.A.; Salazar, M.M.; Drnevich, J.; Holmes, J.R.; Tseng, M.C.; Kolb, F.L.; Mideros, S.X. Field pathogenomics of *Fusarium* head blight reveals pathogen transcriptome differences due to host resistance. *Mycologia* **2019**, *111*, 563–573. [CrossRef] [PubMed]
118. Kubicek, C.P.; Starr, T.L.; Glass, N.L. Plant cell wall–degrading enzymes and their secretion in plant-pathogenic fungi. *Annu. Rev. Phytopathol.* **2014**, *52*, 427–451. [CrossRef] [PubMed]
119. Lo Presti, L.; Lanver, D.; Schweizer, G.; Tanaka, S.; Liang, L.; Tollot, M.; Zuccaro, A.; Reissmann, S.; Kahmann, R. Fungal effectors and plant susceptibility. *Annu. Rev. Plant Biol.* **2015**, *66*, 513–545. [CrossRef] [PubMed]
120. Cumagun, C.J.R.; Bowden, R.L.; Jurgenson, J.E.; Leslie, J.F.; Miedaner, T. Genetic mapping of pathogenicity and aggressiveness of *Gibberella zeae* (*Fusarium graminearum*) toward wheat. *Phytopathology* **2004**, *94*, 520–526. [CrossRef]
121. Kazan, K.; Gardiner, D.M.; Manners, J.M. On the trail of a cereal killer: Recent advances in *Fusarium graminearum* pathogenomics and host resistance. *Mol. Plant Pathol.* **2012**, *13*, 399–413. [CrossRef]
122. Alexander, N.J.; McCormick, S.P.; Waalwijk, C.; van der Lee, T.; Proctor, R.H. The genetic basis for 3-ADON and 15-ADON trichothecene chemotypes in *Fusarium*. *Fungal Genet. Biol.* **2011**, *48*, 485–495. [CrossRef]
123. Buerstmayr, H.; Ban, T.; Anderson, J.A. QTL mapping and marker assisted selection for *Fusarium* head blight resistance in wheat: A review. *Plant Breed.* **2009**, *128*, 1–26. [CrossRef]
124. Dhokane, D.; Karre, S.; Kushalappa, A.C.; McCartney, C. Integrated metabolo-transcriptomics reveals *Fusarium* Head Blight candidate resistance genes in wheat QTL-Fhb2. *PLoS ONE* **2016**, *11*, e0155851. [CrossRef]

125. Sørensen, J.L.; Sondergaard, T.E.; Covarelli, L.; Fuertes, P.R.; Hansen, F.T.; Frandsen, R.J.; Saei, W.; Lukassen, M.B.; Wimmer, R.; Nielsen, K.F.; et al. Identification of the biosynthetic gene clusters for the *lipopeptides fusaristatin* A and W493 B in *Fusarium graminearum* and *F. pseudograminearum*. *J. Nat. Prod.* **2014**, *77*, 2619–2625. [CrossRef]
126. Pan, Y.; Liu, Z.; Rocheleau, H.; Fauteux, F.; Wang, Y.; McCartney, C.; Ouellet, T. Transcriptome dynamics associated with resistance and susceptibility against *Fusarium* head blight in four wheat genotypes. *BMC Genom.* **2018**, *19*, 642. [CrossRef]
127. Zhao, C.; Waalwijk, C.; De Wit, P.J.; van der Lee, T.; Tang, D. EBR1, a novel Zn2Cys6 transcription factor, affects virulence and apical dominance of hyphal tip in *Fusarium graminearum*. *Mol. Plant Microbe Interact.* **2011**, *24*, 1407–1418. [CrossRef] [PubMed]
128. Seong, K.Y.; Zhao, X.; Xu, J.R.; Güldener, U.; Kistler, H.C. Conidial germination in the filamentous fungus *Fusarium graminearum*. *Fungal Genet. Biol.* **2008**, *45*, 389–399. [CrossRef] [PubMed]
129. Dixon, D.C.; Cutt, J.R.; Klessig, D.F. Differential targeting of the tobacco PR-1 pathogenesis-related proteins to the extracellular space and vacuoles of crystal idioblasts. *EMBO J.* **1991**, *10*, 1317–1324. [CrossRef] [PubMed]
130. Wang, Q.; Jiang, C.; Wang, C.; Chen, C.; Xu, J.R.; Liu, H. Characterization of the two-speed subgenomes of *Fusarium graminearum* reveals the fast-speed subgenome specialized for adaption and infection. *Front. Plant Sci.* **2017**, *8*, 140. [CrossRef] [PubMed]
131. Sperschneider, J.; Gardiner, D.M.; Taylor, J.M.; Hane, J.K.; Singh, K.B.; Manners, J.M.A. Comparative hidden Markov model analysis pipeline identifies proteins characteristic of cereal-infecting fungi. *BMC Genom.* **2013**, *14*, 807–829. [CrossRef] [PubMed]
132. Laurent, B.; Moinard, M.; Spataro, C.; Ponts, N.; Barreau, C.; Foulongne-Oriol, M. Landscape of genomic diversity and host adaptation in *Fusarium graminearum*. *BMC Genom.* **2017**, *18*, 203–229. [CrossRef]
133. Schmidt, S.M.; Panstruga, R. Pathogenomics of fungal plant parasites: What have we learnt about pathogenesis? *Curr. Opin. Plant Biol.* **2011**, *14*, 392–399. [CrossRef]
134. Garvey, G.S.; McCormick, S.P.; Alexander, N.J.; Rayment, I. Structural and functional characterization of TRI3 trichothecene 15-O-acetyltransferase from *Fusarium sporotrichioides*. *Protein Sci.* **2009**, *18*, 747–761.
135. Nguyen, L.N.; Bormann, J.; Le, G.T.; Stärkel, C.; Olsson, S.; Nosanchuk, J.D.; Giese, H.; Schäfer, W. Autophagy-related lipase FgATG15 of *Fusarium graminearum* is important for lipid turnover and plant infection. *Fungal Genet. Biol.* **2011**, *48*, 217–224. [CrossRef]
136. Pasquali, M.; Migheli, Q. Genetic approaches to chemotype determination in type B-trichothecene producing *Fusaria*. *Int. J. Food Microbiol.* **2014**, *189*, 164–182. [CrossRef]
137. Taylor, R.D.; Saparno, A.; Blackwell, B.; Anoop, V.; Gleddie, S.; Tinker, N.A.; Harris, L.J. Proteomic analyses of *Fusarium graminearum* grown under mycotoxin-inducing conditions. *Proteomics* **2008**, *8*, 2256–2265. [CrossRef] [PubMed]
138. Yang, F.; Jacobsen, S.; Jørgensen, H.J.; Collinge, D.B.; Svensson, B.; Finnie, C. *Fusarium graminearum* and its interactions with cereal heads: Studies in the proteomics era. *Front. Plant Sci.* **2013**, *4*, 37. [CrossRef] [PubMed]
139. Paper, J.M.; Scott-Craig, J.S.; Adhikari, N.D.; Cuomo, C.A.; Walton, J.D. Comparative proteomics of extracellular proteins *in vitro* and *in planta* from the pathogenic fungus *Fusarium graminearum*. *Proteomics* **2007**, *7*, 3171–3183. [CrossRef] [PubMed]
140. Bollina, V.; Kumaraswamy, G.K.; Kushalappa, A.C.; Choo, T.M.; Dion, Y.; Rioux, S.; Faubert, D.; Hamzehzarghani, H. Mass spectrometry-based metabolomics application to identify quantitative resistance-related metabolites in barley against *Fusarium* head blight. *Mol. Plant Pathol.* **2010**, *11*, 769–782. [CrossRef]
141. Kumaraswamy, G.K.; Bollina, V.; Kushalappa, A.C.; Choo, T.M.; Dion, Y.; Rioux, S.; Mamer, O.; Faubert, D. Metabolomics technology to phenotype resistance in barley against *Gibberella zeae*. *Eur. J. Plant Pathol.* **2011**, *130*, 29–43. [CrossRef]
142. Walkowiak, S.; Bonner, C.T.; Wang, L.; Blackwell, B.; Rowland, O.; Subramaniam, R. Intraspecies interaction of *Fusarium graminearum* contributes to reduced toxin production and virulence. *Mol. Plant Microbe Interact.* **2015**, *28*, 1256–1267. [CrossRef]
143. Boddu, J.; Cho, S.; Muehlbauer, G.J. Transcriptome analysis of trichothecene-induced gene expression in barley. *Mol. Plant Microbe Interact.* **2007**, *20*, 1364–1375. [CrossRef]

144. Jonkers, W.; Dong, Y.; Broz, K.; Kistler, H.C. The Wor1-like protein Fgp1 regulates pathogenicity, toxin synthesis and reproduction in the phytopathogenic fungus *Fusarium graminearum*. *PLoS Pathog.* **2012**, *8*, e1002724. [CrossRef]
145. Hofstad, A.N.; Nussbaumer, T.; Akhunov, E.; Shin, S.; Kugler, K.G.; Kistler, H.C.; Mayer, K.F.; Muehlbauer, G.J. Examining the transcriptional response in wheat Fhb1 near-isogenic lines to *Fusarium graminearum* infection and deoxynivalenol treatment. *Plant Genome* **2016**, *9*. [CrossRef]
146. Brown, N.A.; Antoniw, J.; Hammond-Kosack, K.E. The predicted secretome of the plant pathogenic fungus *Fusarium graminearum*: A refined comparative analysis. *PLoS ONE* **2012**, *7*, e33731. [CrossRef]
147. Marshall, A.G.; Hendrickson, C.L. High-resolution mass spectrometers. *Annu. Rev. Anal. Chem.* **2008**, *1*, 579–599. [CrossRef] [PubMed]
148. Yates, J.R.; Cociorva, D.; Liao, L.J.; Zabrouskov, V. Performance of a linear ion trap-Orbitrap hybrid for peptide analysis. *Anal. Chem.* **2006**, *78*, 493–500. [CrossRef] [PubMed]
149. Makarov, A.; Denisov, E.; Lange, O.; Horning, S. Dynamic range of mass accuracy in LTQ Orbitrap hybrid mass spectrometer. *J. Am. Soc. Mass Spectrom.* **2006**, *17*, 977–982. [CrossRef] [PubMed]
150. Hu, Q.; Noll, R.J.; Li, H.; Makarov, A.; Hardman, M.; Graham Cooks, R. The Orbitrap: A new mass spectrometer. *J. Mass Spectrom.* **2005**, *40*, 430–443. [CrossRef]
151. Mehta, A.; Brasileiro, A.C.; Souza, D.S.; Romano, E.; Campos, M.A.; Grossi-de-Sá, M.F.; Silva, M.S.; Franco, O.L.; Fragoso, R.R.; Bevitori, R.; et al. Plant–pathogen interactions: What is proteomics telling us? *FEBS J.* **2008**, *275*, 3731–3746. [CrossRef]
152. Zhou, W.; Eudes, F.; Laroche, A. Identification of differentially regulated proteins in response to a compatible interaction between the pathogen *Fusarium graminearum* and its host, *Triticum aestivum*. *Proteomics* **2006**, *6*, 4599–4609. [CrossRef]
153. Lu, S.; Edwards, M.C. Genome-wide analysis of small secreted cysteine-rich proteins identifies candidate effector proteins potentially involved in *Fusarium graminearum*–wheat interactions. *Phytopathology* **2016**, *106*, 166–176. [CrossRef]
154. Rampitsch, C.; Day, J.; Subramaniam, R.; Walkowiak, S. Comparative secretome analysis *of Fusarium graminearum* and two of its non-pathogenic mutants upon deoxynivalenol induction in vitro. *Proteomics* **2013**, *13*, 1913–1921. [CrossRef]
155. Geddes, J.; Eudes, F.; Laroche, A.; Selinger, L.B. Differential expression of proteins in response to the interaction between the pathogen *Fusarium graminearum* and its host, *Hordeum vulgare*. *Proteomics* **2008**, *8*, 545–554. [CrossRef]
156. Liñeiro, E.; Chiva, C.; Cantoral, J.M.; Sabidó, E.; Fernández-Acero, F.J. Phosphoproteome profile of *B. cinerea* under different pathogenicity stages by plant based elicitors. *J. Proteom.* **2016**, *139*, 84–94. [CrossRef]
157. Liñeiro, E.; Chiva, C.; Cantoral, J.M.; Sabidó, E.; Fernández-Acero, F.J. Modifications of fungal membrane proteins profile under pathogenicity induction: A proteomic analysis of *Botrytis cinerea* membranome. *Proteomics* **2016**, *16*, 2363–2376. [CrossRef]
158. Liñeiro, E.; Chiva, C.; Cantoral, J.M.; Sabidó, E.; Fernández-Acero, F.J. Dataset of the *Botrytis cinerea* phosphoproteome induced by different plant-based elicitors. *J. Proteom.* **2016**, *7*, 1447–1450. [CrossRef] [PubMed]
159. Yang, F.; Melo-Braga, M.N.; Larsen, M.R.; Jørgensen, H.J.; Palmisano, G. Battle through signaling between wheat and the fungal pathogen *Septoria tritici* revealed by proteomics and phosphoproteomics. *Mol. Cell. Proteom.* **2013**, *12*, 2497–2508. [CrossRef] [PubMed]
160. Machado, A.K.; Brown, N.A.; Urban, M.; Kanyuka, K.; Hammond-Kosack, K.E. RNAi as an emerging approach to control *Fusarium* head blight disease and mycotoxin contamination in cereals. *Pest Manag. Sci.* **2018**, *74*, 790–799. [CrossRef] [PubMed]

microorganisms

MDPI

Article

In Vitro Fumonisin Biosynthesis and Genetic Structure of *Fusarium verticillioides* Strains from Five Mediterranean Countries

Giovanni Beccari [1], Łukasz Stępień [2], Andrea Onofri [1], Veronica M. T. Lattanzio [3], Biancamaria Ciasca [3], Sally I. Abd-El Fatah [4], Francesco Valente [1], Monika Urbaniak [2] and Lorenzo Covarelli [1,*,†]

1 Department of Agricultural, Food and Environmental Sciences, University of Perugia, 06121 Perugia, Italy; giovanni.beccari@unipg.it (G.B.); andrea.onofri@unipg.it (A.O.); fv220@exter.ac.uk (F.V.)
2 Department of Pathogen Genetics and Plant Resistance, Institute of Plant Genetics, Polish Academy of Sciences, 60-479 Poznan, Poland; lste@igr.poznan.pl (Ł.S.); murb@igr.poznan.pl (M.U.)
3 National Research Council of Italy, Institute of Sciences of Food Production (ISPA-CNR), 70126 Bari, Italy; veronica.lattanzio@ispa.cnr.it (V.M.T.L.); biancamaria.ciasca@ispa.cnr.it (B.C.)
4 Food Toxins and Contaminants Department, National Research Centre, Cairo 12622, Egypt; simaged@yahoo.com
* Correspondence: lorenzo.covarelli@unipg.it; Tel.: +39-0755856464
† Current co-address: Centre for Crop and Disease Management, School of Molecular and Life Science, Curtin University, Bentley, Perth 6102, Australia.

Received: 24 January 2020; Accepted: 6 February 2020; Published: 11 February 2020

Abstract: Investigating the in vitro fumonisin biosynthesis and the genetic structure of *Fusarium verticillioides* populations can provide important insights into the relationships between strains originating from various world regions. In this study, 90 *F. verticillioides* strains isolated from maize in five Mediterranean countries (Italy, Spain, Tunisia, Egypt and Iran) were analyzed to investigate their ability to in vitro biosynthesize fumonisin B_1, fumonisin B_2 and fumonisin B_3 and to characterize their genetic profile. In general, 80% of the analyzed strains were able to biosynthesize fumonisins (range 0.03–69.84 μg/g). Populations from Italy, Spain, Tunisia and Iran showed a similar percentage of fumonisin producing strains (>90%); conversely, the Egyptian population showed a lower level of producing strains (46%). Significant differences in fumonisin biosynthesis were detected among strains isolated in the same country and among strains isolated from different countries. A portion of the divergent *FUM1* gene and of intergenic regions *FUM6-FUM7* and *FUM7-FUM8* were sequenced to evaluate strain diversity among populations. A high level of genetic uniformity inside the populations analyzed was detected. Apparently, neither geographical origin nor fumonisin production ability were correlated to the genetic diversity of the strain set. However, four strains from Egypt differed from the remaining strains.

Keywords: *Fusarium*; ear rot; maize; fumonisins; *FUM1*

1. Introduction

Fusarium verticillioides (Sacc.) Nirenberg is a member of the *Gibberella fujikuroi* species complex, also called *Fusarium fujikuroi* species complex (FFSC), a group of 40 closely related *Fusarium* species defined by morphological traits, sexual compatibility and DNA-based phylogenetic analysis [1,2].

In particular, *F. verticillioides* belongs to the "African" clade of the FFSC [3], and it is the main causal agent of *Fusarium* ear rot of maize (*Zea mays* L.) [4,5]. This fungus has been reported worldwide and, in particular, it prevails in drier and warmer climatic regions [6,7] such as those present in temperate, semitropical and tropical regions including European [4], Mediterranean [8], African [9] and Middle

Eastern [10] maize-growing areas. For example, *F. verticillioides* was the species isolated more frequently from maize kernels harvested in Italy [11–13], Spain [14–16], Egypt [17–21] and Iran [22]. This is also one of the species able to biosynthesize the secondary metabolites fumonisins [23]. Specifically, *F. verticillioides* is considered the main fumonisin producer; therefore, this is the most important species associated with fumonisin contamination of maize grains [24]. Fumonisins occur worldwide in maize, including Mediterranean [4,8,24,25] farming areas, where this is one of the most widely cultivated crops [26,27]. Fumonisin accumulation in maize grains can occur in the field, following preharvest infections, and possibly continue during grain storage [28].

Contaminations strongly impair maize grain quality because of the negative impact on animal and human health [29]. Fumonisin mycotoxins can be divided into four main groups, with the most abundant fumonisins found in nature included in the B group: fumonisin B_1 (FB_1), fumonisin B_2 (FB_2) and fumonisin B_3 (FB_3). Among B analogues, FB_1 is the most detected fumonisin in maize as well as the most toxicologically active [24,30]. In fact, after ingestion, fumonisins may cause a wide range of toxic effects, especially towards liver and kidneys [31–35]. For this reason, the European Commission has established maximum limits for the sum of FB_1 and FB_2 in maize for human consumption [36,37].

The amount of fumonisins found in maize kernels is also dependent on the toxigenic ability of the *F. verticillioides* populations occurring in a certain cultivated field or in a specific geographic area [38]. In fact, within the *F. verticillioides* species, fumonisin production commonly varies quantitatively because of the different strain abilities to biosynthesize different levels of these mycotoxins [15,24,39–41]. The amount of fumonisins produced may also vary in quantity depending on substrate [42], biotic and abiotic factors [43] as well as on the relative expression of the genes involved in the biosynthetic pathway [44]. In fact, fumonisin production in *F. verticillioides* is regulated by the *FUM* biosynthetic gene cluster [45], and some of the differences between strains can be explained by *FUM* gene sequence differences [46,47]. Thus, it is very important to determine the variations of fumonisin production by *F. verticillioides* to understand the biosynthetic potential of a certain population in a specific cultivation area.

The characterization of fumonisin biosynthesis by *F. verticillioides* strains isolated from different geographic areas has been often coupled to the study of the genetic structure of these populations to investigate the degree of genetic diversity between the different strains within the same species [44,48–50]. This can provide an important insight on the relationships, the variations and/or the similarities among strains originating from various regions as well as on the possible correlations between genetic variability and different fumonisin production [38,51–54]. Analyses of fumonisin biosynthesis and/or molecular characterization of *F. verticillioides* strains have been conducted in populations from different countries such as Argentina [55], Brazil [38,41,44,49], Italy [50], Iran [22,52], Ethiopia [53] and Nigeria [54].

A similar approach was adopted in the present work to characterize selected *F. verticillioides* strains originating from five Mediterranean countries to simultaneously compare them in a wider geographical context by evaluating their in vitro fumonisin production and genetic profile. Specifically, the main objectives of the present study were to:

(i) investigate the abilities of selected *F. verticillioides* strains isolated from maize kernels in five Mediterranean countries to in vitro biosynthesize FB_1, FB_2 and FB_3;
(ii) characterize the genetic structure of these selected strains to assess for possible variability within strains originating from each of the surveyed countries and between the strains originating from different countries.

2. Materials and Methods

2.1. Fungal Strains

A total of 90 *F. verticillioides* strains (Table 1) isolated from single maize kernels harvested from different fields in five Mediterranean countries (22 from Italy, 9 from Spain, 16 from Tunisia, 28 from

Egypt and 15 from Iran) were used in this study (Figure 1). Isolation operations were carried out in the country of origin where all strains were properly stored in fungal collections. The investigated strains had not been extensively subcultured, thus avoiding possible alterations in fumonisin production. Some of the Italian strains used in this work had been already investigated in a previous study [50] and were included to further characterize them in a wider geographical context (Figure 1).

Table 1. Strain ID, country of origin and fumonisin B_1, fumonisin B_2 and fumonisin B_3 production (μg/g) with standard errors (±SE) by *Fusarium verticillioides* strains isolated from maize kernels harvested in five Mediterranean countries and analyzed in this study.

Strain ID	Origin	Fumonisin Production (μg/g) *								
		Fumonisin B_1		Fumonisin B_2		Fumonisin B_3		Total Fumonisins **,§		
PG 21C	Italy	nd †	-	nd	-	nd	-	nd	-	-
PG 39B	Italy	nd	-	nd	-	nd	-	nd	-	-
ITEM 9313	Italy	0.03	(±0.01)	nd	-	nd	-	0.03	(±0.01)	a
ITEM 9319	Italy	0.16	(±0.08)	0.03	(±0.01)	0.05	(±0.02)	0.24	(±0.11)	ab
PG 60A1	Italy	0.20	(±0.02)	0.04	(±0.01)	0.05	(±0.01)	0.29	(±0.02)	b
ITEM 9330	Italy	0.30	(±0.08)	0.05	(±0.01)	0.06	(±0.01)	0.41	(±0.09)	ab
ITEM 9320	Italy	0.63	(±0.60)	0.10	(±0.10)	0.08	(±0.07)	0.81	(±0.77)	ab
ITEM 9300	Italy	0.65	(±0.37)	0.11	(±0.06)	0.11	(±0.05)	0.87	(±0.48)	ab
PG 28A	Italy	1.01	(±0.40)	0.22	(±0.10)	0.25	(±0.08)	1.49	(±0.58)	ab
ITEM 9318	Italy	1.03	(±0.68)	0.22	(±0.15)	0.35	(±0.24)	1.59	(±1.07)	ab
PG 22A	Italy	1.67	(±1.52)	0.24	(±0.23)	0.25	(±0.22)	2.16	(±1.97)	ab
PG 20A	Italy	2.81	(±1.50)	0.66	(±0.35)	0.40	(±0.16)	3.87	(±2)	abc
ITEM 9310	Italy	6.56	(±3.09)	2.46	(±1.19)	0.68	(±0.29)	9.69	(±4.56)	abcd
PG 5A	Italy	6.99	(±0.89)	2.35	(±0.37)	0.85	(±0.06)	10.19	(±1.27)	cd
ITEM 9309	Italy	7.70	(±3.45)	2.23	(±1)	0.80	(±0.30)	10.74	(±4.74)	abcd
PG 76A1	Italy	8.78	(±4.50)	2.32	(±1.29)	1.24	(±0.60)	12.34	(±6.39)	abcde
PG 30B	Italy	10.36	(±1.25)	2.95	(±0.45)	1.26	(±0.26)	14.57	(±1.92)	d
ITEM 9329	Italy	10.71	(±2.32)	3.04	(±0.71)	0.84	(±0.16)	14.59	(±3.16)	cd
PG 35A	Italy	13.30	(±6.96)	4.39	(±2.26)	1.78	(±0.80)	19.47	(±10)	abcde
PG 58A1	Italy	19.39	(±5.28)	7.51	(±1.73)	2.16	(±0.15)	29.07	(±7.05)	abcde
ITEM 10027	Italy	23.64	(±1.57)	7.22	(±0.44)	2.49	(±0.05)	33.35	(±1.99)	e
PG 36B	Italy	23.87	(±0.44)	5.63	(±1.56)	4.23	(±0.19)	33.73	(±1.49)	e
03-2/A	Spain	0.24	(±0.17)	nd	-	nd	-	0.24	(±0.17)	
FVMM 3-2	Spain	0.78	(±0.29)	0.03	(±0.03)	0.01	(±0.01)	0.82	(±0.33)	a
C1-2 SEV	Spain	2.24	(±1.19)	0.53	(±0.42)	0.01	-	2.77	(±1.61)	ab
FVMM 2-1	Spain	2.60	(±1.60)	0.55	(±0.46)	0.24	(±0.13)	3.38	(±2.17)	ab
FVMM AD 2-4	Spain	6.38	(±3.28)	1.61	(±0.91)	0.20	(±0.05)	8.19	(±4.19)	ab
03-5/B SEV.1	Spain	6.63	(±1.08)	1.31	(±0.31)	0.31	(±0.05)	8.24	(±1.43)	b
03-5/B SEV	Spain	7.70	(±3.57)	1.81	(±0.92)	1.06	(±0.66)	10.57	(±5.01)	ab
FVMM 1-1	Spain	15.63	(±4.19)	4.68	(±1.25)	1.77	(±0.33)	22.08	(±5.74)	ab
0-C-1-3 2/2	Spain	56.12	(±5.31)	10.67	(±1.35)	3.04	(±0.21)	69.84	(±6.57)	c
M16	Tunisia	nd	-	nd	-	nd	-	nd	-	-
M11	Tunisia	0.29	(±0.07)	0.04	(±0.02)	nd	-	0.33	(±0.09)	a
M19	Tunisia	0.30	(±0.07)	0.03	(±0.02)	0.11	(±0.03)	0.45	(±0.11)	a
M12	Tunisia	0.56	(±0.23)	0.12	(±0.05)	0.06	(±0.02)	0.74	(±0.30)	ab
M15	Tunisia	0.47	(±0.17)	0.06	(±0.02)	0.27	(±0.08)	0.80	(±0.28)	ab
M20	Tunisia	0.92	(±0.13)	nd	-	0.01	-	0.93	(±0.13)	b
M17	Tunisia	0.91	(±0.21)	0.12	(±0.03)	0.55	(±0.12)	1.58	(±0.36)	ab
M5	Tunisia	2.55	(±1.43)	0.27	(±0.26)	nd	-	2.83	(±1.69)	ab
M2	Tunisia	3.21	(±1.32)	0.61	(±0.31)	0.01	-	3.82	(±1.63)	ab
M8	Tunisia	3.53	(±1.80)	1.01	(±0.58)	0.01	-	4.55	(±2.39)	abc
M7	Tunisia	3.80	(±3.05)	0.77	(±0.75)	0.40	(±0.32)	4.97	(±4.11)	abc
M22	Tunisia	6.85	(±3.59)	1.15	(±0.40)	2.07	(±0.47)	10.07	(±4.45)	abc
M21	Tunisia	7.10	(±4.93)	1.47	(±1.24)	1.72	(±1.09)	10.29	(±7.24)	abc
M1	Tunisia	8.82	(±1.28)	2.16	(±0.35)	0.68	(±0.23)	11.66	(±1.81)	c
M14	Tunisia	10.50	(±0.10)	1.72	(±0.12)	1.07	(±0.10)	13.28	(±0.18)	c
M10	Tunisia	11.07	(±1.71)	2.48	(±0.55)	0.04	(±0.03)	13.59	(±2.23)	c

Table 1. *Cont.*

Strain ID	Origin	Fumonisin Production (μg/g) *								
		Fumonisin B_1		Fumonisin B_2		Fumonisin B_3		Total Fumonisins **,§		
F2	Egypt	nd	-	nd	-	nd	-	nd	-	-
F6	Egypt	nd	-	nd	-	nd	-	nd	-	-
F7	Egypt	nd	-	nd	-	nd	-	nd	-	-
F10	Egypt	nd	-	nd	-	nd	-	nd	-	-
F12	Egypt	nd	-	nd	-	nd	-	nd	-	-
F19	Egypt	nd	-	nd	-	nd	-	nd	-	-
F22	Egypt	nd	-	nd	-	nd	-	nd	-	-
F23	Egypt	nd	-	nd	-	nd	-	nd	-	-
F25	Egypt	nd	-	nd	-	nd	-	nd	-	-
F26	Egypt	nd	-	nd	-	nd	-	nd	-	-
F27	Egypt	nd	-	nd	-	nd	-	nd	-	-
F30	Egypt	nd	-	nd	-	nd	-	nd	-	-
F36	Egypt	nd	-	nd	-	nd	-	nd	-	-
F38	Egypt	nd	-	nd	-	nd	-	nd	-	-
F41	Egypt	nd	-	nd	-	nd	-	nd	-	-
F39	Egypt	0.22	(±0.02)	nd	-	nd	-	0.22	(±0.02)	a
F29	Egypt	0.81	(±0.05)	0.19	(±0.04)	0.12	(±0.03)	1.12	(±0.11)	b
F8	Egypt	0.96	(±0.90)	0.34	(±0.33)	nd	-	1.29	(±1.23)	ab
F4	Egypt	1.18	(±0.08)	0.10	(±0.02)	0.08	-	1.35	(±0.11)	b
F28	Egypt	1.08	(±0.69)	0.21	(±0.13)	0.09	(±0.05)	1.38	(±0.87)	ab
F9	Egypt	1.14	(±0.79)	0.15	(±0.13)	0.32	(±0.25)	1.61	(±1.17)	ab
F32	Egypt	1.11	(±0.34)	0.72	(±0.27)	0.38	(±0.20)	2.21	(±0.80)	ab
F5	Egypt	4.10	(±2.16)	0.70	(±0.40)	0.05	(±0.03)	4.85	(±2.60)	abc
F11	Egypt	3.56	(±1.88)	0.70	(±0.44)	0.58	(±0.37)	4.85	(±2.68)	abc
F17	Egypt	4.35	(±3.24)	2.03	(±1.57)	nd	-	6.38	(±4.81)	abc
F13	Egypt	6.02	(±1.45)	0.88	(±0.11)	0.33	(±0.12)	7.23	(±1.67)	abc
F15	Egypt	6.32	(±4.25)	1.29	(±0.98)	0.38	(±0.22)	7.99	(±5.45)	abc
F3	Egypt	7.52	(±0.08)	1.95	(±0.15)	1.75	(±0.15)	11.23	(±0.32)	c
35	Iran	nd	-	nd	-	nd	-	nd	-	-
4	Iran	0.03	(±0.02)	nd	-	nd	-	0.03	(±0.02)	a
25	Iran	0.10	(±0.02)	nd	-	nd	-	0.10	(±0.02)	b
2	Iran	0.27	(±0.08)	nd	-	nd	-	0.27	(±0.08)	ab
9	Iran	0.47	(±0.37)	nd	-	nd	-	0.47	(±0.37)	ab
18	Iran	1.21	(±0.25)	0.10	(±0.05)	0.09	(±0.04)	1.40	(±0.35)	abc
39	Iran	1.65	(±0.45)	0.19	(±0.18)	0.42	(±0.12)	2.26	(±0.73)	abc
56	Iran	2.21	(±1.12)	0.34	(±0.18)	0.30	(±0.16)	2.85	(±1.42)	abc
1	Iran	3.94	(±0.76)	0.56	(±0.18)	0.22	(±0.07)	4.72	(±1)	c
3	Iran	4.48	(±1.22)	0.76	(±0.22)	0.47	(±0.16)	5.71	(±1.59)	abc
22	Iran	4.61	(±1.38)	1.65	(±0.53)	nd	-	6.26	(±1.91)	abc
16	Iran	4.66	(±1.63)	1.48	(±0.58)	0.40	(±0.18)	6.55	(±2.39)	abc
5	Iran	9.92	(±5.52)	2.15	(±1.35)	1.17	(±0.71)	13.25	(±7.59)	abcd
7	Iran	13.65	(±4.74)	3.23	(±1.15)	1.45	(±0.50)	18.33	(±6.40)	abcd
89	Iran	30.81	(±4.39)	7.23	(±1.01)	1.75	(±0.28)	39.79	(±5.25)	d

* values represent the average (±SE) of three biological replicates. ** sum of fumonisin B_1, fumonisin B_2 and fumonisin B_3. † nd: not detected (<0.002 μg/g for fumonisin B_1 and <0.001 μg/g for fumonisin B_2 and fumonisin B_3). § within the same country of origin, means followed by different letters are significantly different ($p < 0.05$).

Figure 1. Countries of origin (red dots) of the *Fusarium verticillioides* strains used in this study. Map downloaded from www.google.com/maps and modified by the authors.

2.2. Confirmation of *F. verticillioides* Identity by PCR Assays

To preliminarily confirm the identity of the 90 *F. verticillioides* strains used in this study, species-specific PCR assays were conducted. All strains were grown on Potato Dextrose Agar (PDA (Biolife Italiana, Milan, Italy)) at 22 °C for 14 d in the dark. DNA was extracted as described by Beccari et al. [56,57]. PCR assays were carried out with the specific primers VERT1 (GTCAGAATCCATGCCAGAACG) and VERT2 (CACCCGCAGCAATCCATCAG) [58]. A single PCR protocol was optimized using a total reaction of 20 µL. Each reaction contained 9.2 µL of sterile water for molecular biology (5prime, Hilden, Germany), 1.5 µL of cresol red (Sigma-Aldrich, Saint Louis, MO, USA), 2 µL of 10X PCR buffer (Microtech, Pozzuoli, Naples, Italy), 1.2 µL of magnesium chloride (Microtech), 2 µL of 10 mM DNTP mix (Microtech), 1 µL of 10 µM forward and reverse primers, 0.1 µL of 5 U/µL Taq polymerase (Microtech) and 2 µL of template DNA. The PCR cycle consisted of an initial denaturation step at 94 °C for 2 min, followed by 30 cycles of denaturation (94 °C for 35 s), annealing (60 °C for 30 s), extension (72 °C for 2 min) and a final extension at 72 °C for 5 min. PCR assays contained a positive control (template DNA of *F. verticillioides*) and a negative control with no DNA added. The amplification was performed in a T-100 thermal cycler (Bio Rad, Hercules, CA, USA). All PCR fragments were separated by electrophoresis by applying a tension of 110 V for about 1 h. Electrophoretic runs were visualized using an UV Image analyzer (Euroclone, Pero, Milan, Italy).

2.3. Determiantion of Fumonisin Biosynthesis by *F. verticillioides* In Vitro

2.3.1. *F. verticillioides* Cultures

To determine in vitro fumonisin biosynthesis, cultures of *F. verticillioides* strains were obtained following the protocol indicated by Covarelli et al. [50] with slight modifications. In brief, 15 g of finely ground maize grains and 15 mL of deionized sterile water were added into 100 mL glass flasks (Duran, Mainz, Germany) to obtain the right moisture for allowing fungal development and then autoclaved three times at alternate days. Three flasks (replicates) for each *F. verticillioides* strain were then inoculated with a mycelium plug (0.6 cm diameter) taken from the growing edge of one-week-old pure fungal cultures of each strain developed on PDA at 22 °C in the dark. Three flasks (replicates) were used as controls by adding only a PDA plug. Flasks were incubated in the dark at 22 °C for 4 w, and developed cultures were then freeze-dried for 24 h using a lyophilizer instrument (Heto Powder Dry LL3000, Thermo Fisher Scientific, Waltham, MA, USA), ground with mortar and pestle and stored at −80 °C until further analysis.

2.3.2. Fumonisin Extraction and LC-MS/MS Analysis

Each fungal culture was extracted and analyzed in triplicate according to the validated and routine procedure also described by Covarelli et al. [50] with slight modifications. In brief, 1 g of ground sample was extracted with 5 mL of methanol/water (75:25, *v*/*v*) following 60 min shaking. The extract was filtered through filter paper. Prior to liquid chromatography, tandem Mass Spectrometry (LC-MS/MS) analysis, the extract was diluted by default 1:50 with a mixture of methanol/water (60:40), then filtered through 0.45 µm syringe filter. Twenty microliters were injected into the LC-MS/MS apparatus. If fumonisin levels were out of the calibration range, a further dilution (1:500 or 1:5000) was applied to the raw extract and then re-analyzed.

LC-MS/MS analyses were performed on a QTrap MS/MS system, from Applied Biosystems (Foster City, CA, USA), equipped with an Electrospray Ionization (ESI) interface and a 1100 series micro-Liquid Chromatography system comprising a binary pump and a micro-autosampler from Agilent Technologies (Waldbronn, Germany). The analytical column was a Gemini® C18 column (150 × 2 mm, 5 µm particles) (Phenomenex, Torrance, CA, USA), preceded by a Gemini® C18 guard column (4 × 2 mm, 5 µm particles). The column oven was set at 40 °C. The flow rate of the mobile phase was 200 µL/min, and the injection volume was 20 µL.

The column effluent was directly transferred into the ESI interface, without splitting. Eluent A was water and eluent B was methanol, both containing 0.5% acetic acid. A gradient elution was performed as follows. The percentage of eluent B was increased from 40% to 80% in 10 min, kept constant 3 min, then increased to 100% in 1 min, and kept constant for 4 min. The column was re-equilibrated with 40% eluent B for 7 min. The ESI interface was used in positive ion mode with the following settings: temperature 350 °C; curtain gas, nitrogen, 30 psi; nebulizer gas, air, 10 psi; heater gas, air, 30 psi; ion spray voltage +4500 V. The mass spectrometer operated in Multiple Reaction Monitoring (MRM) mode. Mycotoxin quantification was performed by external calibration in neat solvent. The identity of fumonisins was confirmed by comparison with the analytical standard considering chromatography retention time and MRM transitions (ion ratios) in agreement with the official guidelines for mycotoxin identification by Mass Spectrometry [59]. Detection limits in maize fungal cultures were 0.002 μg/g for FB_1 and 0.001 μg/g for FB_2 and FB_3.

Methanol (HPLC grade) and glacial acetic acid were purchased from Mallinckrodt Baker (Milan, Italy). Ultrapure water was produced by a Millipore Milli-Q system (Millipore, Bedford, MA, USA). Filter papers (Whatman no. 4) were obtained from Whatman International Ltd. (Maidstone, UK). HPLC syringe filters (regenerated cellulose, 0.45 μm) were from Alltech (Deerfield, IL, USA).

2.4. Genetic Structure of Different *F. verticillioides* Populations

For genetic diversity assessment, all *F. verticillioides* strains were cultured on PDA for 7 d. Mycelia were harvested, homogenized in liquid nitrogen, and genomic DNA was extracted using the method already described by Stępień et al. [60]. A pre-validated *FUM1*-specific marker that showed intraspecific polymorphism in *F. verticillioides* and *F. proliferatum* in previous studies [61,62] was used. Briefly, Fum1F1 (CACATCTGTGGGCGATCC)/Fum1R2 (ATATGGCCCCAGCTGCATA) primers were used for *FUM1* gene fragment PCR-based amplification and sequencing according to Waśkiewicz et al. [61]. Additionally, *FUM6-FUM7* and *FUM7-FUM8* intergenic regions were amplified using the primers Fum6eF (AGATTTCCCAACAGTGGCAG)/Fum7bR (GTTTGCTTGGTGGAACTGGT) and Fum7eF (ATCCGGTTGAGTTGGACAAG)/Fum8eR (GGAACAGATGCCCATACCAT) according to Stępień et al. [47].

The BigDye Terminator kit v. 3.1 (Life Technologies, Carlsbad, CA, USA) was used for fluorescent labeling according to the manufacturer's instructions. DNA fragments were purified using alkaline phosphatase and exonuclease I (Thermo Fisher Scientific)) and precipitated using ice-cold 96% ethanol (Sigma Aldrich, St. Louis, MO, USA). Sequence reading was performed using Applied Biosystems equipment. Sequence reads were analyzed using BioEdit software [63] and aligned using MEGA5 software package [64] using Maximum Parsimony heuristics with standard settings. Based on *FUM1* sequences, the most parsimonious tree was calculated (bootstrap test with 1000 replications).

Sequences were compared to the NCBI GenBank-deposited sequence (*FUM* cluster NCBI (AF155733)) and, in addition, a total of five *F. verticillioides FUM1* sequences (F.v.F1.8.I.I, F.v.10I3 (*Pisum sativum*, Wiatrowo, Poland); F.v.KF3477, F.v.F1M1.1 (*Z. mays*, Poland); F.v.KF3537 (*Ananas comosus*, Costa Rica)) were used as references. A total of four *Fusarium proliferatum FUM1* sequences (15 *F. proliferatum* (*Z. mays*, Iran); *F. proliferatum* Gar3.2, Gar1 and Gar3.0 (*Allium sativum*, Poznan, Poland)) were used as outgroup.

2.5. Statistical Analysis

To analyze the in vitro fumonisin biosynthesis within each country of origin, total fumonisin content was submitted to ANOVA by allowing a different standard deviation per strain to comply with heteroscedasticity. Generalized least-squares were used for model fitting, as implemented in the gls() function of the nlme package [65] within the R statistical environment [66]. Heteroscedastic Welch's *t*-tests were used for pairwise comparisons of strains, within country [67].

3. Results

3.1. Identity Confirmation of *F. verticillioides*

DNA extracted from the 90 *F. verticillioides* strains was subject to PCR assays using the species-specific primer pair VERT1/VERT2. As expected, a single fragment of 800 bp amplified in all the samples, thus confirming their identity as *F. verticillioides*.

3.2. Fumonisin Biosynthesis by *F. verticillioides* In Vitro

Data on the in vitro biosynthesis of FB_1, FB_2 and FB_3 with the calculation of total fumonisins (sum of FB_1, FB_2 and FB_3) by the 90 *F. verticillioides* strains are summarized in Table 1.

In general, this analysis revealed that 80% (n = 71) of the *F. verticillioides* strains investigated in this study were able to produce fumonisins at variable levels, while the remaining 20% (n = 19) showed undetectable levels (not detected; nd) of fumonisins and were considered, in this experimental condition, as non-producing strains.

Total fumonisins biosynthesized by all positive strains (n = 71) varied from 0.03 to 69.84 μg/g (average 7.88 μg/g), with FB_1 being the most abundant analogue followed by FB_2 and FB_3. All positive strains (100%, n = 71) produced FB_1 in levels ranging from 0.03–56.12 μg/g (average 5.9 μg/g), while 64 of 71 strains (90%) produced FB_2 in levels ranging from 0.03–10.67 μg/g (average 1.6 μg/g). Finally, 59 of 71 strains (83%) biosynthesized FB_3 in a range from 0.01–4.23 μg/g (average 0.7 μg/g). The average ratios of FB_1:total fumonisins, FB_2:total fumonisins and FB_3:total fumonisins were 0.77, 0.13 and 0.05, respectively. The three fumonisin analogues analyzed in this study (FB_1, FB_2 and FB_3) were simultaneously produced by 81% of positive strains (n = 58), while two analogues, FB_1 and FB_2 as well as FB_1 and FB_3, were simultaneously biosynthesized by 7% (n = 5) and 1% (n = 1) of positive strains, respectively. Finally, 7 out of 71 strains (10%) producerd only FB_1. No strains biosynthesized FB_2 or FB_3 only.

In most cases, considering all producing strains (n = 71), differences in fumonisin production were detected among the strains isolated in the same country.

In detail, 20 out of 22 strains (91%; Figure 2) isolated from maize grains in Italy and analyzed in this study showed the ability to biosynthesize fumonisins in variable levels (Table 1). Total fumonisins biosynthesized by the Italian positive strains (n = 20) varied from 0.03 to 33.73 μg/g (average 9.98 μg/g). All fumonisin-producing Italian strains (100%, n = 20) biosynthesized FB_1 in levels ranging from 0.03–23.87 μg/g (average 5.7 μg/g), while 19 out of 20 strains (95%) produced FB_2 and FB_3 in levels ranging from 0.03–5.63 μg/g (average 2.20 μg/g) and 0.05–4.23 μg/g (average 0.94 μg/g), respectively. The average ratios of FB_1:total fumonisins, FB_2:total fumonisins and FB_3:total fumonisins were 0.71, 0.18 and 0.10, respectively. The three fumonisin analogues (FB_1, FB_2 and FB_3) were simultaneously produced by 95% of positive Italian strains (n = 20), while 1 out of 20 strains (5%) produced only FB_1. Strains ITEM 10027 and PG 36B showed a significantly higher biosynthesis of total fumonisins with respect to the other Italian strains ($p < 0.02$), with the exception of strains PG 58A1, PG 35A and PG 76A1 ($p > 0.07$).

Considering the Spanish strains analyzed in this study, all of them (100%, n = 9; Figure 2) were able to in vitro biosynthesize different levels of fumonisins. Total fumonisins produced by these strains ranged from 0.24 to 69.84 μg/g (average 14.01 μg/g) with FB_1 being the most abundant (range 0.24–56.12 μg/g; average 10.9 μg/g), followed by FB_2 (range 0.03–10.67 μg/g; average 2.4 μg/g) and FB_3 (range 0.01–3.04 μg/g; average 0.7 μg/g). The average ratios of FB_1:total fumonisins, FB_2:total fumonisins and FB_3:total fumonisins were 0.81, 0.15 and 0.04, respectively. Eight out of 9 strains (89%) simultaneously biosynthesized all three fumonisin analogues, while in 1 out of 9 strains (11%) only FB_1 was detected. Strain 0-C-1–3 2/2 showed a significantly higher ($p < 0.008$) production of total fumonisins with respect to the other Spanish strains analyzed in this study.

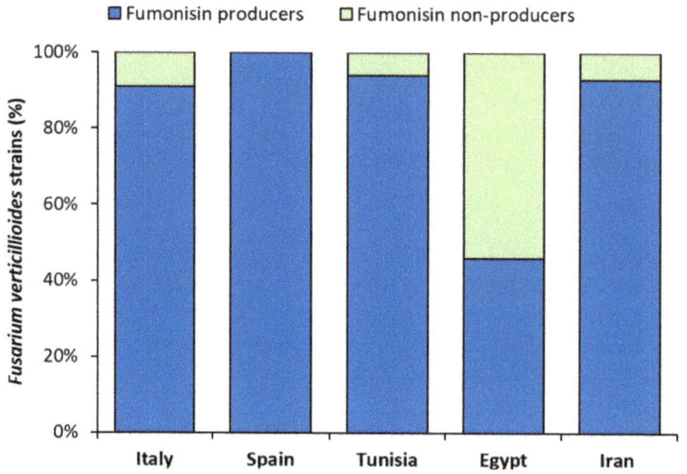

Figure 2. *Fusarium verticillioides* strains (%) isolated from maize kernels harvested in five Mediterranean countries that showed in vitro production of detectable (fumonisin producers) and non-detectable levels (fumonisin non-producers) of total fumonisins. Italy, $n = 22$; Spain, $n = 9$; Tunisia, $n = 16$; Egypt, $n = 28$; Iran, $n = 15$.

Focusing on the Tunisian strains analyzed in this study, 15 out of 16 strains (94%; Figure 2) produced detectable amounts of fumonisins *in vitro*. Total fumonisin levels ranged from 0.33 to 13.59 μg/g, with an average production equal to 5.36 μg/g. Twelve out of 15 strains (80%) biosynthesized all the analogues, while 2 out of 15 strains (13%) produced FB_1 and FB_2, and the remaining strain (7%; $n = 1$) produced FB_1 and FB_3. The gradient of production did not differ from that detected for the other strains: FB_1 (average 4.01 μg/g) > FB_2 (average 0.86 μg/g) > FB_3 (average 0.54 μg/g). The average ratios of FB_1:total fumonisins, FB_2:total fumonisins and FB_3:total fumonisins were 0.76, 0.13 and 0.11, respectively. Strains M10, M14 and M1 showed significantly higher total fumonisin biosynthesis with respect to the other Tunisian strains ($p < 0.02$), with the exception of strains M21, M22, M7 and M8 ($p > 0.05$).

The *F. verticillioides* population isolated from maize kernels in Egypt and analyzed in this study showed a low percentage of fumonisin-producing strains (46%, $n = 13$; Figure 2) with an average total fumonisin production of 3.98 μg/g (range 0.22–11.23 μg/g). All producing strains biosynthesized FB_1 (range 0.22–7.52 μg/g; average 2.95 μg/g), while 12 out of 13 strains (92%; average 0.77 μg/g) and 10 out of 13 strains (77%; average 0.40 μg/g) showed the ability to biosynthesize FB_2 and FB_3, respectively. In other words, 77% of producing strains were able to simultaneously produce all three fumonisin analogues, while 15% ($n = 2$) and 8% ($n = 1$) of the Egyptian strains showed the ability to biosynthesize FB_1 and FB_2 or FB_1 alone, respectively. The average ratios of FB_1: total fumonisins, FB_2:total fumonisins and FB_3:total fumonisins were 0.76, 0.17 and 0.09, respectively. The Egyptian strain F3 showed a significantly higher ($p < 0.01$) production of total fumonisins than F39, F29, F8, F4, F28, F9 and F32 strains.

In the *F. verticillioides* population isolated from maize kernels in Iran and anlyized in this study, a total of 14 fumonisin-producing strains were recovered (93%; Figure 2). Total fumonisins biosynthesized by all positive strains ($n = 14$) varied from 0.03 to 39.79 μg/g (average 7.28 μg/g). All producing Iranian strains (100%, $n = 14$) biosynthesized FB_1 in levels ranging from 0.03–30.81 μg/g (average 5.57 μg/g), while 11 out of 14 strains (71%) produced FB_2 in levels ranging from 0.1–7.23 μg/g (average 0.70 μg/g), and 10 out of 14 strains (64%) biosynthesized FB_3 in levels ranging from 0.09–1.75 μg/g (average 0.70 μg/g), respectively. The average ratios of FB_1:total fumonisins, FB_2:total fumonisins and FB_3:total fumonisins were 0.83, 0.14 and 0.07, respectively. The three fumonisin analogues (FB_1, FB_2 and FB_3)

were simultaneously produced by 64% of positive Iranian strains ($n = 9$), while 4 out of 14 strains (29%) produced only FB_1, and 1 out of 14 strains (7%) biosynthesized FB_1 and FB_2. The Iranian strain 89 showed a significantly higher total fumonisin biosynthesis than the other strains from the same country ($p < 0.01$), with the exception of strains 5 and 7 ($p > 0.05$).

Taking into account all fumonisin-producing strains of each country analyzed in this study, differences in total fumonisin biosynthesis among countries were also detected (Figure 3). In particular, the Spanish strains used in this study showed a significantly higher total fumonisin production (average 14.01 µg/g) than the Egyptian ones (average 3.98 µg/g) ($p = 0.02$). Also, the total fumonisin productions detected for the Italian (average 9.98 µg/g), Tunisian (average 5.36 µg/g) and Iranian (average 6.79 µg/g) strains were higher than the Egyptian ones and lower than the Spanish ones, even if no significant differences were recorded ($p > 0.46$ and $p > 0.47$, respectively) (Figure 3).

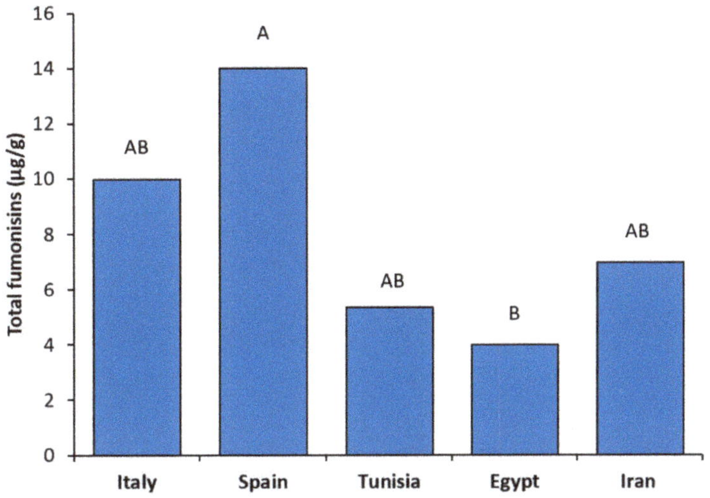

Figure 3. Average of total fumonisins (µg/g) biosynthesized by *Fusarium verticillioides* fumonisin-producing strains isolated from maize kernels harvested in each of the five countries analyzed in this study. Means with different letters are significantly different ($p < 0.05$).

3.3. *Genetic Structure and Variability of F. verticillioides Populations*

We sequenced a portion of a divergent *FUM1* gene to evaluate the diversity among the five populations of *F. verticillioides* originating from various countries. All strains amplified DNA fragments of about 1100 bp in length. Additionally, the *FUM6-FUM7* (ca. 550 bp) and *FUM7-FUM8* (ca. 500 bp) intergenic regions were sequenced using the primers described previously [47].

The sequences were aligned, the ends trimmed manually using MEGA 5 software, and dendrograms of similarities were calculated. Interestingly, the intergenic regions did not show polymorphisms, which was rather unexpected, since these regions normally accumulated more point mutations than the coding regions. However, this means that the *F. verticillioides* strains characterized in this study, even if originating from different countries, were basically uniform (results not shown).

Therefore, only slightly more polymorphic *FUM1* sequences were analyzed and shown (Figure 4). Apparently, neither geographical origin nor fumonisin production ability were correlated to the genetic diversity of the strain set, as almost all of them grouped together. Only four strains from Egypt (F10, F12, F13 and F36) were distinguished from the remaining strains at a bootstrap value of 60, including our five reference sequences [61] and NCBI GenBank-deposited *FUM* cluster sequences (AF155773) reported by Proctor et al. [45].

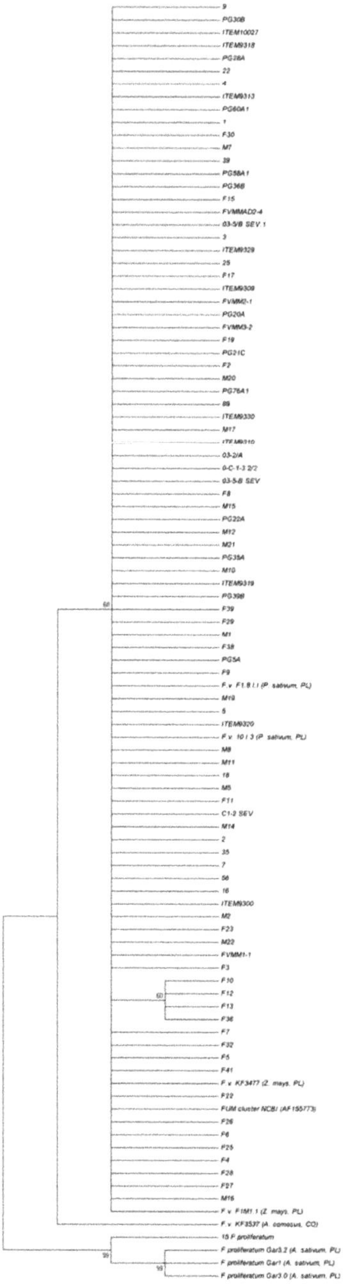

Figure 4. A most parsimonious tree calculated based on the partial *FUM1* sequences of 90 *Fusarium verticillioides* strains isolated from *Zea mays* of different origins using the maximum parsimony setting, bootstrap set to 50%, and 1000 replications were done. Five reference strains isolated from *Pisum sativum* (F.v. F1.8.I.I; F.v. 10 I 3), *Z. mays* (F.v. KF3477; F.v. F1M1.1) and *Ananas comosus* (F.v. KF3537) were added to the analysis, as well as the NCBI GenBank-deposited *FUM* cluster sequences (AF155773). Four *Fusarium proliferatum* sequences were also included as outgroup (15; Gar3.2; Gar1; Gar3.0).

4. Discussion

This study was aimed at investigating the different ability of selected *F. verticillioides* strains isolated from maize kernels harvested in five Mediterranean countries to in vitro biosynthesize fumonisins as well as at characterizing their genetic structure to assess possible variabilities among them. So far, various studies have been conducted to analyze the ability of different *F. verticillioides* strains from diverse geographic areas to biosynthesize fumonisins. In several investigations, a large percentage of strains able to produce detectable amounts of these mycotoxins were usually found. However, the presence of strains that were not able to biosynthesize measurable levels of fumonisins was also reported. In this research, the majority of the strains isolated from maize grains in Italy, Spain, Tunisia and Iran, analyzed in this study, produced detectable levels of fumonisins (91%, 100%, 94% and 94% respectively; Figure 2), while the remaining part showed a lack of ability to produce measurable amounts of these mycotoxins. Similar percentages of fumonisin-producing strains (> 80%) were also detected in other *F. verticillioides* populations isolated from maize in Croatia [68], Spain [15,69], Italy [50], Iran [22], Egypt [17], Brazil [41,44,49], Korea [70], USA [71], Argentina [55,72] and from durum wheat in Argentina [2].

Conversely, in this study, only 46% of the analyzed Egyptian strains showed the ability to biosynthesize detectable amounts of fumonisins (Figure 2). Similarly to other studies, low incidences of producing strains were also recorded in other *F. verticillioides* populations such as those isolated from maize in Croatia (55%) [73], Taiwan (66%) [74] and Spain (36%) [14].

In general, the producing strains analyzed in this study biosynthesized fumonisin analogues following the "typical" gradient: $FB_1 > FB_2 > FB_3$. A predominance of FB_1 compared to the other analyzed fumonisin analogues was recovered also in other *F. verticillioides* populations such as those isolated from maize in Spain [15,75], Italy [76], Iran [22], Brazil [44,49], Argentina [55,72], Egypt [17], South Korea and South Africa [39]. In this study, no *F. verticillioides* strains producing more FB_2 or FB_3 than FB_1 were recorded. Conversely, these types of strains were observed in *F. verticillioides* populations isolated from durum wheat in Argentina [2] and from maize and sorghum cultivated in the United States [77].

As known, fumonisin production within the *F. verticillioides* species could quantitatively vary due to the different biosynthetic ability of the different strains [24,40]. Also in this study, variability of fumonisin production among strains isolated in the same country was found, highlighting that mycotoxigenic diversity occurred within the five investigated *F. verticillioides* populations. Variability among *F. verticillioides* strains isolated from maize in the same country was commonly detected in many surveys in other parts of the world [2,8,15,17,22,44,49,55,73–75].

Variability in fumonisin production was also recorded among *F. verticillioides* strains isolated from different countries [30,39,71]. Also in this study, differences in fumonisin production among strains of different geographic origin were detected. In particular, the Spanish and Egyptian strains analyzed in this study showed a high level of mycotoxigenic variability, being the populations with the highest and the lowest fumonisin productions, respectively.

Interestingly, these two populations were also those with the highest and lowest percentages of fumonisin-producing (Spain) and non-producing (Egypt) strains. Conversely, the other three investigated populations of *F. verticillioides* (isolated from Italy, Tunisia and Iran) considered in this study did not show a significant variability of fumonisin production. In agreement with the results of Vogelgsang et al. [78], it is important to consider that in vitro results cannot be fully extrapolated to in vivo conditions because there are several factors influencing *Fusarium* infections and secondary metabolite production in the field. However, in vitro results could provide important information, which may be useful to understand intra-population variability within a single country as well as inter-population variability among different countries.

In this study, the mycotoxigenic characterization of *F. verticillioides* strains from different geographic origins was coupled to the study of the genetic structure of these populations. The genetic diversity of *F. verticillioides* has been studied using multiple techniques, including AFLP and RAPD methods [50,53,79].

Recently, however, direct sequencing of specific genomic regions has become more popular because of its high discrimination power and accuracy. The *FUM1* gene has already beeeen proven to be useful to assess species diversity inside the FFSC, serving as a source of phylogenetic and chemotypic markers [47], showing often higher levels of polymorphisms than constitutively expressed genes [e.g., *beta tubulin* (*tub2*) or *translation elongation factor 1α* (*tef-1α*)].

Our previous studies suggested there might be high levels of intraspecific genetic uniformity inside *F. verticillioides* populations, particularly when compared to the high diversity of the closely related species *F. proliferatum* [61,62,80,81]. The use of the *FUM1* gene sequence analysis allowed for discrimination of subpopulations likely related to the host species of origin. We assumed that a similar rule would be valid for *F. verticillioides*; therefore, we added some pea- and pineapple-derived strains to the analysis (Figure 4). It was also possible that geographical differences between populations would become visible.

However, in the present study we could not confirm this hypothesis. In fact, this was in accordance to previous findings, which did not reveal significant differences between *F. verticillioides* strains from different hosts [61]. This was also confirmed by the sequence analysis of the intergenic regions between *FUM6* and *FUM7* as well as *FUM7* and *FUM8* genes (results not shown), which were previously used for polymorphism screening [47]. The most likely explanation for this situation may be the endophytic type of growth observed for this pathogen in maize, which combined with the extensive seed material transfer between countries and continents made the population uniform across the world. Another possibility is that *FUM* cluster integrity and structure undergoes much more strict selection pressure in *F. verticillioides* than in *F. proliferatum*. This may implicate that fumonisin production by *F. verticillioides* is more essential to complete its life cycle than it is for *F. proliferatum*. This issue was already reported by Glenn et al. [82] but never confirmed for *F. proliferatum*.

The only outlier obtained in this study was a group of four strains (F10, F12, F13 and F36) isolated from Egypt (Figure 4), which was distinct from the remaining strains. Only one of these strains (F13) produced fumonisins in detectable amounts (Table 1). They should be further studied to explain their genetic diversity.

5. Conclusions

In this study, we analyzed fumonisin production as well as genetic structures of five *F. verticillioides* populations isolated from maize kernels in five Mediterranean countries.

The characterization of a selected number of strains per country does not allow a general conclusion to be drawn at the country level; however, the results obtained in these experimental conditions highlighted:

(i) the presence of an Egyptian population which differed from the others for its low percentage of fumonisin-producing strains;
(ii) the presence of significant differences in fumonisin production within the strains isolated in each of the surveyed countries and, in some cases, also among populations isolated from different countries;
(iii) the high level of genetic uniformity inside the populations analyzed;
(iv) the general absence of correlation between geographical origin and/or fumonisin production ability with the genetic diversity of the strain set;
(v) the presence of four Egyptian strains that were distinguished from the other strains at a bootstrap value of 60.

Author Contributions: Conceptualization, L.C.; validation, G.B. and Ł.S.; formal analysis, Ł.S., A.O. and V.M.T.L.; investigation, G.B., Ł.S., V.M.T.L., B.C., S.I.A.-E.F., F.V. and M.U.; resources, Ł.S., V.M.T.L. and L.C.; data curation, G.B. and Ł.S.; writing—original draft preparation, G.B.; writing—review and editing, Ł.S., V.M.T.L. and L.C.; visualization, G.B.; supervision, L.C. All authors have read and agree to the published version of the manuscript.

Funding: This research received no external funding.

Acknowledgments: The authors wish to thank Antonio Moretti (National Research Council, ISPA-CNR, Bari, Italy) for providing the Italian strains from the ITEM collection, Vicente Sanchis (University of Lleida, Spain) for providing the strains from Spain, Souheib Oueslati (Université Libre de Tunis, Tunisia) for providing the strains from Tunisia and Younes Rezaee Danesh (Urmia University, Urmia, Iran) for providing the strains from Iran.

Conflicts of Interest: The authors declare no conflicts of interest.

References

1. O'Donnell, K.; Nirenberg, H.I.; Aoki, T.; Cigelnik, E. A multigene phylogeny of the *Gibberella fujikuroi* species complex: detection of additional phylogenetic distinct species. *Mycoscience* **2000**, *41*, 61–78. [CrossRef]
2. Palacios, S.A.; Susca, A.; Haidukowski, M.; Stea, G.; Cendoya, E.; Ramírez, M.L.; Chulze, S.N.; Farnochi, M.C.; Moretti, A.; Torres, A.M. Genetic variability and fumonisin production by *Fusarium proliferatum* isolated from durum wheat grains in Argentina. *Int. J. Food Microbiol.* **2015**, *18*, 35–41. [CrossRef]
3. Kvas, M.; Marasas, W.F.O.; Wingfield, B.D.; Wingfield, M.J.; Steenkamp, E.T. Diversity and evolution of *Fusarium* species in the *Gibberella fujikuroi* complex. *Fungal Divers.* **2009**, *34*, 1–21.
4. Logrieco, A.; Mulè, G.; Moretti, A.; Bottalico, A. Toxigenic *Fusarium* species and mycotoxins associated with maize ear rot in Europe. *Eur. J. Plant Pathol.* **2002**, *108*, 597–609. [CrossRef]
5. Folcher, L.; Jarry, M.; Weissenberger, A.; Gérault, F.; Eychenne, N.; Delos, M.; Regnault-Roger, C. Comparative activity of agrochemical treatments on mycotoxin levels with regard to corn borers and *Fusarium* mycoflora in maize (*Zea mays* L.) fields. *Crop Prot.* **2009**, *28*, 302–308. [CrossRef]
6. Bottalico, A. *Fusarium* diseases of cereals: Species complex and related mycotoxins profiles, in Europe. *J. Plant Pathol.* **1998**, *80*, 85–103.
7. Oldenburg, E.; Höppner, F.; Ellner, F.; Weinert, J. *Fusarium* disease of maize associated with mycotoxin contamination of agricultural products intended to be used for food and feed. *Mycotoxin Res.* **2017**, *33*, 167–182. [CrossRef]
8. Marín, S.; Ramos, A.J.; Cano-Sancho, G.; Sanchis, V. Reduction of mycotoxins and toxigenic fungi in the Mediterranean basin maize chain. *Phytopathol. Mediterr.* **2012**, *51*, 93–118.
9. Fandohan, P.; Hell, K.; Marasas, W.F.O.; Wingfield, M.J. Infection of maize by *Fusarium* species and contamination with fumonisin in africa. *Afr. J. Biotechnol.* **2003**, *2*, 570–579.
10. Mirzadi Gohari, A.M.; Javan-Nikkhah, M.; Hedjaroude, G.A.; Abbasi, M.; Rahjoo, V.; Sedaghat, N. Genetic diversity of *Fusarium verticillioides* isolates from maize in Iran based on vegetative compatibility grouping. *J. Plant Pathol.* **2008**, *90*, 113–116.
11. Covarelli, L.; Beccari, G.; Salvi, S. Infection by mycotoxigenic fungal species and mycotoxin contamination of maize grain in Umbria, central Italy. *Food Chem. Toxicol.* **2011**, *49*, 2365–2369. [CrossRef] [PubMed]
12. Venturini, G.; Assante, G.; Vercesi, A. *Fusarium verticillioides* contamination patterns in northern Italian maize during the growing season. *Phytopathologia Mediterr.* **2011**, *50*, 110–120.
13. Lazzaro, I.; Moretti, A.; Giorni, P.; Brera, C.; Battilani, P. Organic vs conventional farming: differences in infection by mycotoxin-producing fungi on maize and wheat in Northern and Central Italy. *Crop Prot.* **2015**, *72*, 22–30. [CrossRef]
14. Sala, N.; Sanchis, V.; Vilaro, P.; Viladrich, R.; Torres, M.; Viñas, I.; Canela, R. Fumonisin producing capacity of *Fusarium* strains isolated from cereals in Spain. *J. Food Prot.* **1994**, *57*, 915–917. [CrossRef]
15. Ariño, A.; Juan, T.; Estopañan, G.; González-Cabo, J.F. Natural occurrence of *Fusarium* species, fumonisin production by toxigenic strains, and concentrations of fumonisins B_1 and B_2 in conventional and organic maize grown in Spain. *J. Food Protect.* **2007**, *70*, 151–156. [CrossRef]
16. Aguín, O.; Cao, A.; Pintos, C.; Santiago, R.; Mansilla, P.; Butrón, A. Occurence of *Fusarium* species in maize kernels grown in northwestern Spain. *Plant Pathol.* **2014**, *63*, 946–951. [CrossRef]
17. Fadl Allah, E.M. Occurrence and toxigenicity of *Fusarium moniliforme* from freshly harvested maize ears with special references to fumonisin production in Egypt. *Mycopathologia* **1998**, *140*, 99–103. [CrossRef]
18. Aboul-Nasr, M.B.; Obied-Allah, M.R.A. Biological and chemical detection of fumonisins produced on agar medium by *Fusarium verticillioides* isolates collected from corn in Sohag, Egypt. *Microbiology* **2013**, *159*, 1720–1724. [CrossRef]

19. Abd El-Fatah, S.I.; Naguib, M.M.; El-Hossiny, E.N.; Sultan, Y.Y. Occurrence of *Fusarium* species and the potential accumulation of its toxins in Egyptian maize grains. *Int. J. Adv. Res.* **2015**, *3*, 1435–1444.
20. Abd-El Fatah, S.I.; Naguib, M.M.; El-Hossiny, E.N.; Sultan, Y.Y.; Abodalam, T.H.; Yli-Mattila, T. Molecular versus morphological identification of *Fusarium* spp. isolated from Egyptian corn. *Res. J. Pharm. Biol. Chem. Sci.* **2015**, *6*, 1813–1822.
21. Hussien, T.; Carlobos-Lopez, A.L.; Cumagun, C.J.R.; Yli-Mattila, T. Identification and quantification of fumonisin-producing *Fusarium* species in grain and soil samples from Egypt and the Philippines. *Phytopathol. Mediterr.* **2017**, *56*, 146–153.
22. Ghiasian, S.A.; Rezayat, S.M.; Kord-Bacheh, P.; Hossein, M.; Yazdanpanah, H.; Shephard, G.S.; van der Westhuizen, L.; Vismer, H.; Marasas, W.F.O. Fumonisin production by *Fusarium* species isolated from freshly harvested corn in Iran. *Mycopathologia* **2005**, *159*, 31–40. [CrossRef]
23. Gelderblom, W.C.A.; Jaskiewicz, J.; Marasas, W.F.O.; Thiel, P.G.; Horak, R.M.; Vleggar, R.; Kriek, N.P.J. Fumonisins—Novel mycotoxins with cancer-promoting activity produced by *Fusarium moniliforme*. *Appl. Environ. Microbiol.* **1988**, *54*, 1806–1811. [CrossRef]
24. Ferrigo, D.; Raiola, A.; Causin, R. *Fusarium* toxins in cereals: Occurrence, legislation, factors promoting the appearance and their management. *Molecules* **2016**, *21*, 627. [CrossRef]
25. Shephard, G.S.; Marasas, W.F.O.; Leggott, N.L.; Yazdanpanah, H.; Rahimian, H. Natural occurrence of fumonisins in corn from Iran. *J. Agric. Food Chem.* **2000**, *48*, 1860–1864. [CrossRef]
26. African Development Bank Group. Annual Core Data. Available online: http://high5.opendataforafrica.org (accessed on 28 June 2018).
27. Food and Agriculture Organization of the United Nations. Statistic Division Database 2014. Available online: http://faostat.fao.org (accessed on 3 January 2018).
28. Lanubile, A.; Maschietto, V.; Borrelli, V.M.; Stagnati, L.; Logrieco, A.F.; Marocco, A. Molecular basis of resistance to *Fusarium* ear rot in maize. *Front. Plant Sci.* **2017**, *8*, 1774. [CrossRef]
29. Marasas, W.F.O. Discovery and occurrence of the fumonisins: A historical perspective. *Environ. Health Perspect.* **2011**, *109*, 239–243.
30. Rheeder, J.P.; Marasas, W.O.; Vismer, H.F. Production of fumonisin analogs by *Fusarium* species. *Appl. Environ. Microbiol.* **2002**, *68*, 2101–2105. [CrossRef]
31. Summary and Conclusions. In Proceedings of the Eighty-Third Meeting, Joint FAO/WHO Expert Committee on Food Additives, Rome, Italy, 8–17 November 2016; p. 15.
32. Bondy, G.; Mehta, R.; Caldwell, D.; Coady, L.; Armstrong, C.; Savard, M.; Miller, J.D.; Chomyshyn, E.; Bronson, R.; Zitomer, N.; et al. Effect of long term exposure to the mycotoxin fumonisin B_1 in p53 heterozygous and p53 homozygous transgenic mice. *Food Chem. Toxicol.* **2012**, *50*, 3604–3613. [CrossRef]
33. Escriva, L.; Font, G.; Manyes, L. In vivo studies of fusarium mycotoxins in the last decade: A review. *Food Chem. Toxicol.* **2015**, *78*, 185–206. [CrossRef]
34. Müller, S.; Dekant, W.; Mally, A. Fumonisin B1 and the kidney: Modes of action for renal tumor formation by fumonisin B1 in rodents. *Food Chem. Toxicol.* **2012**, *50*, 3833–3846.
35. Missmer, S.A.; Suarez, L.; Felkner, M.; Wang, E.; Merril, A.H., Jr.; Rothman, K.J.; Hendricks, K.A. Exposure to fumonisins and the occurrence of neural tube defects along the Texas-Mexico border. *Environ. Health Perspect.* **2006**, *114*, 237–241. [CrossRef] [PubMed]
36. European Commission. Commission Recommendation (EC) 2006/576/CE on the presence of deoxynivalenol, zearalenone, ocharatoxin A, T-2 and HT-2 and fumonisins in products intended for animal feeding. *Off. J. Eur. Union* **2006**, *L229*, 7–9.
37. European Commission. Commission Regulation (EC) No. 1126/2007 on maximum levels for certain contaminants in foodstuffs as regards *Fusarium* toxins in maize and maize products. *Off. J. Eur. Union* **2007**, *L255*, 14–17.
38. Silva, J.J.; Viaro, H.P.; Ferranti, L.S.; Oliveira, A.L.M.; Ferreira, J.M.; Ruas, C.F.; Ono, E.Y.S.; Fungaro, M.H.P. Genetic structure of *Fusarium verticillioides* populations and occurrence of fumonisins in maize grown in Southern Brazil. *Crop Prot.* **2017**, *99*, 160–167. [CrossRef]

39. Sewram, V.; Mshicileli, N.; Shephard, G.S.; Vismer, H.F.; Rheeder, J.P.; Lee, Y.; Leslie, J.F.; Marasas, W.F.O. Production of fumonisin B and C analogues by several *Fusarium* species. *J. Agric. Food Chem.* **2005**, *53*, 4861–4866. [CrossRef]
40. Logrieco, A.; Moretti, A.; Perrone, G.; Mulè, G. Biodiversity of complexes of mycotoxigenic fungal species associated with *Fusarium* ear rot of maize and *Aspergillus* rot of grape. *Int. J. Food Microbiol.* **2007**, *119*, 11–16. [CrossRef]
41. Lanza, F.E.; Zambolim, L.; da Costa, R.V.; Queiroz, V.A.V.; Cota, L.V.; da Silva, D.D.; de Souza, A.G.C.; Figueiredo, J.E.F. Prevalence of fumonisin-producing *Fusarium* species in Brazilian corn grains. *Crop Prot.* **2014**, *65*, 232–237. [CrossRef]
42. Falavigna, C.; Lazzaro, I.; Galaverna, G.; Dall'Asta, C.; Battilani, P. Oleoyl and linoleoyl esters of fumonisin B1 are differently produced by *Fusarium verticillioides* on maize and rice based media. *Int. J. Food Microbiol.* **2016**, *217*, 79–84. [CrossRef]
43. Marín, S.; Magan, N.; Ramos, A.J.; Sanchis, V. Fumonisin-producing strains of *Fusarium*: A review of their ecophysiology. *J. Food Prot.* **2004**, *67*, 1792–1805.
44. Rocha, L.O.; Barroso, V.M.; Andrade, L.J.; Pereira, G.H.A.; Ferreira-Castro, F.L.; Duarte, A.P.; Michelotto, M.D.; Correa, B. *FUM* gene expression profile and fumonisin production by *Fusarium verticillioides* inoculated in *Bt* and non-*Bt* Maize. *Front. Microbiol.* **2015**, *6*, 1503. [CrossRef] [PubMed]
45. Proctor, R.H.; Brown, D.W.; Plattner, R.D.; Desjardins, A.E. Co-expression of 15 contiguous genes delineates a fumonisin biosynthetic gene cluster in *Gibberella moniliformis*. *Fungal Genet. Biol.* **2003**, *38*, 237–249. [CrossRef]
46. Proctor, R.H.; Plattner, R.D.; Desjardins, A.E.; Busman, M.; Butchko, R.A.E. Fumonisin production in the maize pathogen *Fusarium verticillioides*: Genetic basis of naturally occurring chemical variation. *J. Agric. Food Chem.* **2006**, *54*, 2424–2430. [CrossRef] [PubMed]
47. Stępień, Ł.; Koczyk, G.; Waśkiewicz, A. FUM cluster divergence in fumonisins-producing *Fusarium* species. *Fungal Biol.* **2011**, *115*, 112–123. [CrossRef]
48. Moretti, A.; Mulè, G.; Susca, A.; González-Jaén, M.T.; Logrieco, A. Toxin profile, fertility and AFLP analysis of *Fusarium verticillioides* from banana fruits. *Eur. J. Plant Pathol.* **2004**, *110*, 601–609. [CrossRef]
49. da Silva, V.N.; Fernandes, F.M.C.; Cortez, A.; Ribeiro, D.H.B.; de Almeida, A.P.; Hassegawa, R.H.; Corrêa, B. Characterization and genetic variability of *Fusarium verticillioides* strains isolated from corn and sorghum in Brazil based on fumonisins production, microsatellites, mating type locus, and mating cross. *Can. J. Microbiol.* **2006**, *52*, 798–804. [CrossRef]
50. Covarelli, L.; Stifano, S.; Beccari, G.; Raggi, L.; Lattanzio, V.M.T.; Albertini, E. Characterization of *Fusarium verticillioides* strains isolated from maize in Italy: Fumonisins production, pathogenicity and genetic variability. *Food Microbiol.* **2012**, *31*, 17–24. [CrossRef]
51. Reynoso, M.M.; Chulze, S.N.; Zeller, K.A.; Torres, A.M.; Leslie, J.F. Genetic structure of *Fusarium verticillioides* populations isolated from maize in Argentina. *Eur. J. Plant Pathol.* **2009**, *123*, 207–215. [CrossRef]
52. Momeni, H.; Nazari, F. Population genetic structure among Iranian isolates of *Fusarium verticillioides*. *J. Plant Pathol. Microbiol.* **2016**, *7*, 6. [CrossRef]
53. Tsehaye, H.; Elameen, A.; Tronsmo, A.M.; Sundheim, L.; Tronsmo, A.; Assefa, D.; Brurberg, M.B. Genetic variation among *Fusarium verticillioides* isolates associated with Ethiopian maize kernels as revealed by AFLP analysis. *Eur. J. Plant Pathol.* **2016**, *146*, 807–816. [CrossRef]
54. Olowe, O.M.; Odebode, A.C.; Olawuyi, O.J.; Sobowale, A.A. Molecular variability of *Fusarium verticillioides* (Sacc.) in maize from three agro-ecological zones of southwest Nigeria. *Am. J. Mol. Biol.* **2017**, *7*, 30–40. [CrossRef]
55. Reynoso, M.M.; Torres, A.M.; Chulze, S.N. Fusaproliferin, beauvericin and fumonisin production by different mating populations among *Gibberella fujikuroi* complex isolated from maize. *Mycol. Res.* **2004**, *108*, 154–160. [CrossRef]
56. Beccari, G.; Caproni, L.; Tini, F.; Uhlig, S.; Covarelli, L. Presence of *Fusarium* species and other toxigenic fungi in malting barley and multi-mycotoxin analysis by liquid chromatography-high-resolution mass spectrometry. *J. Agric. Food Chem.* **2016**, *64*, 4390–4399. [CrossRef]
57. Beccari, G.; Colasante, V.; Tini, F.; Senatore, M.T.; Prodi, A.; Sulyok, M.; Covarelli, L. Causal agents of *Fusarium* head blight of durum wheat (*Triticum durum* Desf.) in central Italy and their in vitro biosynthesis of secondary metabolites. *Food Microbiol.* **2018**, *70*, 17–27. [CrossRef]

58. Patiño, B.; Mirete, S.; González-Jaén, M.T.; Mulé, G.; Rodríguez, M.T.; Vázquez, C. PCR detection assay of fumonisin-producing *Fusarium verticillioides* strains. *J. Food Prot.* **2004**, *67*, 1278–1283. [CrossRef]
59. SANTE/12089/2016. *Guidance Document on Identification of Mycotoxins in Food and Feed*; SANTE: Warszawa, Poland, 2016.
60. Stępień, Ł.; Jestoi, M.; Chełkowski, J. Cyclic hexadepsipeptides in wheat field samples and *esyn1* gene divergence among enniatin producing *Fusarium avenaceum* strains. *World Mycotoxin J.* **2013**, *6*, 399–409. [CrossRef]
61. Waśkiewicz, A.; Stępień, Ł.; Wilman, K.; Kachlicki, P. Diversity of pea-associated *F. proliferatum* and *F. verticillioides* populations revealed by *FUM1* sequence analysis and fumonisin biosynthesis. *Toxins* **2013**, *5*, 488–503. [CrossRef]
62. Stępień, Ł.; Waśkiewicz, A.; Wilman, K. Host extract modulates metabolism and fumonisin biosynthesis by the plant-pathogenic fungus *Fusarium proliferatum*. *Int. J. Food Microbiol.* **2015**, *193*, 74–81. [CrossRef]
63. Hall, T. BioEdit: An important software for molecular biology. *GERF Bull. Biosci.* **2011**, *2*, 60–61.
64. Tamura, K.; Peterson, D.; Peterson, N.; Stecher, G.; Nei, M.; Kumar, S. MEGA5: Molecular evolutionary genetics analysis using maximum likelihood, evolutionary distance, and maximum parsimony methods. *Mol. Biol. Evol.* **2011**, *28*, 2731–2739. [CrossRef]
65. Pinheiro, J.C.; Bates, D.M. *Mixed-Effects Models in S and S-Plus*; Springer: New York, NY, USA, 2000.
66. R Core Team. *R: A Language and Environment for Statistical Computing*; R Foundation for Statistical Computing: Vienna, Austria, 2017; Available online: https://www.R-project.org (accessed on 10 September 2018).
67. Welch, B.L. The generalization of "Student's" problem when several different population variances are involved. *Biometrika* **1947**, *34*, 28–35. [CrossRef]
68. Segvic, M.; Pepeljnjak, S. Distribution and fumonisin B1 production capacity of *Fusarium moniliforme* isolated from corn in Croatia. *Period. Biol.* **2003**, *105*, 275–279.
69. Castella, G.; Bragulat, M.R.; Cabañes, F.J. Mycoflora and fumonisin-producing strains of *Fusarium moniliforme* in mixed poultry and component raw material. *Mycopathologia* **1996**, *133*, 181–184. [CrossRef]
70. Lee, U.S.; Lee, M.Y.; Shin, W.S.; Min, Y.S.; Cho, C.M.; Yoshio, U. Production of fumonisin B_1 and B_2 by *Fusarium moniliforme* isolated from Korean corn kernels for feed. *Mycotoxin Res.* **1994**, *10*, 67–72.
71. Nelson, P.E.; Plattner, R.D.; Shackelford, D.D.; Desjardins, A.E. Production of fumonisins by *Fusarium moniliforme* strains from various substrates and geographic areas. *Appl. Environ. Microbiol.* **1991**, *57*, 2410–2412. [CrossRef]
72. Chulze, S.; Ramirez, M.L.; Pascale, M.; Visconti, A. Fumonisin production by, and mating type population of, *Fusarium* section *Liseola* isolates from maize in Argentina. *Mycol. Res.* **1998**, *102*, 141–144. [CrossRef]
73. Cvetnić, Z.; Pepeljkjak, S.; šegvić, M. Toxigenic potential of *Fusarium* species isolated from non-harvested maize. *Arh. Hig. Rada. Toksikol.* **2005**, *56*, 275–280.
74. Tseng, T.C.; Lee, K.L.; Deng, T.S.; Liu, T.S.; Liu, C.Y.; Huang, J.W. Production of fumonisins by *Fusarium* species of Taiwan. *Mycopathologia* **1995**, *130*, 117–121. [CrossRef]
75. Sanchis, V.; Abadias, M.; Oncins, L.; Sala, N.; Viñas, I.; Canela, R. Fumonisins B1 and B2 and toxigenic *Fusarium* strains in feeds from the Spanish market. *Int. J. Food Microbiol.* **1995**, *27*, 37–44. [CrossRef]
76. Moretti, A.; Bennett, G.A.; Logrieco, A.; Bottalico, A.; Beremand, M.N. Fertility of *Fusarium moniliforme* from maize and sorghum related to fumonisin production in Italy. *Mycopathologia* **1995**, *131*, 25–29. [CrossRef]
77. Plattner, R.D.; Desjardins, A.E.; Leslie, J.F.; Nelson, P.E. Identification and characterization of strains of *Gibberella fujikuroi* mating population A with rare fumonisin production phenotypes. *Mycologia* **1996**, *88*, 416–424. [CrossRef]
78. Vogelgsang, S.; Sulyok, M.; Bäzinger, I.; Krska, R.; Schuhmacher, R.; Forrer, H.R. Effect of fungal strain and cereal susbtrate on in vitro mycotoxin production by *Fusarium poae* and *Fusarium avenaceum*. *Food Add. Contam.* **2008**, *25*, 745–757. [CrossRef] [PubMed]
79. Ortiz, C.S.; Richards, C.; Terry, A.; Parra, J.; Won-Bo, S. Genetic variability and geographical distribution of mycotoxigenic *Fusarium verticillioides* strains isolated from maize fields in Texas. *Plant Pathol. J.* **2015**, *31*, 203–211. [CrossRef] [PubMed]
80. Gálvez, L.; Urbaniak, M.; Waśkiewicz, A.; Stępień, Ł.; Palmero, D. *Fusarium proliferatum* – Causal agent of garlic bulb rot in Spain: Genetic variability and mycotoxin production. *Food Microbiol.* **2017**, *67*, 41–48. [CrossRef]

81. Stępień, Ł.; Koczyk, G.; Waśkiewicz, A. Genetic and phenotypic variation of *Fusarium proliferatum* isolates from different host species. *J. Appl. Genet.* **2011**, *52*, 487–496. [CrossRef]
82. Glenn, A.E.; Zitomer, N.C.; Zimeri, A.M.; Williams, L.D.; Riley, R.T.; Proctor, R.H. Transformation-mediated complementation of a *FUM* gene cluster deletion in *Fusarium verticillioides* restores both fumonisin production and pathogenicity on maize seedlings. *Mol. Plant Microbe Interact.* **2008**, *21*, 87–97. [CrossRef]

microorganisms

MDPI

Article

Pro-Inflammatory Effects of NX-3 Toxin Are Comparable to Deoxynivalenol and not Modulated by the Co-Occurring Pro-Oxidant Aurofusarin

Lydia Woelflingseder [1], Nadia Gruber [1], Gerhard Adam [2] and Doris Marko [1,*]

1 Department of Food Chemistry and Toxicology, Faculty of Chemistry, University of Vienna, 1090 Vienna, Austria; lydia.woelflingseder@univie.ac.at (L.W.); nadia.gruber@gmx.at (N.G.)

2 Institute of Microbial Genetics, Department of Applied Genetics and Cell Biology (DAGZ), University of Natural Resources and Life Sciences, Vienna (BOKU), 3430 Tulln, Austria; gerhard.adam@boku.ac.at

* Correspondence: doris.marko@univie.ac.at

Received: 9 March 2020; Accepted: 20 April 2020; Published: 21 April 2020

Abstract: The type A trichothecene NX-3, produced by certain *Fusarium graminearum* strains, is similar to the mycotoxin deoxynivalenol (DON), with the exception that it lacks the carbonyl moiety at the C-8 position. NX-3 inhibits protein biosynthesis and induces cytotoxicity to a similar extent as DON, but so far, immunomodulatory effects have not been assessed. In the present study, we investigated the impact of NX-3 on the activity of the nuclear factor kappa B (NF-κB) signaling pathway in direct comparison to DON. Under pro-inflammatory conditions (IL-1β treatment), the impact on cytokine mRNA levels of NF-κB downstream genes was studied in human colon cell lines, comparing noncancer (HCEC-1CT) and cancer cells (HT-29). In addition, potential combinatory effects with the co-occurring *Fusarium* secondary metabolite aurofusarin (AURO), a dimeric naphthoquinone known to induce oxidative stress, were investigated. NX-3 and DON (1 μM, 20 h) significantly activated a NF-κB regulated reporter gene to a similar extent. Both trichothecenes also enhanced transcript levels of the known NF-κB-dependent pro-inflammatory cytokines IL-8, IL-6, TNF-α and IL-1β. Comparing the colon cancer HT-29 and noncancer HCEC-1CT cells, significant differences in cytokine signaling were identified. In contrast, AURO did not affect NF-κB pathway activity and respective cytokine expression levels at the tested concentration. Despite its pro-oxidant potency, the combination with AURO did not significantly affect the immunomodulatory effects of the tested trichothecenes. Taken together, the present study reveals comparable potency of DON and NX-3 with respect to immunomodulatory and pro-inflammatory potential. Consequently, not only DON but also NX-3 should be considered as factors contributing to intestinal inflammatory processes.

Keywords: mycotoxin; trichothecene; NF-κB; intestinal inflammation; combinatory effects; food safety

1. Introduction

Mycotoxins are toxic, secondary metabolites produced by certain filamentous fungi, mainly belonging to the genera of *Fusarium, Aspergillus, Penicillium* and *Alternaria.* Contaminating food and feed pre- and postharvest, mycotoxins pose a potential risk to food safety and thus to human and animal health. After ingestion, the intestinal tract represents the first barrier of the host against food contaminants such as mycotoxins and is therefore the first line for many defense mechanisms. In order to protect the entire organism from the entrance of unwanted solutes, microorganisms and luminal antigens, a proper function of the intestinal barrier and its innate immune response is crucial. As a defense mechanism against external stressors, intestinal epithelial cells are able to secrete cytokines and chemokines, including transforming growth factor-α (TGF-α), interleukin-1 (IL-1), interleukin-6 (IL-6), interleukin-8 (IL-8) or interleukin-10 (IL-10) in order to activate the immune response and to recruit respective immune cells [1]. However, if not controlled properly, excessive immune response and cytokine release can lead to chronic intestinal inflammation and contribute to the progression of inflammatory disorders, such as inflammatory bowel diseases (IBDs) [2].

One of the most prevalent mycotoxins in temperate climate regions is the trichothecene mycotoxin deoxynivalenol (DON, vomitoxin, Figure 1A), which frequently contaminates grain- and cereal-based products [3–5]. Human biomonitoring studies revealed that, due to its ubiquitous occurrence, consumers are chronically exposed to low levels of DON [6–8]. The epoxide moiety at position C12-C13 is considered a key factor in the toxin's main mechanism of action, the inhibition of eukaryotic protein synthesis. By binding to the 60S ribosomal subunit, DON causes a ribotoxic stress response [9,10], resulting in the activation of mitogen-activated protein kinases (MAPKs) and furthermore in the induction of apoptosis and inflammation [11]. Depending on dose, exposure frequency and duration, DON can induce both immunostimulatory and immunosuppressive effects [11–13]. Low dose exposure caused transcriptionally and post-transcriptionally upregulation of immunostimulating cytokines and chemokines, whereas high dose exposure was shown to promote apoptosis with concomitant immunosuppressive effects [11,13]. Modulatory effects of DON on cytokine production in intestinal tissue and intestinal epithelial cancer cell lines have been studied and reviewed extensively during the last decades [11,14,15]. DON was reported to induce the secretion of IL-8 via a nuclear factor "kappa-light-chain-enhancer of activated B cells"-mediated (NF-κB) mechanism in various human intestinal epithelial cells such as Caco-2 or HT-29 [16–18]. When mimicking an inflamed intestinal epithelium, co-exposure experiments with various pro-inflammatory stimuli including interleukin-1β (IL-1β), the tumor necrosis factor-α (TNF-α) or lipopolysaccharide (LPS) resulted in additive and synergistic effects regarding cytokine secretion and NF-κB activation [16]. However, the effects of DON on the inflammatory response have so far only been reported in cancer cell models, but not in noncancer intestinal epithelial cells.

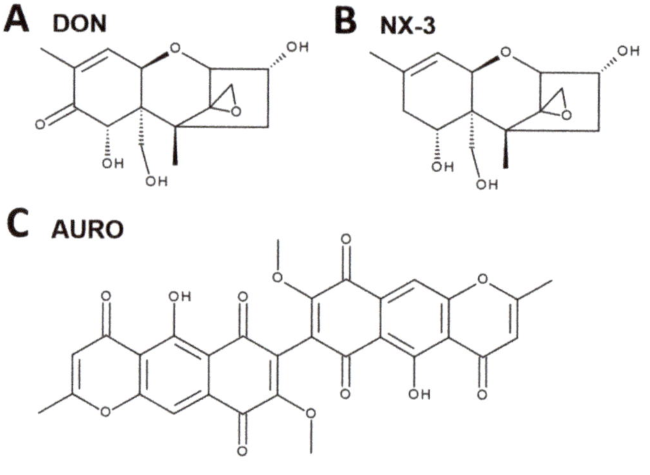

Figure 1. Chemical structures of the investigated *Fusarium* secondary metabolites: (**A**) deoxynivalenol (DON), (**B**) type A trichothecene (NX-3) and (**C**) aurofusarin (AURO).

Moreover, no toxicological characterization of the recently discovered type A trichothecene NX-3 (Figure 1B) regarding immunomodulatory effects has been performed to date. Structurally identified by Varga et al. [19] as an analogue of DON lacking the carbonyl moiety at the C8-position, this toxin was found to be produced by *F. graminearum* [20,21]. In vitro translation assays showed a similar inhibitory potency on protein biosynthesis as DON [19]. Likewise, cytotoxicity studies revealed comparable effects of both trichothecenes in human liver cancer (HepG2), nontransformed colon cells (HCEC-1CT) and colon cancer cells (HT-29) [22,23]. However, potential inflammatory properties of NX-3 have not been investigated yet.

Among many other aspects, oxidative stress plays an essential role in the pathogenesis and progression of IBDs [24]. Another *Fusarium* secondary metabolite reported to induce oxidative stress in human colon adenocarcinoma cells HT-29, is the coloring pigment aurofusarin (AURO, Figure 1C) [25]. Even though AURO was already isolated in 1937 from the mycelium of *Fusarium culmorum* [26], toxicological studies are still limited. In HT-29 and HCEC-1CT cells, Jarolim, Wolters, Woelflingseder, Pahlke, Beisl, Puntscher, Braun, Sulyok, Warth and Marko [25] reported AURO to be cytotoxic at concentrations ≥ 1 μM. Although AURO is frequently found at high concentrations in various food commodities [27–29], immunomodulatory effects have not been assessed yet. As *Fusarium* fungi can produce several mycotoxins simultaneously, infested food and feed may be contaminated by a high number of toxins. In recent years, human biomonitoring studies confirmed that humans are exposed to a variety of toxins [30–32]. Thus, the assessment of potential interactions of mycotoxins is crucial for proper risk assessment.

In the present study, we addressed the question whether the *Fusarium* metabolites NX-3 and AURO affect the NF-κB signaling pathway as individual compounds and as binary mixtures. Furthermore, we characterized alterations in the gene expression profiles of the pro-inflammatory cytokines IL-8, IL-6, TNF-α and IL-1β in a cancer and a noncancer colon cell line, mimicking an inflamed intestinal epithelium by IL-1β stimulation.

2. Materials and Methods

2.1. Chemicals and Reagents

DON was purchased from Romer Labs (Tulln, Austria). NX-3 was produced and purified by preparative HPLC from NX-2 (purity >99% according to LC–UV at 200 nm) as published by Varga, Wiesenberger, Hametner, Ward, Dong, Schofbeck, McCormick, Broz, Stuckler, Schuhmacher, Krska, Kistler, Berthiller and Adam [19]. AURO was purchased from Biovitica (purity: 97.5%; Biovitica Naturstoffe GmbH, Dransfeld, Germany). DON and NX-3 were dissolved in water (LC–MS grade) to obtain stock solutions of 10 mM, which were further dissolved, aliquoted and stored at −20 °C. AURO was dissolved in dimethyl sulfoxide (DMSO). Stock solutions of 1 mM were ultrasonicated for 5 min, aliquoted and stored at −80 °C. For each incubation, a new aliquot was thawed.

2.2. Cell Culture and Treatment

The human monocytic cell line THP1-Lucia™ NF-κB, deriving from the human THP-1 monocyte cell line by stable integration of an NF-κB-inducible luciferase reporter construct, was purchased from Invivogen (San Diego, CA, USA) and HT-29, a human colorectal adenocarcinoma cell line from the German Collection of Microorganisms and Cell Cultures (DSMZ, Braunschweig, Germany). THP1-Lucia™ were cultured in RPMI 1640 medium, HT-29 in Dulbecco's Modified Eagle's Medium (DMEM), both supplemented with 10% (*v/v*) heat-inactivated fetal bovine serum and 1% (*v/v*) penicillin–streptomycin (100 U/mL). THP1-Lucia™ were treated alternately with zeocin and normocin (100 μg/mL; Invivogen, San Diego, CA, USA). Noncancer human colon epithelial cells HCEC-1CT [33,34] were kindly provided by Prof. Jerry W. Shay (UT Southwestern Medical Center, Dallas, TX, USA). HCEC-1CT cells were cultivated in high glucose DMEM. Basal medium was supplemented with the following components: Medium 199 (10×; 2% (*v/v*)), HyClone™ Cosmic Calf™ Serum (2% (*v/v*)), gentamicin (50 μg/mL), 4-(2-hydroxyethyl)-1-piperazineethanesulfonic acid (20 mM), insulin-transferrin-selenium-G (10 μg/mL; 5.5 μg/mL; 6.7 ng/mL), hydrocortisone (1 μg/mL) and recombinant human epidermal growth factor (18.7 ng/mL). All three cell lines were subcultured every 3–4 d, maintained in humidified incubators at 37 °C and 5% CO_2 and routinely tested for the absence of mycoplasm contamination. Cell culture media, supplements and material were purchased from GIBCO Invitrogen (Karlsruhe, Germany), Sigma-Aldrich (Munich, Germany) and Sarstedt AG

& Co. (Nuembrecht, Germany). DON, NX-3, AURO and their combinations were added to the incubation solutions, resulting in a final solvent concentration of 1% (*v/v*) DMSO. In order to ensure data comparability, combinatory effects were always assessed on the same culture plate, in parallel to the individual substance.

2.3. NF-κB Reporter Gene Assay

DON, NX-3, AURO and their combinations were prepared in reaction tubes to be further diluted 1:100 by the addition of the THP1-Lucia™ cell suspension. Cells were counted, centrifuged for 5 min at 250 × *g* and resuspended at 1×10^6 cells/mL in fresh, prewarmed growth medium. Cell suspension was added to the respective toxin preparations, mixed gently and transferred into a 96-well plate (100 μL/well). After 2 h of incubation at 37 °C and 5% CO_2, cells were treated with LPS (10 ng/mL) and incubated for an additional 18 h. 1% (*v/v*) DMSO with and without LPS treatment were used as solvent control. Heat-killed *Listeria monocytogenes* (HKLM; 20×10^6 cells/well) were used as positive control for Toll-like receptor-mediated activation of the NF-κB pathway. Following treatment, cells were centrifuged (250 × *g*, 5 min) and 10 μL of each supernatant were collected and the reporter gene assay was performed according to the manufacturer's protocol. For luminescence measurements QUANTI-Luc™ (Invivogen, San Diego, CA, USA), containing the luciferase substrate coelenterazine, was used.

In parallel cellular metabolic activity was monitored by the alamarBlue® assay (Invitrogen, Carlsbad, CA, USA). After supernatant collection for luciferase activity measurements, 10 μL of alamarBlue® reagent (Invitrogen, Carlsbad, CA, USA) were added to the cells and incubated for 2 h. Subsequently, 50 μL/well were transferred to a black 96-well plate and fluorescence intensity was measured at 530/560 nm (excitation/emission). Both luciferase activity and fluorescence intensity measurements were performed on a Synergy™ H1 hybrid multimode reader (BioTek, Bad Friedrichshall, Germany), assessing at least five independent experiments in duplicates.

2.4. Quantitative Analysis of Cytokine Gene Transcription

Gene transcription levels of up to four pro-inflammatory cytokines in two colon cell lines (HT-29: TNF-α, IL-1β, IL-8; HCEC-1CT: TNF-α, IL-1β, IL-8, IL-6) were analyzed by quantitative real-time PCR (qRT-PCR). Cells were seeded in 12-well plates (HT-29: 150,000 cells/well; HCEC-1CT: 50,000 cells/well) and allowed to grow for 48 h. Cells were incubated for 5 h, consisting of 2 h preincubation with the test compounds (DON, NX-3, AURO and the respective combinations) followed by IL-1β co-treatment (25 ng/mL) for additional 3 h. Total RNA was extracted using Maxwell® 16 Cell LEV Total RNA Purification Kits (Promega, Madison, WI, USA) and reversed transcribed into complementary DNA (cDNA) by QuantiTect® Reverse Transcription Kit (Qiagen, Hilden, Germany) according to the manufacturer's protocols. cDNA samples were amplified in duplicates in presence of gene specific primers (QuantiTect® Primer Assays, Qiagen, Hilden, Germany) and QuantiTect® SYBR Green Master Mix (Qiagen, Hilden, Germany) using a StepOnePlus™ System (Applied Biosystems, Foster City, CA, USA). The following primer assays were used: β-actin (ACTB1, Hs_ACTB1_1_SG, QT00095431); glyceraldehyde 3-phosphate dehydrogenase (GAPDH, Hs_GAPDH_1_SG, QT00079247); TNF-α (Hs_TNF_1_SG; QT00029162); IL-1β (Hs_IL1B_1_SG, QT00021385); IL-8 (Hs_CXCL8_1_SG; QT00000322); IL-6 (Hs_IL6_1_SG, QT00083720). The applied PCR protocol included 15 min enzyme activation at 95 °C, 45 cycles of 15 s at 94 °C, 30 s at 55 °C and 30 s at 72 °C. For the quantification of the fluorescence signal and further data analysis, StepOnePlus® software (Applied Biosystems, Foster City, CA, USA) was used. For each tested sample, at least five independent experiments were performed. Presented transcript data were normalized to the mean of transcript levels of endogenous control genes (ACTB1, GAPDH) by applying the ΔΔCt-method [35] for relative quantification. In relation to the unchallenged solvent control, IL-1β-stimulationenhanced cytokine transcription levels of TNF-α, IL-1β, IL-8 and IL-6 already 4200-, 500-, 4700- and 1600-fold, respectively.

2.5. Determination of Cellular Protein Content and Metabolic Activity

To determine effects on the cellular protein content, the metabolic viability, and to preclude cytotoxic effects for qRT-PCR experiments, the sulforhodamine B (SRB) assay according to Skehan, et al. [36] and the alamarBlue® assay were performed in HT-29 and HCEC-1CT cells. HT-29 (5500 cells/well) and HCEC-1CT cells (2000 cells/well) were seeded into 96-well plates and allowed to grow for 48 h. Cells were incubated for 5 h, including preincubation for 2 h with the substances (DON, NX-3, AURO and the respective combinations) followed by 3 h IL-1β co-treatment (25 ng/mL). Following 4 h of incubation, 10 µL alamarBlue® reagent were added to the incubation solution. After 75 min, 70 µL of the supernatant were transferred into a black 96-well plate and fluorescence intensity was measured at 530/560 nm (excitation/emission) using a Synergy™ H1 hybrid multimode reader (BioTek, Bad Friedrichshall, Germany). Subsequent to the fluorescence readout, cells were rinsed once with prewarmed PBS, fixed by the addition of 5% (*v/v*) trichloroacetic acid and incubated at 4 °C for 30 min. After the fixation, plates were washed four times with water, dried overnight at room temperature and stained for 1 h by adding a solution of 0.4% (*w/v*) SRB in 1% (*v/v*) acetic acid. Plates were washed twice with water and 1% acetic acid solution and dried at room temperature in the dark. Finally, 10 mM Tris buffer (pH 10; 100 µL/well) was added to dissolve the dye, and single wavelength absorbance (570 nm) was measured using a Synergy™ H1 hybrid multimode reader (BioTek, Bad Friedrichshall, Germany). One percent (*v/v*) water (LC–MS grade) and 1% (*v/v*) DMSO with and without IL-1β stimulation served as solvent control, whereas 1% (*v/v*) triton X-100 was used as positive control. Cell-free blank values were subtracted and measured data were referred to the respective solvent control. Each cell line was tested in duplicate with a minimum of five independent experiments.

2.6. Statistical Analysis

Normal distribution of data was tested with the Shapiro–Wilk test. Correction of outliers was performed according to Nalimov. Statistical significances were calculated using OriginPro 2018G (Origin Lab, Northampton, MA, USA) applying one-way ANOVA followed by Bonferroni post hoc testing or one- and two-sample Student's *t*-test.

3. Results

3.1. Impact of Fusarium Secondary Metabolites on LPS-Induced NF-κB Activation

Modulatory effects of DON and NX-3 and potential combinatory effects with the pro-oxidant co-contaminant AURO on LPS-induced NF-κB pathway activation were assessed in THP-1 NF-κB Luc Reporter Monocytes (Figure 2A). DON (1 µM), as well as NX-3, significantly increased the luciferase signal up to 241% ± 35% (DON) and 207% ± 29% of the LPS-induced signal (solid line, 100%). AURO caused an increase of luciferase signal limited to 139% ± 6% of the LPS-induced signal at 0.01 µM, whereas the other concentrations did not significantly modulate NF-κB activity. Combined incubations of AURO and DON or NX-3 resulted in a slightly, but not significantly reduced luminescence intensity compared to effects caused by DON and NX-3 alone.

Cell viability was monitored using the alamarBlue® assay and all data were normalized to the evaluated metabolic activity (Figure 2B). A pronounced decrease of the fluorescence signal was determined in cells incubated with 5 and 10 µM DON and NX-3 and the respective combinations with 0.5 and 1 µM AURO, in line with a substantially decreased NF-κB activity. In most tested conditions, similar effects of NX-3 and DON could be observed. Only in the case of 5 µM NX-3 and the combination with 0.5 µM AURO, which showed similar effects on cell viability as the respective DON-treated samples, significant differences in NF-κB activity were determined.

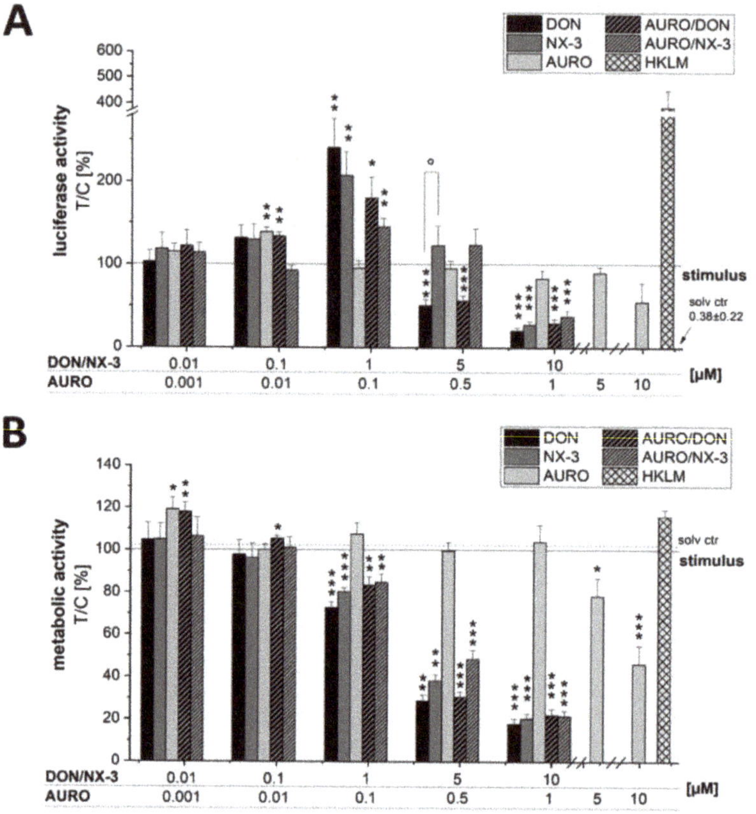

Figure 2. Activity of nuclear factor kappa B (NF-κB) in (**A**) lipopolysaccharide (LPS)-stimulated human monocytic THP1-Lucia™ NF κB cells. THP1-Lucia NF-κB cells were preincubated with the compounds (DON, NX-3, AURO and their combinations) for 2 h followed by an 18 h LPS challenge (10 ng/mL). Heat-killed *Listeria monocytogenes* (HKLM; 20 × 10^6 cells/well) served as positive control for Toll-like receptor-mediated activation of the NF-κB pathway. Luminescence intensity data are expressed as mean values ± SE normalized to LPS-treated solvent control and to the respective cell viability data, assessed in (**B**) the alamarBlue® cell viability assay of at least five independent experiments. One percent DMSO and 1% water (LC–MS grade) served as solvent control (dotted line). Significant differences to LPS, calculated with one-sample *t*-test, are indicated with * ($p < 0.05$), ** ($p < 0.01$) and *** ($p < 0.001$), whereas differences between DON and NX-3, calculated with a two-sample *t*-test, are indicated with ° ($p < 0.05$).

3.2. Modulation of Cytokine Gene Transcription by Fusarium Secondary Metabolites

In order to assess the effects of DON and NX-3 (1 µM) and also their combination with AURO (0.1 µM) on NF-κB-dependent cytokine transcription, two colon cell lines, the cancer cells HT-29 and the noncancer cells HCEC-1CT, were exposed to the *Fusarium* secondary metabolites in the presence of the pro-inflammatory stimulus IL-1β (25 ng/mL). In HT-29 cells (Figure 3A), both trichothecene mycotoxins DON and NX-3 significantly increased TNF-α, IL-1β and IL-8 mRNA levels, whereas AURO did not lead to alterations of the analyzed cytokine transcription levels. Combinatory treatments of DON/AURO or NX-3/AURO resulted again in a slightly decreased signal when compared to the effects of the respective trichothecene single treatments.

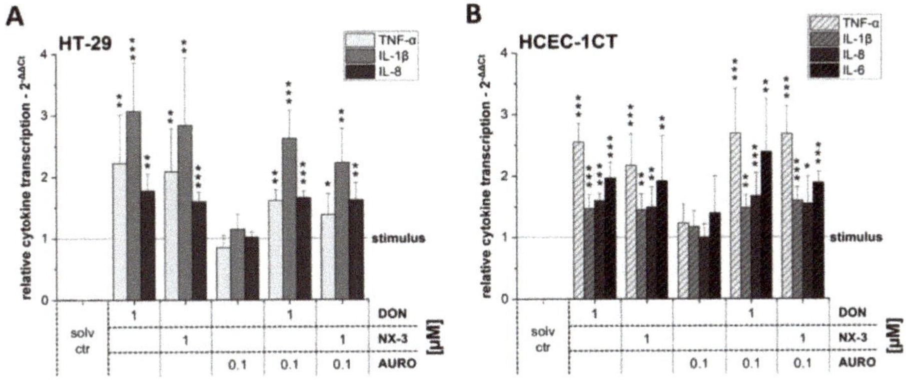

Figure 3. Relative gene transcription levels of TNF-α, IL-1β, IL-8 and IL-6 in (**A**) HT-29 and (**B**) HCEC-1CT cells (calibrator was IL-1β-treated solvent control, which was set to 1). Cells were preincubated with the compounds (DON, NX-3, AURO and their combinations) for 2 h followed by a 3 h IL-1β challenge (25 ng/mL). Relative transcript levels were measured with qRT-PCR. Data are expressed as mean values ± SD normalized to IL-1β-treated solvent control samples of at least five independent experiments. One percent DMSO and 1% water (LC–MS grade) served as solvent control. Significant differences to IL-1β-stimulation, calculated with two-sample *t*-test, are indicated with * ($p < 0.05$), ** ($p < 0.01$) and *** ($p < 0.001$).

In the noncancer cell line HCEC-1CT, a different cytokine pattern could be identified (Figure 3B). While in HT-29 the strongest induction was found for IL-1β transcription, followed by TNF-α and IL-8, in HCEC-1CT more TNF-α mRNA was present compared to the transcript levels of the other tested cytokines. In addition to IL-8 and IL-1β, IL-6 mRNA also could be identified in HCEC-1CT samples. While DON and NX-3 enhanced significantly the cytokine mRNA levels of all four target genes tested, respective cytokine transcripts were only marginally modulated after AURO treatment. Upon co-incubation of DON or NX-3 with AURO, increased mRNA levels similar to those following DON and NX-3 single substance treatment were identified.

When comparing the TNF-α transcription levels of the two colon cell lines, significant differences were observed (Figure 4A). While in the samples exposed to DON or NX-3 as single compounds no significant differences in TNF-α gene transcription were determined, combination with AURO decreased the TNF-α mRNA levels in HT-29 (DON/AURO: 1.6 ± 0.2 and NX-3/AURO: 1.3 ± 0.3 rel. transcription). These differences reached statistical significance in comparison to levels detected in HCEC-1CT cells. Combinations with AURO in the noncancer cell line resulted namely even in an increase in TNF-α gene transcription compared to the single compound treatments (DON/AURO: 2.7 ± 0.7 and NX-3/AURO: 2.7 ± 0.5 rel. transcription). Regarding IL-1β gene transcription (Figure 4B), significant differences between the two cell lines were already present in the samples exposed to DON and NX-3 as single compounds (DON: 3.1 ± 0.8 in HT-29 and 1.5 ± 0.2 in HCEC-1CT rel. transcription; NX-3: 2.8 ± 1.1 in HT-29 and 1.4 ± 0.3 in HCEC-1CT rel. transcription). Comparable to the effects in HT-29 cells, a slight decrease in IL-1β mRNA levels was observed, reaching statistical significance in the case of co-incubation with DON and AURO. IL-8 transcription levels did not differ between the two cell lines (Figure 4C).

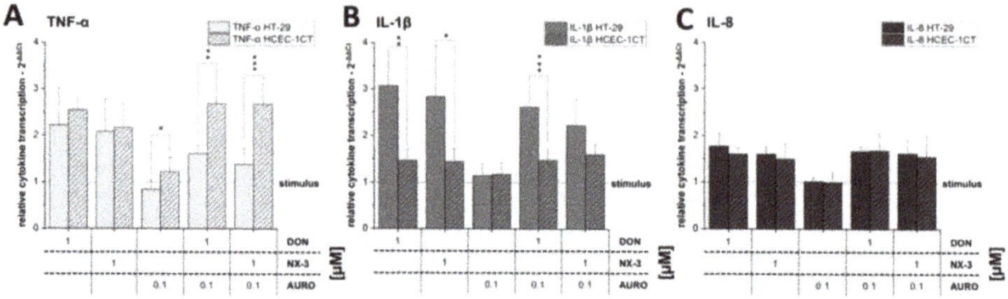

Figure 4. Relative gene transcription levels of (**A**) TNF-α, (**B**) IL-1β and (**C**) IL-8 in HT-29 and HCEC-1CT cells (calibrator was IL-1β-treated solvent control, which was set to 1). Cells were preincubated with the fungal metabolites (DON, NX-3, AURO and their combinations) for 2 h followed by a 3 h IL-1β challenge (25 ng/mL). Relative transcript levels were measured with qRT-PCR. Data are expressed as mean values ± SD normalized to IL-1β-treated solvent control samples of at least five independent experiments. One percent DMSO and 1% water (LC–MS grade) served as solvent control. Significant differences between the two cell lines, calculated with two-sample *t*-test, are indicated with * ($p < 0.05$), ** ($p < 0.01$) and *** ($p < 0.001$).

3.3. Effects of *Fusarium* Secondary Metabolites on Cell Viability

In order to rule out cytotoxicity potentially compromising the analysis of immunomodulatory effects, the impact of the tested concentrations on cell viability was determined by the SRB (Figure 5A,B) and the alamarBlue® assay (Figure 5C,D). In both cell lines, pronounced effects on the cellular protein content and on the metabolic activity after DON and NX-3 treatment for 5 h (last 3 h co-exposed to 25 ng/mL IL-1β) could be identified at concentrations ≥ 5 µM. AURO did not trigger any significant effects on cell viability except for the highest tested concentration (10 µM), which caused a pronounced decrease of the fluorescence signal in the alamarBlue® assay in both cell lines. Partly significant differences between the samples treated in combination with AURO and the DON- or NX-3-single incubations were determined (Figure 5A–D, highlighted with ° symbols). However, due to the fact that the observed effects were of rather limited nature, no appropriate mathematical model for a correct evaluation of the combinatory interactions (e.g., the model of independent joint action [37] or the multiple drug effect equation [38]) could be applied.

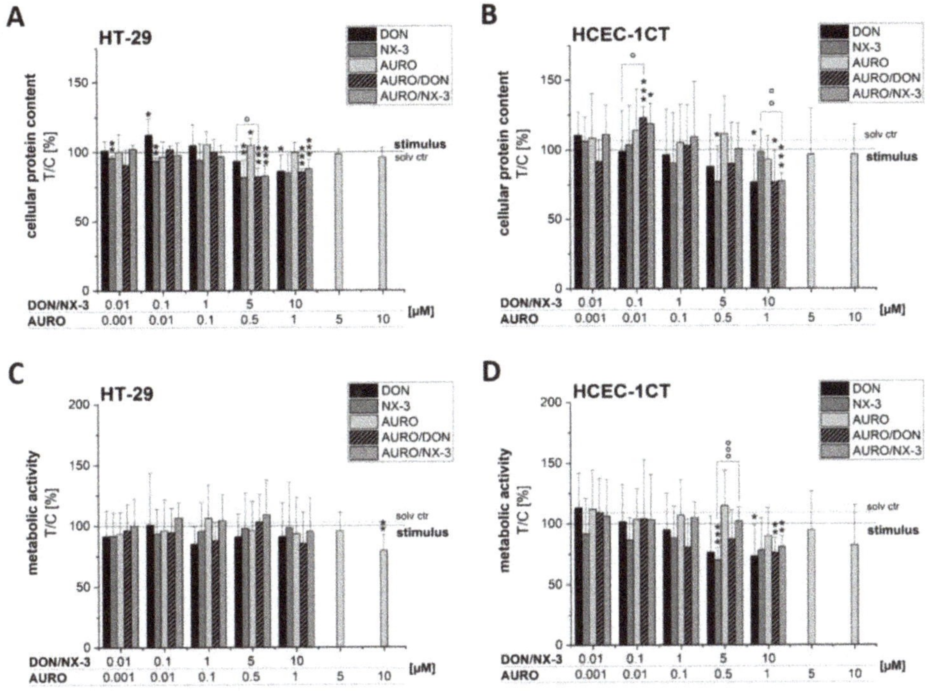

Figure 5. Effects of DON, NX-3, AURO and their combinations on the cellular protein content (**A**,**B**) and viability (**C**,**D**) of the two human colon cell lines HT-29 and HCEC-1CT determined in the sulforhodamine B (HT-29: **A**; HCEC-1CT: **B**) and alamarBlue® assay (HT-29: **C**; HCEC-1CT: **D**). Cells were preincubated with the compounds (DON, NX-3, AURO and their combinations) for 2 h followed by a 3 h IL-1β challenge (25 ng/mL). Data are expressed as mean values ± SD normalized to IL-1β-treated solvent control samples of at least five independent experiments. One percent DMSO and 1% water (LC–MS grade) served as solvent control (dotted line). Significant differences to IL-1β-treated solvent control, calculated with one-sample *t*-test, are indicated with * ($p < 0.05$), ** ($p < 0.01$) and *** ($p < 0.001$), whereas differences between DON, NX-3 and their combinations with AURO, calculated with a two-sample *t*-test, are indicated with ° ($p < 0.05$), °° ($p < 0.01$) and °°° ($p < 0.001$).

4. Discussion

The novel type A trichothecene NX-3 was recently reported to possess comparable inhibitory potency on protein biosynthesis and similar cytotoxic potential as the well-known *Fusarium* mycotoxin DON [22,23]. In the present study, we explored in direct comparison to DON the immunomodulatory effects of NX-3, including its impact on NF-κB signaling pathway activation and on the expression of NF-κB target cytokines in two different intestinal cell lines, comparing the impact on a tumor cell line to noncancer cells. Furthermore, we investigated combinatory effects with AURO, a frequently co-occurring *Fusarium* secondary metabolite, which so far has not been assessed in any cell system regarding its immunomodulatory effects.

Activation of the NF-κB pathway plays a crucial role in inflammatory processes through its ability to induce the expression of various pro-inflammatory genes, including cytokines, chemokines, and adhesion molecules [39,40]. Numerous studies have reported a dysregulation of NF-κB signaling in patients suffering from irritable bowel syndrome [41–43] and identified this pathway as one of the major regulatory components in the complex pathogenesis and progression of chronic intestinal inflammatory disorders like Crohn's disease and ulcerative colitis [44,45]. In order to assess the impact of the three

Fusarium secondary metabolites DON, NX-3 and AURO on this important inflammatory signaling pathway, THP-1 monocytes carrying a NF-κB-inducible luciferase reporter construct were used. NX-3, as well as DON, activated the NF-κB pathway at a concentration of 1 μM (Figure 2A), whereas at higher concentrations, in line with a substantial decrease in cell viability (Figure 2B), a significantly reduced luciferase signal was observed. While this is the first report of NX-3-induced NF-κB activation, the respective effects of DON were already extensively studied during the last decades [12,46]. In Caco-2 cells, DON at concentrations between 1.6 and 16 μM slightly induced NF-κB pathway activity observed due to an increased phosphorylation of its inhibitor IκB and IL-8 secretion, whereas co-exposure to IL-1β or LPS resulted in a more pronounced pathway induction [16]. In HT-29 cells, microscopic localization of NF-κB p65 revealed a nuclear translocation within 15 min after DON treatment at a concentration of 0.8 μM, still active after 60 min [47]. Similar concentrations increased NF-κB p65 expression in HT-29 cells [48] and NF-κB binding in RAW 264.7 murine macrophage cells after 2 and 8 h of DON treatment in the presence and absence of LPS [49].

Due to the ability to influence the amount of intracellular reactive oxygen species (ROS), the transcription factor NF-κB and the regulation of downstream transcriptional targets play a crucial role in cell survival and in the prevention of cellular oxidative damage [50]. Low or transient levels of ROS are reported to trigger an inflammatory response through activation of the NF-κB signaling pathway [51–53]. Recently, the dimeric naphthoquinone AURO was shown to enhance intracellular ROS levels causing significant pro-oxidative DNA damage in HT-29 cells [25]. High levels of AURO contamination were reported in occurrence studies analyzing various food and feed components [4,5,29]. However, the impact of AURO on the NF-κB signaling pathway and potential combinatory interactions with co-occurring trichothecenes such as DON and NX-3 on the inflammatory response have not been addressed yet. AURO is known to be rather unstable and concentrations of about 10 μM were already reported to induce pronounced cytotoxic effects in the used cell systems [25,54]. We therefore used the low level of 1 μM to limit cytotoxicity and to be able to observe possible combinatory effects with the trichothecenes. Despite its reported pro-oxidative properties [25], AURO modulated NF-κB activity only marginally (Figure 2A). At 5 and 10 μM significant cytotoxic effects were detected, concomitantly with a decrease in NF-κB activity. Beside intracellular ROS formation, AURO was previously reported to enhance the ratio of GSSG/GSH and to induce significant oxidative DNA damage in HT-29 cells [25]. However, this pro-oxidative effect seems not sufficient to activate the NF-κB signaling cascade. Accordingly, no significant interactions were observed in the combinatory treatments of AURO with the trichothecenes DON and NX-3 (Figure 2A).

As a consequence of increased NF-κB transcription factor activity, enhanced expression levels of downstream target genes, including interleukins (IL-1β, IL-8 and IL-6) or the tumor necrosis factor (TNF-α) are expected [55]. Upregulation of cytokine mRNA expression can be triggered either transcriptionally or post-transcriptionally via increase of mRNA stability [56]. Since differences in cellular response between tumor and nontumor cells cannot be excluded, the impact of the tested mycotoxins on mRNA levels of pro-inflammatory cytokines were assessed using two intestinal epithelial cell models, the adenocarcinoma cell line HT-29 and the extended primary cell line HCEC-1CT (Figure 3A,B). In addition, co-treatments with IL-1β were used to mimic a potentially inflamed, pathologic IBD disordered intestinal epithelium [16,18,57]. In both cell lines, NX-3-enhanced mRNA levels of the assessed pro-inflammatory cytokines were comparable to DON treatment.

Several studies reported effects of DON on IL-secretion, focusing mainly on IL-8 [16,18], a cytokine acting as an early marker in inflammatory processes, mediating the activation and migration of neutrophils [58]. Six hours after DON exposure, Maresca, Yahi, Younes-Sakr, Boyron, Caporiccio and Fantini [18] reported a dose-dependent increase in IL-8 mRNA levels in Caco-2 cells at concentrations between 1 and 100 μM. Respective IL-8 protein levels were enhanced only at DON concentrations up to 25 μM, whereas higher concentrations resulted in a decrease of IL-8 secretion [18]. Similar effects on NF-kB-dependent IL-8 secretion were determined by Van De Walle, Romier, Larondelle and Schneider [16], revealing that IL-8 induction was potentiated upon pro-inflammatory stimulation by

IL-1β and LPS. NX-3 was found to induce comparable effects as its type B trichothecene derivative DON, regarding cytotoxicity, induction of oxidative stress and GSH modulation [22,23]. In our study, both trichothecenes induced significantly enhanced transcript levels of different pro-inflammatory cytokines to a similar extent. Accordingly, a concomitant increase in cytokine secretion levels, as previously reported by Van De Walle, Romier, Larondelle and Schneider [16] and Maresca, Yahi, Younes-Sakr, Boyron, Caporiccio and Fantini [18], in Caco-2 cells after DON-treatment, is expected after DON as well as after NX-3 exposure in HT-29 and HCEC-1CT cells.

The effects observed on the NF-κB signaling pathway after AURO treatment indicate the conclusion that this *Fusarium* secondary metabolite lacks immunomodulatory potency, at least with respect to the spectrum of cytokines tested so far. These results argue for the fact that additional pro-inflammatory signaling pathways, such as the Toll-like receptors or retinoic acid-inducible gene-I-like receptors [59], are not substantially affected by low AURO concentration (0.1 μM, 5 h).

Studies focusing on the immunomodulatory effects of DON in normal, noncancer intestinal epithelial cells, to the best of our knowledge, are still limited to nontransformed porcine intestinal epithelial cells [60]. In that model, a concentration of 10 μM DON induced a pro-inflammatory response resulting in significantly increased transcription levels of mRNAs encoding for IL-8, IL-1α, IL-1β and TNF-α, reaching their maximum levels after 4 h of DON exposure. Similar results were reported using an ex vivo model of porcine jejunal explants [60].

Noteworthy, analysis of the cytokine mRNA levels in HT-29 and HCEC-1CT cells showed substantial differences in the cytokine transcription pattern (Figure 4). Under the applied experimental conditions, IL-6 transcript levels in HT-29 were below the detection limit, whereas in HCEC-1CT cells substantially higher IL-6 mRNA concentrations were observed (Figure 3). While in the case of IL-8 no significant differences between the two cell lines were noted, transcript analyses for TNF-α and IL-1β revealed significantly different expression patterns (Figure 4A,B). IL-1β transcription levels were much higher in HT-29 cells, compared to the levels in HCEC-1CT. Again, in both cell models, combinatory treatment with AURO did not substantially modulate IL-1β mRNA levels, compared to the effects caused by 1 μM DON or NX-3 alone. However, in the case of TNF-α significant differences between the two intestinal cell lines were determined only in combination with AURO. In HT-29 reduced amounts of TNF-α mRNA were found while HCEC-1CT cells showed increased transcript levels. As described in literature, HCEC-1CT cells are more susceptible to the toxic effects and stress induced by mycotoxins [22,25,61,62].

Since to-date no study evaluated the cytotoxic effects of NX-3 after short-term exposure and in order to rule out potential cytotoxicity affecting the analysis of cytokine transcripts by qRT-PCR, respective experiments were performed as part of this study (Figure 5A–D). A slight decrease in cell viability, more pronounced in HCEC-1CT cells, caused by the highest tested concentrations of DON and NX-3 could be determined. Varga, Wiesenberger, Woelflingseder, Twaruschek, Hametner, Vaclavikova, Malachova, Marko, Berthiller and Adam [22] reported NX-3 to induce pronounced cytotoxic effects in HT-29 and HCEC-1CT cells at concentrations ≥ 10 μM. Similar results were observed in the human hepatocyte carcinoma cell line HepG2 [23]. AURO affected cell viability only marginally at the highest concentration tested (10 μM). Jarolim, Wolters, Woelflingseder, Pahlke, Beisl, Puntscher, Braun, Sulyok, Warth and Marko [25] reported AURO to induce only minor cytotoxic effects in HT-29 and HCEC-1CT cells after 1 h of incubation at a concentration of 10 μM, whereas after 24 h, 5 μM AURO had already caused a statistically significant decrease in cell viability.

Taken together, despite the pro-oxidative properties of the potentially co-occurring bisnaphthoquinone derivative AURO, no immunomodulatory effects, neither alone nor in combination with NX-3 or DON were observed. The present study shows that the recently discovered type A trichothecene NX-3 can be seen as equipotent to DON in its potency to activate the NF-κB signaling pathway. Thereby, respective pro-inflammatory response was found not only in tumor cells but also in nontumorigenic intestinal cells. Altogether, this study underlines the importance to continuously explore the complex interaction between food contaminants and the intestinal inflammatory system.

In order to allow proper risk assessment beyond healthy intestinal epithelia, known pathologic gastrointestinal tracts, e.g., from patients suffering IBDs that might be more sensitive to the effects of individual food contaminants and their mixtures, need to be taken into account.

Author Contributions: Conceptualization, L.W. and D.M.; methodology, L.W.; validation, L.W. and N.G.; formal analysis, L.W. and N.G.; investigation, L.W. and N.G.; resources, D.M.; data curation, L.W. and N.G.; writing—original draft preparation, L.W.; writing—review and editing, G.A. and D.M.; visualization, L.W.; supervision, L.W. and D.M.; project administration, D.M. and G.A.; funding acquisition, D.M. and G.A. All authors have read and agreed to the published version of the manuscript.

Funding: This research was supported by the Austrian Science Fund (FWF) via the special research project Fusarium (F3701, F3702 and F3718).

Acknowledgments: Open Access Funding by the Austrian Science Fund (FWF).

Conflicts of Interest: The authors declare no conflict of interest. The funders had no role in the design of the study; in the collection, analyses, or interpretation of data; in the writing of the manuscript, or in the decision to publish the results.

References

1. Stadnyk, A.W. Intestinal epithelial cells as a source of inflammatory cytokines and chemokines. *J. Can. Gastroenterol.* **2002**, *16*, 241–246. [CrossRef]
2. Takeuchi, O.; Akira, S. Pattern recognition receptors and inflammation. *Cell* **2010**, *140*, 805–820. [CrossRef]
3. Kovalsky, P.; Kos, G.; Nahrer, K.; Schwab, C.; Jenkins, T.; Schatzmayr, G.; Sulyok, M.; Krska, R. Co-Occurrence of Regulated, Masked and Emerging Mycotoxins and Secondary Metabolites in Finished Feed and Maize-An Extensive Survey. *Toxins* **2016**, *8*, 363. [CrossRef] [PubMed]
4. Streit, E.; Schwab, C.; Sulyok, M.; Naehrer, K.; Krska, R.; Schatzmayr, G. Multi-mycotoxin screening reveals the occurrence of 139 different secondary metabolites in feed and feed ingredients. *Toxins* **2013**, *5*, 504–523. [CrossRef] [PubMed]
5. Uhlig, S.; Eriksen, G.S.; Hofgaard, I.S.; Krska, R.; Beltran, E.; Sulyok, M. Faces of a changing climate: Semi-quantitative multi-mycotoxin analysis of grain grown in exceptional climatic conditions in Norway. *Toxins* **2013**, *5*, 1682–1697. [CrossRef]
6. Warth, B.; Sulyok, M.; Fruhmann, P.; Berthiller, F.; Schuhmacher, R.; Hametner, C.; Adam, G.; Frohlich, J.; Krska, R. Assessment of human deoxynivalenol exposure using an LC-MS/MS based biomarker method. *Toxicol. Lett.* **2012**, *211*, 85–90. [CrossRef] [PubMed]
7. Turner, P.C.; Rothwell, J.A.; White, K.L.; Gong, Y.; Cade, J.E.; Wild, C.P. Urinary deoxynivalenol is correlated with cereal intake in individuals from the United kingdom. *Environ. Health Perspect.* **2008**, *116*, 21–25. [CrossRef]
8. Ali, N.; Blaszkewicz, M.; Degen, G.H. Assessment of deoxynivalenol exposure among Bangladeshi and German adults by a biomarker-based approach. *Toxicol. Lett.* **2016**, *258*, 20–28. [CrossRef] [PubMed]
9. Garreau de Loubresse, N.; Prokhorova, I.; Holtkamp, W.; Rodnina, M.V.; Yusupova, G.; Yusupov, M. Structural basis for the inhibition of the eukaryotic ribosome. *Nature* **2014**, *513*, 517–522. [CrossRef] [PubMed]
10. Ueno, Y. Mode of action of trichothecenes. *Ann. Nutr. Aliment.* **1977**, *31*, 885–900. [CrossRef]
11. Pestka, J.J.; Zhou, H.R.; Moon, Y.; Chung, Y.J. Cellular and molecular mechanisms for immune modulation by deoxynivalenol and other trichothecenes: Unraveling a paradox. *Toxicol. Lett.* **2004**, *153*, 61–73. [CrossRef] [PubMed]
12. Pestka, J.J. Mechanisms of deoxynivalenol-induced gene expression and apoptosis. *Food Addit. Contam.* **2008**, *25*, 1128–1140. [CrossRef]
13. Pestka, J.J. Deoxynivalenol: Mechanisms of action, human exposure, and toxicological relevance. *Arch. Toxicol.* **2010**, *84*, 663–679. [CrossRef] [PubMed]
14. Pinton, P.; Oswald, I.P. Effect of deoxynivalenol and other Type B trichothecenes on the intestine: A review. *Toxins* **2014**, *6*, 1615–1643. [CrossRef] [PubMed]
15. Pestka, J.J. Deoxynivalenol-induced proinflammatory gene expression: Mechanisms and pathological sequelae. *Toxins* **2010**, *2*, 1300–1317. [CrossRef] [PubMed]

16. Van De Walle, J.; Romier, B.; Larondelle, Y.; Schneider, Y.J. Influence of deoxynivalenol on NF-kappaB activation and IL-8 secretion in human intestinal Caco-2 cells. *Toxicol. Lett.* **2008**, *177*, 205–214. [CrossRef] [PubMed]
17. Del Favero, G.; Woelflingseder, L.; Braun, D.; Puntscher, H.; Kutt, M.L.; Dellafiora, L.; Warth, B.; Pahlke, G.; Dall'Asta, C.; Adam, G.; et al. Response of intestinal HT-29 cells to the trichothecene mycotoxin deoxynivalenol and its sulfated conjugates. *Toxicol. Lett.* **2018**, *295*, 424–437. [CrossRef]
18. Maresca, M.; Yahi, N.; Younes-Sakr, L.; Boyron, M.; Caporiccio, B.; Fantini, J. Both direct and indirect effects account for the pro-inflammatory activity of enteropathogenic mycotoxins on the human intestinal epithelium: Stimulation of interleukin-8 secretion, potentiation of interleukin-1beta effect and increase in the transepithelial passage of commensal bacteria. *Toxicol. Appl. Pharmacol.* **2008**, *228*, 84–92. [CrossRef]
19. Varga, E.; Wiesenberger, G.; Hametner, C.; Ward, T.J.; Dong, Y.; Schofbeck, D.; McCormick, S.; Broz, K.; Stuckler, R.; Schuhmacher, R.; et al. New tricks of an old enemy: Isolates of Fusarium graminearum produce a type A trichothecene mycotoxin. *Environ. Microbiol.* **2015**, *17*, 2588–2600. [CrossRef]
20. Kelly, A.; Proctor, R.H.; Belzile, F.; Chulze, S.N.; Clear, R.M.; Cowger, C.; Elmer, W.; Lee, T.; Obanor, F.; Waalwijk, C.; et al. The geographic distribution and complex evolutionary history of the NX-2 trichothecene chemotype from Fusarium graminearum. *Fungal Genet. Biol.* **2016**, *95*, 39–48. [CrossRef]
21. Lofgren, L.; Riddle, J.; Dong, Y.; Kuhnem, P.R.; Cummings, J.A.; Del Ponte, E.M.; Bergstrom, G.C.; Kistler, H.C. A high proportion of NX-2 genotype strains are found among Fusarium graminearum isolates from northeastern New York State. *Eur. J. Plant Pathol.* **2018**, *150*, 791–796. [CrossRef]
22. Varga, E.; Wiesenberger, G.; Woelflingseder, L.; Twaruschek, K.; Hametner, C.; Vaclavikova, M.; Malachova, A.; Marko, D.; Berthiller, F.; Adam, G. Less-toxic rearrangement products of NX-toxins are formed during storage and food processing. *Toxicol. Lett.* **2018**, *284*, 205–212. [CrossRef] [PubMed]
23. Woelflingseder, L.; Del Favero, G.; Blazevic, T.; Heiss, E.H.; Haider, M.; Warth, B.; Adam, G.; Marko, D. Impact of glutathione modulation on the toxicity of the Fusarium mycotoxins deoxynivalenol (DON), NX-3 and butenolide in human liver cells. *Toxicol. Lett.* **2018**, *299*, 104–117. [CrossRef] [PubMed]
24. Tian, T.; Wang, Z.; Zhang, J. Pathomechanisms of Oxidative Stress in Inflammatory Bowel Disease and Potential Antioxidant Therapies. *Oxidative Med. Cell. Longev.* **2017**, *2017*, 4535194. [CrossRef]
25. Jarolim, K.; Wolters, K.; Woelflingseder, L.; Pahlke, G.; Beisl, J.; Puntscher, H.; Braun, D.; Sulyok, M.; Warth, B.; Marko, D. The secondary Fusarium metabolite aurofusarin induces oxidative stress, cytotoxicity and genotoxicity in human colon cells. *Toxicol. Lett.* **2018**, *284*, 170–183. [CrossRef]
26. Ashley, J.N.; Hobbs, B.C.; Raistrick, H. Studies in the biochemistry of micro-organisms: The crystalline colouring matters of Fusarium culmorum (W. G. Smith) Sacc. and related forms. *Biochem. J.* **1937**, *31*, 385–397.
27. Beccari, G.; Colasante, V.; Tini, F.; Senatore, M.T.; Prodi, A.; Sulyok, M.; Covarelli, L. Causal agents of Fusarium head blight of durum wheat (Triticum durum Desf.) in central Italy and their in vitro biosynthesis of secondary metabolites. *Food Microbiol.* **2018**, *70*, 17–27. [CrossRef]
28. Ezekiel, C.N.; Bandyopadhyay, R.; Sulyok, M.; Warth, B.; Krska, R. Fungal and bacterial metabolites in commercial poultry feed from Nigeria. *Food Addit. Contam.* **2012**, *29*, 1288–1299. [CrossRef]
29. Nichea, M.J.; Palacios, S.A.; Chiacchiera, S.M.; Sulyok, M.; Krska, R.; Chulze, S.N.; Torres, A.M.; Ramirez, M.L. Presence of Multiple Mycotoxins and Other Fungal Metabolites in Native Grasses from a Wetland Ecosystem in Argentina Intended for Grazing Cattle. *Toxins* **2015**, *7*, 3309–3329. [CrossRef]
30. Marin, S.; Cano-Sancho, G.; Sanchis, V.; Ramos, A.J. The role of mycotoxins in the human exposome: Application of mycotoxin biomarkers in exposome-health studies. *Food Chem. Toxicol.* **2018**, *121*, 504–518. [CrossRef]
31. Warth, B.; Sulyok, M.; Krska, R. LC-MS/MS-based multibiomarker approaches for the assessment of human exposure to mycotoxins. *Anal. Bioanal. Chem.* **2013**, *405*, 5687–5695. [CrossRef] [PubMed]
32. Abia, W.A.; Warth, B.; Sulyok, M.; Krska, R.; Tchana, A.; Njobeh, P.B.; Turner, P.C.; Kouanfack, C.; Eyongetah, M.; Dutton, M.; et al. Bio-monitoring of mycotoxin exposure in Cameroon using a urinary multi-biomarker approach. *Food Chem. Toxicol.* **2013**, *62*, 927–934. [CrossRef] [PubMed]
33. Roig, A.I.; Eskiocak, U.; Hight, S.K.; Kim, S.B.; Delgado, O.; Souza, R.F.; Spechler, S.J.; Wright, W.E.; Shay, J.W. Immortalized epithelial cells derived from human colon biopsies express stem cell markers and differentiate in vitro. *Gastroenterology* **2010**, *138*, 1011–1015. [CrossRef]
34. Roig, A.I.; Shay, J.W. Immortalization of adult human colonic epithelial cells extracted from normal tissues obtained via colonoscopy. *Protoc. Exch.* **2010**. [CrossRef]

35. Schmittgen, T.D.; Livak, K.J. Analyzing real-time PCR data by the comparative C(T) method. *Nat. Protoc.* **2008**, *3*, 1101–1108. [CrossRef] [PubMed]
36. Skehan, P.; Storeng, R.; Scudiero, D.; Monks, A.; McMahon, J.; Vistica, D.; Warren, J.T.; Bokesch, H.; Kenney, S.; Boyd, M.R. New colorimetric cytotoxicity assay for anticancer-drug screening. *J. Natl. Cancer Inst.* **1990**, *82*, 1107–1112. [CrossRef] [PubMed]
37. Webb, J.L. *Enzyme and Metabolic Inhibitors*; Academic Press: New York, NY, USA, 1963.
38. Chou, T.C. Theoretical basis, experimental design, and computerized simulation of synergism and antagonism in drug combination studies. *Pharmacol. Rev.* **2006**, *58*, 621–681. [CrossRef] [PubMed]
39. Lawrence, T. The nuclear factor NF-kappaB pathway in inflammation. *Cold Spring Harb. Perspect. Biol.* **2009**, *1*, a001651. [CrossRef]
40. Tak, P.P.; Firestein, G.S. NF-kappaB: A key role in inflammatory diseases. *J. Clin. Investig.* **2001**, *107*, 7–11. [CrossRef]
41. Rogler, G.; Brand, K.; Vogl, D.; Page, S.; Hofmeister, R.; Andus, T.; Knuechel, R.; Baeuerle, P.A.; Scholmerich, J.; Gross, V. Nuclear factor kappaB is activated in macrophages and epithelial cells of inflamed intestinal mucosa. *Gastroenterology* **1998**, *115*, 357–369. [CrossRef]
42. Schreiber, S.; Nikolaus, S.; Hampe, J. Activation of nuclear factor kappa B inflammatory bowel disease. *Gut* **1998**, *42*, 477–484. [CrossRef] [PubMed]
43. Neurath, M.F.; Pettersson, S.; Meyer zum Buschenfelde, K.H.; Strober, W. Local administration of antisense phosphorothioate oligonucleotides to the p65 subunit of NF-kappa B abrogates established experimental colitis in mice. *Nat. Med.* **1996**, *2*, 998–1004. [CrossRef] [PubMed]
44. Atreya, I.; Atreya, R.; Neurath, M.F. NF-kappaB in inflammatory bowel disease. *J. Intern. Med.* **2008**, *263*, 591–596. [CrossRef] [PubMed]
45. Abraham, C.; Cho, J.H. Inflammatory bowel disease. *New Engl. J. Med.* **2009**, *361*, 2066–2078. [CrossRef] [PubMed]
46. Katika, M.R.; Hendriksen, P.J.; van Loveren, H.; Pijnenburg, A.C.M. Characterization of the modes of action of deoxynivalenol (DON) in the human Jurkat T-cell line. *J. Immunotoxicol.* **2015**, *12*, 206–216. [CrossRef] [PubMed]
47. Krishnaswamy, R.; Devaraj, S.N.; Padma, V.V. Lutein protects HT-29 cells against Deoxynivalenol-induced oxidative stress and apoptosis: Prevention of NF-kappaB nuclear localization and down regulation of NF-kappaB and Cyclo-Oxygenase-2 expression. *Free Radic. Biol. Med.* **2010**, *49*, 50–60. [CrossRef]
48. Kalaiselvi, P.; Rajashree, K.; Bharathi Priya, L.; Padma, V.V. Cytoprotective effect of epigallocatechin-3-gallate against deoxynivalenol-induced toxicity through anti-oxidative and anti-inflammatory mechanisms in HT-29 cells. *Food Chem. Toxicol.* **2013**, *56*, 110–118. [CrossRef]
49. Wong, S.S.; Zhou, H.R.; Pestka, J.J. Effects of vomitoxin (deoxynivalenol) on the binding of transcription factors AP-1, NF-kappaB, and NF-IL6 in raw 264.7 macrophage cells. *J. Toxicol. Environ. Health* **2002**, *65*, 1161–1180. [CrossRef]
50. Morgan, M.J.; Liu, Z.-g. Crosstalk of reactive oxygen species and NF-κB signaling. *Cell Res.* **2011**, *21*, 103–115. [CrossRef]
51. Reuter, S.; Gupta, S.C.; Chaturvedi, M.M.; Aggarwal, B.B. Oxidative stress, inflammation, and cancer: How are they linked? *Free radical biology & medicine* **2010**, *49*, 1603–1616. [CrossRef]
52. Gloire, G.; Legrand-Poels, S.; Piette, J. NF-kappaB activation by reactive oxygen species: Fifteen years later. *Biochem. Pharmacol.* **2006**, *72*, 1493–1505. [CrossRef] [PubMed]
53. Mittal, M.; Siddiqui, M.R.; Tran, K.; Reddy, S.P.; Malik, A.B. Reactive oxygen species in inflammation and tissue injury. *Antioxid. Redox Signal.* **2014**, *20*, 1126–1167. [CrossRef] [PubMed]
54. Vejdovszky, K.; Warth, B.; Sulyok, M.; Marko, D. Non-synergistic cytotoxic effects of Fusarium and Alternaria toxin combinations in Caco-2 cells. *Toxicol. Lett.* **2016**, *241*, 1–8. [CrossRef] [PubMed]
55. Oeckinghaus, A.; Ghosh, S. The NF-kappaB family of transcription factors and its regulation. *Cold Spring Harb. Perspect. Biol.* **2009**, *1*, a000034. [CrossRef]
56. Wong, S.; Schwartz, R.C.; Pestka, J.J. Superinduction of TNF-alpha and IL-6 in macrophages by vomitoxin (deoxynivalenol) modulated by mRNA stabilization. *Toxicology* **2001**, *161*, 139–149. [CrossRef]
57. Schuerer-Maly, C.C.; Eckmann, L.; Kagnoff, M.F.; Falco, M.T.; Maly, F.E. Colonic epithelial cell lines as a source of interleukin-8: Stimulation by inflammatory cytokines and bacterial lipopolysaccharide. *Immunology* **1994**, *81*, 85–91.

58. Műzes, G.; Molnár, B.; Tulassay, Z.; Sipos, F. Changes of the cytokine profile in inflammatory bowel diseases. *World J. Gastroenterol.* **2012**, *18*, 5848–5861. [CrossRef]
59. Newton, K.; Dixit, V.M. Signaling in innate immunity and inflammation. *Cold Spring Harb. Perspect. Biol.* **2012**, *4*, a006049. [CrossRef]
60. Cano, P.M.; Seeboth, J.; Meurens, F.; Cognie, J.; Abrami, R.; Oswald, I.P.; Guzylack-Piriou, L. Deoxynivalenol as a new factor in the persistence of intestinal inflammatory diseases: An emerging hypothesis through possible modulation of Th17-mediated response. *PLoS ONE* **2013**, *8*, e53647. [CrossRef]
61. Howells, L.M.; Mitra, A.; Manson, M.M. Comparison of oxaliplatin- and curcumin-mediated antiproliferative effects in colorectal cell lines. *Int. J. Cancer* **2007**, *121*, 175–183. [CrossRef]
62. Warth, B.; Del Favero, G.; Wiesenberger, G.; Puntscher, H.; Woelflingseder, L.; Fruhmann, P.; Sarkanj, B.; Krska, R.; Schuhmacher, R.; Adam, G.; et al. Identification of a novel human deoxynivalenol metabolite enhancing proliferation of intestinal and urinary bladder cells. *Sci. Rep.* **2016**, *6*, 33854. [CrossRef] [PubMed]

microorganisms

MDPI

Communication

First Demonstration of Clinical *Fusarium* Strains Causing Cross-Kingdom Infections from Humans to Plants

Thuluz Meza-Menchaca [1,†], Rupesh Kumar Singh [2,†], Jesús Quiroz-Chávez [3], Luz María García-Pérez [3], Norma Rodríguez-Mora [3], Manuel Soto-Luna [3], Guadalupe Gastélum-Contreras [3], Virginia Vanzzini-Zago [4], Lav Sharma [5] and Francisco Roberto Quiroz-Figueroa [3,*]

1 Laboratorio de Genómica Humana, Facultad de Medicina, Universidad Veracruzana, Médicos y Odontólogos S/N, Col. Unidad del Bosque, C.P. 91010 Xalapa, Veracruz, Mexico; thuluz@gmail.com
2 Centro de Química de Vila Real (CQ-VR), Universidade de Trás-os-Montes e Alto Douro, 5000-801 Vila Real, Portugal; rupesh@utad.pt
3 Instituto Politécnico Nacional, Centro Interdisciplinario de Investigación para el Desarrollo Integral Regional Unidad Sinaloa (CIIDIR-IPN Unidad Sinaloa), Laboratorio de Fitomejoramiento Molecular, Blvd. Juan de Dios Bátiz Paredes no. 250, Col. San Joachín, C.P. 81101 Guasave, Sinaloa, Mexico; yesck_pbs@hotmail.com (J.Q.-C.); luzmgp1@hotmail.com (L.M.G.-P.); rguezmora@hotmail.com (N.R.-M.); che_eucla@hotmail.com (M.S.-L.); gpegastelum21@gmail.com (G.G.-C.)
4 Hospital Para Evitar la Ceguera en México, "Luis Sánchez Bulnes" 11850 Mexico, Mexico; vivanzzini@yahoo.com
5 Syngenta Ghent Innovation Center, Devgen NV, Technologiepark 30, 9052 Gent-Zwijnaarde, Belgium; lavhere@gmail.com
* Correspondence: labfitomol@hotmail.com
† These authors contributed equally to this work.

Received: 11 February 2020; Accepted: 1 March 2020; Published: 23 June 2020

Abstract: Mycotoxins from the *Fusarium* genus are widely known to cause economic losses in crops, as well as high mortalities rates among immunocompromised humans. However, to date, no correlation has been established for the ability of *Fusarium* to cause cross-kingdom infection between plants and humans. The present investigation aims to fill this gap in the literature by examining cross-kingdom infection caused by *Furasium* strains isolated from non-immunocompromised or non-immunosuppressed humans, which were subsequently reinfected in plants and on human tissue. The findings document for the first time cross-kingdom infective events in *Fusarium* species, thus enhancing our existing knowledge of how mycopathogens continue to thrive in different hosts.

Keywords: keratomycosis; onychomycosis; pathogenicity; horizontal cross-kingdom

1. Introduction

The biotic components of any ecosystem are diverse in constitution and dependent on limited and specific resources in order to subsist, adapt, and evolve. Species interact on a broad spectrum, ranging from neutral interaction to lethal parasitism. Elucidating the network of eco-epidemiology is not only vital for understanding disease emergence, but also how it was established and escalated. Other viruses such as bird/swine flu, Ebola, SARS, and recently, nCoV-2019 have been discussed in relation to how cross-species transmission behavior could lead to viral evolution in a new host [1]. Thus, the capacity of a pathogenic organism to be a multi-host infective agent is not restricted to viruses, bacteria, and helminths. Indeed, fungi such as *Aspergillus*, *Penicillium*, and *Fusarium* spp. are able to infect multiple host species [2–5]. Although infective mechanisms have been observed significantly across species, inter-kingdom infective patterns are very rare. Inter-kingdom pathogens like *Fusarium* as well as others with their cognate hosts are well-known. However, to our knowledge, there is no evidence that the infective pattern is sequentially maintained across the inter-kingdom jump. Interestingly, other diverse parasites in humans such as *Pseudomonas* spp. show the same cross-infecting behavior from plant to animal [6].

The genus *Fusarium* is comprised of diverse and ubiquitous hyaline filamentous fungi that are adaptable to any habitable niche, making them the quintessential opportunistic pathogens [7]. These fungi and their mycotoxins adversely affect approximately 80 economically important crops [8]. However, over the last 30 years, they have emerged as an opportunistic human pathogen, producing lethal systemic infections with a wide range of morbidities in superficial infections [9,10]. This change in epidemiology is likely due to a number of complex factors. Certain strains infect a broad spectrum of host organisms, ranging from plants and insects to humans [3,5,11]. In humans, *Fusarium* can produce fungal keratitis, also known as ocular keratomycosis, which results in severe vision complications. This condition is also a significant cause of surgical intervention in 15%–27% of cases, leading to corneal transplantation, enucleation, removal of eye contents, or even treatment for vision deterioration caused by non-effective drug treatment [12]. Various fungi species can cause this illness, but *Fusarium* is the main causative agent in 37%–50% of fungal keratitis cases [13,14]. Onychomycosis is another human infection caused by *Fusarium* fungi [14]. In plants, *Fusarium* is the most persistent fungi isolated from soil that is associated with vascular invasive mycoses, and its conidia can infect aerial tissue such as corn ears. In maize, *Fusarium* spp. can provoke rots and blight that affect stalk, grain, roots, and seedlings [15–17]. Mycotoxins produced by *Fusarium* are a prominent economic issue since they can cause crop loss, in addition to having important animal and human health repercussions [18,19]. *Fusarium* species and their respective strains are rapidly becoming multidrug-resistant [20]. From a health perspective, one of the most harmful *Fusarium*-related diseases is ocular keratomycosis, given the fact that ophthalmic fungal infections represent one of the main etiologic factors of blindness in humans. Although the precise reasons are unknown, the incidence of fungal keratitis has dramatically increased in the last two decades, particularly in countries such as China, India, Brazil, and Mexico [9].

Transmission of pathogens between plants and humans has been hypothesized in the past, but a specific mechanism could not be detected [21]. Fungi are capable of both direct and indirect transmission. Inter-kingdom infective patterns across species is rare. The present study elucidates the ability of clinical samples of *Fusarium* to infect monocotyledonous (corn) or dicotyledonous (*Arabidopsis*) plants as well as human tissue by testing 13 specific fungal keratitis samples from four *Fusarium* species. The findings in the present study are a step forward in clarifying whether there is a cyclical pattern of infection between plants and humans, or whether the infection is only oriented in a plant-to-human direction.

2. Materials and Methods

2.1. Sample Collection

The human-pathogenic fungal strains initially originated from 13 non-immunocompromised or non-immunosuppressed individuals with keratomycosis, and were acquired by courtesy of the hospital staff at "El Hospital para Prevenir la Ceguera en México, Luis Sánchez Bulnes" between January 2013 and August 2016. Patients were registered from nine states of Mexico with variable age and economic activity (Table 1). In addition, to isolate the potential causative agent of keratitis, eye rub from infected eyes were taken with sterile hyssop and cultured on Sabouraud Dextrose Emmons agar medium agar (SGA; Difco, Detroit, MI, USA). Monoconidia cultures were obtained by serial dilution in Spezieller Nährstoffarmer agar medium (SNA) with a 1-cm^2 filter paper at 37°C, as described earlier [8]. Colonies were observed growing during the first 48–72 h. The species were identified morphologically on microculture, as in a previous study [22].

Table 1. Profiles of keratitis patients.

Date	Lab ID	Symptomatology	Gender/Age	Patient Location	Occupation	Species
10/01/2013	21564	Ocular trauma	F/75	Tuzamapan, Puebla	Housewife	*Fusarium solani*
19/02/2013	21791	DM 10 years, cornea trauma	M/73	Zacatecas	Farmer	*Fusarium dimerum*
20/02/2013	21797	Ocular trauma	M/28	Puebla	Farmer	*Fusarium solani*
06/03/2013	21890	Ocular trauma	F/32	Quintana Roo	Housewife	*Fusarium solani*
08/04/2013	22083	Ocular trauma	M/76	Mexico city	Worker	*Fusarium solani*
18/06/2013	22503	Ocular trauma	M/7	Veracruz, Ver	Student	*Fusarium solani*
19/08/2013	22869	Insidious, pain+++, immune ring	F/41	La trinitaria Chiapas	Housewife	*Fusarium dimerum*
23/12/2013	23544	Insidious, pain++	M/47	AltamiraTamaulipas	Worker	*Fusarium solani*
17/02/2014	23813	Ocular trauma	M/41	Mexico city	Builder	*Fusarium solani*
20/08/2014	24810	Ocular trauma	M/49	Durango Durango	Farmer	*Fusarium solani*
30/01/2015	25704	Insidious, pain	F/34	Acapulco Guerrero	Housewife	*Fusarium solani*
27/04/2015	26256	Ocular trauma	F/30	Mexico City	Housewife	*Fusarium dimerum*
18/08/2016	28615	Ocular trauma	M/62	Puebla	Farmer	*Fusarium oxysporum*

Abbreviations used: DM: diabetes mellitus; pain+++: intense pain; Insidious: patient did not feel any trauma or damage. Patients did not have any immunodeficiencies. F: female; M: male.

2.2. *Detached in-vitro Tissue Assay*

2.2.1. Surface Disinfection of Seeds

Maize seeds were surface disinfected by sonication (ultrasonic bath 2.8L, Fisher Scientific) in sterile distilled water with Tween 20 (0.1% *v/v*) for 5 min. Subsequently, seeds were immersed in 1.5% (*v/v*) sodium hypochlorite (NaOCl) at 52°C for 20 min (Thermobath FE-377, Felisa), followed by rinsing three times in sterile distilled water, and air dried in a Class II Type A2 Biological Safety Cabinet (Herasafe KS, Thermo Scientific, Langenselbold, Germany) and grown in culture room in Magenta™ vessel (no. cat V8505, Sigma, MO, USA) with sterile sand.

2.2.2. Conidia Suspensions and Inoculation of Tissue by *Fusarium*

The *Fusarium* strains (Table 1) were cultivated in the SNA medium with a 1-cm^2 filter paper [8] supplemented with neomycin (0.12 mg/mL) and streptomycin (1 µg/mL), and cultured at 25 ± 2°C for 7 days [23]. Conidia were harvested by adding 5 mL of sterile saline solution (0.8% (*w/v*) sodium chloride) to the culture medium with gentle shaking. The conidia quantification was performed using a Neubauer chamber (Hausser Scientific, Horsham, PA, USA) and a light microscope (B-383-M11, Optika, Ponteranica, Italy) and by CFU on PDA plates. The conidia suspension was prepared with the final concentration of inoculation of 1 × 10^6 CFU/mL. Leaves and roots were collected from 2 weeks old in-vitro grown maize plants with no visible fungi contamination [24] for detached tissue assays. The tissues were inoculated with 200 µL of conidia solution and incubated at 25°C in a wet chamber and photographed after 5 days post-inoculation (dpi) (Figure 1).

2.3. *In vitro Seedling Assay*

2.3.1. Surface Disinfection of Seeds

The procedure for maize seed surface disinfection was similar to the detached tissue assay. Columbia-0 *Arabidopsis* seeds were surface sterilized for 5 min with 1.25% (*v/v*) NaOCl containing 0.1% (*v/v*) Triton X-100 for four min then rinsed four times with sterile water and sown in 0.1% (*w/v*) sterile agar. All seeds were stratified at 4°C for 3 days and then planted on Petri dishes containing 0.5× Murashigue and Skoog medium [25] supplemented with 0.5% (*w/v*) sucrose and 1% (*w/v*) agar (pH 5.8).

2.3.2. Inoculation of Maize and *Arabidopsis* Seeds by *Fusarium*

Surface-disinfected maize seeds were immersed in the conidia suspension for five minutes, and planted in in vitro culture container with sand, and cultivated at 25 ± 2°C for 5 days. In the case of *Arabidopsis*, in vitro plants 2 weeks old were transferred into in vitro culture container with fertilized sand and inoculated with 200 uL of work suspensions and incubated at 25 ± 2°C for 7 days. Before every assay, both sand and fertilizer were autoclaved at 120°C for 60 min. Photographs were recorded at 5 and 7 dpi for maize and *Arabidopsis*, respectively (Figure 1).

2.4. Human Onychomycosis Assay

Small pieces of nails (ca. 5 mm) without polish were washed with 1.25% (*v/v*) NaOCl containing 0.1% (*v/v*) Triton X-100 for 5 min. Nails were rinsed three times with sterile water and incubated for 3 days on PDA at 25 ± 2°C. The non-infected nails were washed three times with sterile water to eliminate any adhered PDA fragments. Nails were placed in a wet chamber (Petri dishes) and the nail edge was infected with 100 μL of work conidia solution. Nails were observed for visible fungal growth from the fourth day onwards until the ninth day (Figure 2).

2.5. Fusarium Detection by Confocal Microscopy

In order to visualize *Fusarium* spp. on the nails, samples were stained with WGA-Alexa Fluor® 488 conjugate (W11261, Life Technologies; CA USA), which binds to the chitin molecules on the fungal cell wall [26]. Samples were incubated for 30 min at room temperature in 1× PBS buffer (137 mM NaCl, 10 mM phosphate, 2.7 mM KCl, pH 7.4) supplemented with 1 ng/μL WGA. For visualization, stained nails were placed on a microscope slide and covered with a glass coverslip. Confocal microscopy (Leica TCS SP5 X) was used with the white laser for 499 nm excitation wavelength and emission ranges of 512–526 nm for WGA (green fluorescence) and 632–739 nm to get the autofluorescence signal (red fluorescence).

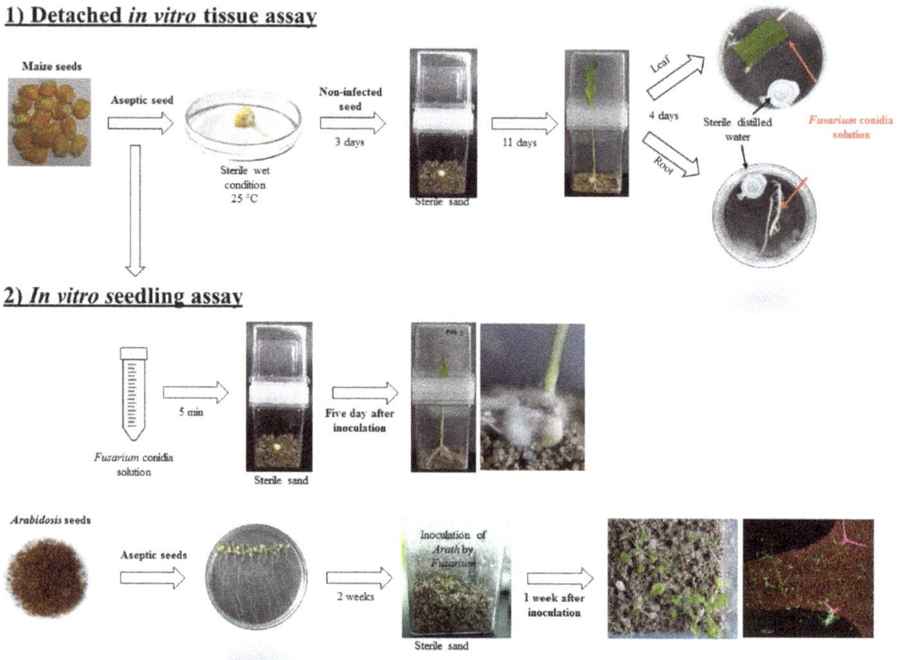

Figure 1. Diagrammatic representation of strategy to evaluate the infective capacity of fungi on plants.

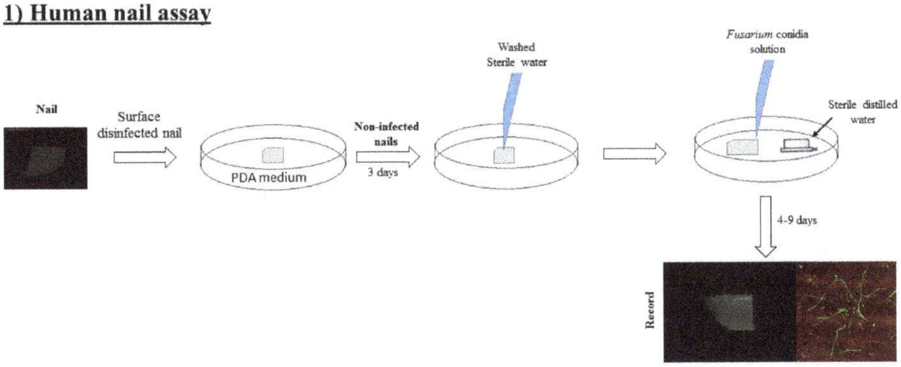

Figure 2. Diagrammatic presentation of strategy to evaluate the infective capacity of fungi to cause onychomycosis.

3. Results

Although the cross-kingdom infective capacity of *Fusarium* has been hypothesized, there is no evidence of any sequential human–plant–human tissue reinfection. In order to better understand the broad infectious properties of *Fusarium* spp., the present study assessed the ability of *Fusarium* isolates from human keratomycosis to infect monocot (maize) and dicot (*Arabidopsis*) plants, as well as human tissue in the form of nails. First, the *Fusarium* genotypes were isolated from 15 fungal keratitis samples, initially taken from patients at the "El Hospital para Prevenir la Ceguera en México, Luis Sánchez Bulnes" between January 2013 and August 2016. Each sample was taken from the eye of a non-immunologically compromised patient. Patients came from nine Mexican states and had diverse occupations. Patient age was not restricted to any specific range, but 87.5% were middle-aged adults or older (Figure S1).

In the first stage, we evaluated the infective capacity of *Fusarium* isolates from patients with keratitis to infect plants using two assays. (1) The first assay consisted of the inoculated tissue of detached leaves and root maize seedlings, placed in wet chamber Petri dishes. (2) In the second assay, whole maize and *Arabidopsis* seedlings were inoculated in vitro while still alive. *Fusarium verticillioides*, which infects corn plants was used as a positive control [23,24]. In both experiments, *Fusarium* conidia germinated, colonized the detached tissues, and deterred the growth of in-vivo seedlings (Figure 3). These results demonstrate that *Fusarium* spp. from keratitis patients conserve their infective capacity during the cross-kingdom reinfection jump from humans to plants.

In the second stage, we evaluated the capacity of *Fusarium* strains to back-infect human tissue. In this experiment, the capacity of the isolates to cause onychomycosis was assessed, using human nail samples (Figure 4). Observations suggested that every strain that infected maize and *Arabidopsis* was also able to back-infect human nails. After 4 days of inoculation, *Fusarium* was well established (Figure 4A) and exudates were observed (arrowheads, Figure 4A). Due to the rough surface of the nails, *Fusarium* was able to colonize and use the keratin in them as a growth substrate (arrow, Figure 4B). These results, together with the literature [14], indicate that *Fusarium* might cause onychomycosis in humans. Even though *Fusarium* has been sampled and well-studied in immune-compromised patients [12], the species found in this study were obtained from non-immune-compromised patients, suggesting that the pathogen reached the patients' eyes by ocular injury.

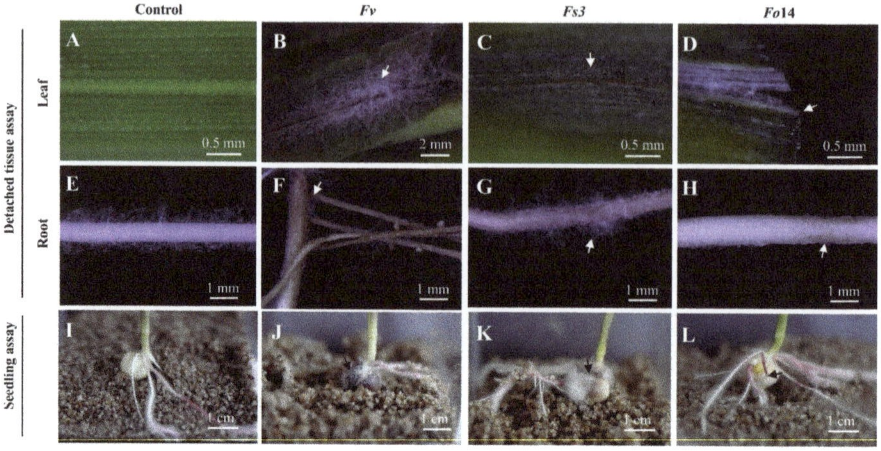

Figure 3. *Fusarium* spp. isolated from human keratitis conserve their plant infective capacity. Detached leaf assay: (**A**) control non-infected leaf; (**B–D**) leaves infected by isolated *Fusarium* strains *Fv*, *Fs*3, and *Fo*14, respectively. Detached root assay: (**E**) control non-infected root; (**F–H**) roots infected by isolated *Fusarium* strains *Fv*, *Fs*3, and *Fo*14, respectively. In both detached tissues, leaves and roots were imaged 12 days after infection. Seedling assay: (**I**) control non-infected seedling; (**J–L**) seedlings from free-infected seeds inoculated with isolated *Fusarium* strains *Fv*, *Fs*3, and *Fo*14, respectively. Seedlings were imaged 5 days after infection. Samples in the detached tissue and seedling growth assays were under sterile in vitro conditions. Arrows show the sites with abundance fungi growth or damaged tissue. *Fv*: *Fusarium verticillioides*; *Fs*3: *Fusarium solani* isolate no. 3; *Fo*14: *Fusarium oxysporum* isolate no. 14.

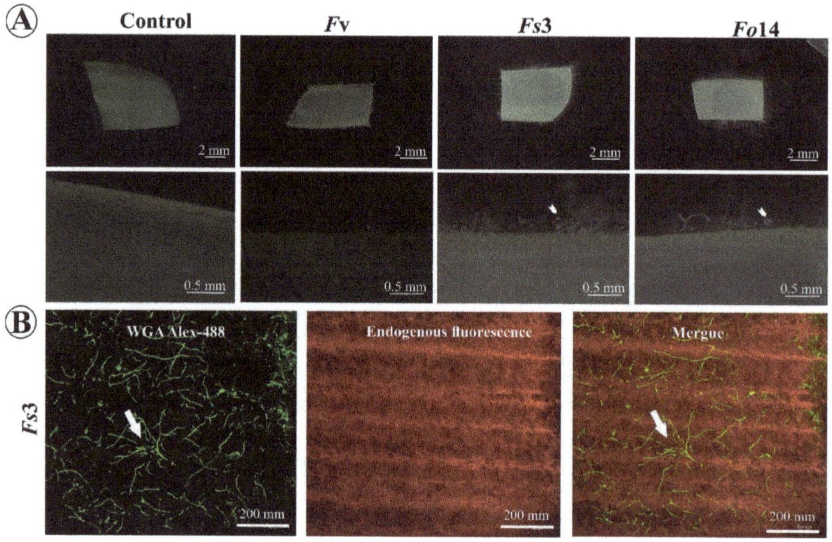

Figure 4. *Fusarium* isolates are a potential causative agent of onychomycosis in human tissue. (**A**) Macroscopic (upper panels) and microscopic (lower panels) imaging of human nails inoculated with *Fv*, *Fs*3, or *Fo*14. (**B**) Colonization of a human nail (which emits endogenous red fluorescence) by *Fusarium Fs*3 isolate (green fluorescence). WGA Alexa-488 was used to detect fungal hyphae in the colonized nail (see 'Merge' image). All photographs (**A**) and scans (**B**) were taken 4 days after inoculation in a wet chamber. *Fv*: *Fusarium verticillioides*; *Fs*3: *Fusarium solani* isolate no. 3; *Fo*14: *Fusarium oxysporum* isolate no. 14.

This study's findings show that *Fusarium*-infective agents may alter human health, and that their infective capacity to colonize human tissue and plants and to back-infect human tissue (nails) is intact.

4. Discussion

There have been several species identified to date that infect both plants and animals. These species range from the extreme, such as *Agrobacterium tumefaciens*, which infects fungi, plants, and animals [27], to the more common species that co-infect plants and animals, including *Aspergillus fumigatus*, *Pleurostomophora richardsiae*, *Pseudomonas aeruginosa*, *Pythium insidiosum*, *Rhizopus oryzae*, *Sporothrix schenckii*, *Staphylococcus aureus*, and *Trichoderma longibrachiatum* [28]. Studies on the infection mechanism in plant hosts, particularly at the immune system level, have revealed that certain features are shared within animal hosts [29]. It is commonly accepted that in plant–animal co-infections, the pathogen is dependent on its ability to recruit iron from the host or environment [30]. Previous studies on *Fusarium* have tried to establish the cross-kingdom pathogenicity between plants, mammals, and insects [31–33]; however, to our knowledge, the present study is the first report to demonstrate this infective property between plants and humans.

Fusarium is one cause of disease in a wide variety of crops as maize, although it does not produce the same lethality in animals as it does in plants [34]. This might be due to differences between the two kingdoms during pattern recognition by the receptors of innate immune cells [35], which could allow *Fusarium* to evade the host's immune defenses. One important difference is that the main components of a blood circulatory system, i.e., macrophages, neutrophils, and dendritic cells are not found in plants. In this sense, the pathogenic outcome in both kingdoms can range from lethal to non-lethal outcomes. This may also cause a different clinical result, resulting in a distinct disease with completely different consequences. Taking this into account, there may be two different diseases produced by the same microbial pathogen, which intriguingly appears to contradict to Koch's third postulate [36]. Just as the well-known evidence of exemptions to Koch's postulates, it shows that not all parasites can be isolated in artificial media, single isolated pathogens could also produce different symptoms depending on the nature of the host and tissue. A very clear example of this is the fact that *Salmonella* spp. can live within plants without producing any lethal effects, while in animals it is fatal. Although *Fusarium* rarely causes a disease harmful enough to lead to mortality in healthy humans, it can be lethal for plant species. It will be interesting in the future to investigate the similarities between how pathogens escape in order to establish disease. The present research may also facilitate our understanding of how some host species survive an infection while others perish from the same pathogen.

In summary, we investigated whether *Fusarium* species isolated from keratomycosis human patients, which are normally pathogens of plants, conserve their infective capacity to re-infect plants and other human tissue. We demonstrated that *Fusarium* spp. conserves their infective mechanism for colonizing human tissue and plants and for back-infecting other human tissue, such as nails. Our results also found a new exemption to Koch's third postulate, as the same fungal pathogen was seen to produce two different diseases. This work could serve as a reference for demonstrating cyclic *Fusarium* reinfections between plants and humans. It thus suggests that the infective mechanism of *Fusarium* could be conserved. Further, Omics studies will help to elucidate how *Fusarium* changes its infective mechanism (gene expression and physiology) to adapt to hosts from a different kingdom.

Supplementary Materials: The following are available online at http://www.mdpi.com/2076-2607/8/6/947/s1, Figure S1. Summary of patient characteristics infected by *Fusarium keratitis*. (A) The Mexican states involved in this study, with percentages of keratitis for each. (B) Distribution of *Fusarium* keratitis by gender. (C) Distribution of patient occupations. (D) Distribution of patient age groups.

Author Contributions: Conceptualization, F.R.Q.-F.; investigation, T.M.-M., J.Q.-C., L.M.G.-P., N.R.-M., V.V.-Z., L.S., M.S.-L., G.G.-C. and R.K.S.; formal analysis, T.M.-M. and F.R.Q.-F.; writing and data curation, T.M.-M., R.K.S., L.S. and F.R.Q.-F.; validation, F.R.Q.-F. All authors have read and agreed to the published version of the manuscript.

Funding: The authors are grateful to the Mexican National Council on Science and Technology (CONACyT) for financial support from Infraestructura no. 250738, Frontera de la Ciencia no. 1070 and SIP 20195805.

Acknowledgments: We wish to thank H. Cámara de Diputados for the Insignia Project "Proyecto de equipamiento en medio ambiente". We are grateful to Brandon Loveall of IMPROVENCE for proofreading the manuscript.

Conflicts of Interest: The authors declare no conflict of interest.

References

1. Streicker, D.G.; Turmelle, A.S.; Vonhof, M.J.; Kuzmin, I.V.; McCracken, G.F.; Rupprecht, C.E. Host Phylogeny Constrains Cross-Species Emergence and Establishment of Rabies Virus in Bats. *Science* **2010**, *329*, 676–679. [CrossRef] [PubMed]
2. Egbuta, M.A.; Mwanza, M.; Babalola, O.O. Health Risks Associated with Exposure to Filamentous Fungi. *Int. J. Env. Res. Public Health* **2017**, *14*. [CrossRef] [PubMed]
3. Sharma, L.; Marques, G. Fusarium, an Entomopathogen-A Myth or Reality? *Pathogens* **2018**, *7*. [CrossRef] [PubMed]
4. Sharma, L.; Oliveira, I.; Raimundo, F.; Torres, L.; Marques, G. Soil Chemical Properties Barely Perturb the Abundance of Entomopathogenic Fusarium oxysporum: A Case Study Using a Generalized Linear Mixed Model for Microbial Pathogen Occurrence Count Data. *Pathogens* **2018**, *7*. [CrossRef]
5. Sharma, L.; Goncalves, F.; Oliveira, I.; Torres, L.; Marques, G. Insect-associated fungi from naturally mycosed vine mealybug Planococcus ficus (Signoret) (Hemiptera: Pseudococcidae). *Biocontrol Sci. Technol.* **2018**, *28*, 122–141. [CrossRef]
6. Sitaraman, R. Pseudomonas spp. as models for plant-microbe interactions. *Front. Plant Sci.* **2015**, *6*. [CrossRef]
7. Ma, L.J.; Geiser, D.M.; Proctor, R.H.; Rooney, A.P.; O'Donnell, K.; Trail, F.; Gardiner, D.M.; Manners, J.M.; Kazan, K. Fusarium Pathogenomics. *Annu. Rev. Microbiol* **2013**, *67*, 399–416. [CrossRef]
8. Leslie, J.F.; Summerell, B.A. *The Fusarium Laboratory Manual*; Blackwell Publishing Ltd.: Hoboken, NJ, USA, 2007; Volume 2, pp. 1–369.
9. Al-Hatmi, A.M.S.; Hagen, F.; Menken, S.B.J.; Meis, J.F.; de Hoog, G.S. Global molecular epidemiology and genetic diversity of Fusarium, a significant emerging group of human opportunists from 1958 to 2015. *Emerg. Microbes. Infec.* **2016**, *5*. [CrossRef]
10. Hayashida, M.Z.; Seque, C.A.; Enokihara, M.; Porro, A.M. Disseminated fusariosis with cutaneous involvement in hematologic malignancies: Report of six cases with high mortality rate. *Bras Derm.* **2018**, *93*, 726–729. [CrossRef]
11. Sun, S.T.; Lui, Q.X.; Han, L.; Ma, Q.F.; He, S.Y.; Li, X.H.; Zhang, H.M.; Zhang, J.J.; Liu, X.H.; Wang, L.Y. Identification and Characterization of Fusarium proliferatum, a New Species of Fungi that Cause Fungal Keratitis. *Sci. Rep.* **2018**, *8*. [CrossRef]
12. Nucci, M.; Anaissie, E. Fusarium infections in immunocompromised patients. *Clin. Microbiol. Rev.* **2007**, *20*, 695–704. [CrossRef]
13. Vanzinni Zago, V.; Manzano-Gayosso, P.; Hernández-Hernández, F. Queratomicosis en un centro de atención oftalmológica en la Ciudad de México. *Rev. Iberoam. Micol.* **2010**, *27*, 57–61. [CrossRef]
14. Monod, M.; Mehul, B. Recent Findings in Onychomycosis and Their Application for Appropriate Treatment. *J. Fungi.* **2019**, *5*. [CrossRef]
15. Gai, X.; Dong, H.; Wang, S.; Liu, B.; Zhang, Z.; Li, X.; Gao, Z. Infection cycle of maize stalk rot and ear rot caused by Fusarium verticillioides. *PLoS ONE* **2018**, *13*, e0201588. [CrossRef] [PubMed]
16. Okello, P.N.; Petrović, K.; Kontz, B.; Mathew, F.M. Eight Species of Fusarium Cause Root Rot of Corn (Zea mays) in South Dakota. *Plant Health Prog.* **2019**, *20*, 38–43. [CrossRef]
17. Baldwin, T.T.; Zitomer, N.C.; Mitchell, T.R.; Zimeri, A.M.; Bacon, C.W.; Riley, R.T.; Glenn, A.E. Maize Seedling Blight Induced by Fusarium verticillioides: Accumulation of Fumonisin B-1 in Leaves without Colonization of the Leaves. *J. Agric. Food Chem.* **2014**, *62*, 2118–2125. [CrossRef] [PubMed]
18. Riley, R.T.; Voss, K.A.; Speer, M.; Stevens, V.L.; Waes, J.G.-V. Fumonisin inhibition of ceramide synthease: A possible risk factor for human nueral tube defects. In *Sphingolipid Biology*, 1st ed.; Hirabayashi, Y., Igarashi, Y., Merrill, A.H., Eds.; Springer: Tokyo, Japan, 2006; pp. 345–361.
19. Antonissen, G.; Martel, A.; Pasmans, F.; Ducatelle, R.; Verbrugghe, E.; Vandenbroucke, V.; Li, S.J.; Haesebrouck, F.; Van Immerseel, F.; Croubels, S. The Impact of Fusarium Mycotoxins on Human and Animal Host Susceptibility to Infectious Diseases. *Toxins* **2014**, *6*, 430–452. [CrossRef]

20. Al-Hatmi, A.M.S.; Meis, J.F.; de Hoog, G.S. Fusarium: Molecular Diversity and Intrinsic Drug Resistance. *PLoS Pathog.* **2016**, *12*. [CrossRef]
21. Zhang, N.; O'Donnell, K.; Sutton, D.A.; Nalim, F.A.; Summerbell, R.C.; Padhye, A.A.; Geiser, D.M. Members of the Fusarium solani species complex that cause infections in both humans and plants are common in the environment. *J. Clin. Microbiol.* **2006**, *44*, 2186–2190. [CrossRef] [PubMed]
22. de Hoog, G.S.; Guarro, J.; Gené, J.; Figueras, M.J. *Atlas of Clinical Fungi*, 2nd ed.; Universatat Rovira i Virgili: Reus, Spain, 2000; p. 1126.
23. Leyva-Madrigal, K.Y.; Larralde-Corona, C.P.; Apodaca-Sanchez, M.A.; Quiroz-Figueroa, F.R.; Mexia-Bolanos, P.A.; Portillo-Valenzuela, S.; Ordaz-Ochoa, J.; Maldonado-Mendoza, I.E. Fusarium Species from the Fusarium fujikuroi Species Complex Involved in Mixed Infections of Maize in Northern Sinaloa, Mexico. *J. Phytopathol.* **2015**, *163*, 486–497. [CrossRef]
24. Román, S.G. *Caracterización de genotipos de maíz (Zea mays L.) a la infección de Fusarium verticillioides en diferentes fases del ciclo de vida de la planta y su correlación con marcadores moleculares de tipo SNPs*; Instituto Politécnico Nacional: Guasave, México, 2017.
25. Murashige, T.; Skoog, F. A revised medium for rapid growth and bioassays with tobacco tissue cultures. *Physiol. Plant.* **1962**, *15*, 473–497. [CrossRef]
26. Figueroa-López, A.M.; Cordero-Ramírez, J.D.; Quiroz-Figueroa, F.R.; Maldonado-Mendoza, I.E. A high-throughput screening assay to identify bacterial antagonists against Fusarium verticillioides. *J. Basic Microbiol.* **2014**, *54*, S125–S133. [CrossRef] [PubMed]
27. Hwang, H.H.; Yu, M.; Lai, E.M. Agrobacterium-mediated plant transformation: Biology and applications. *Arab. Book* **2017**, *15*, e0186. [CrossRef] [PubMed]
28. van Baarlen, P.; van Belkum, A.; Summerbell, R.C.; Crous, P.W.; Thomma, B.P.H.J. Molecular mechanisms of pathogenicity: How do pathogenic microorganisms develop cross-kingdom host jumps? *Fems. Microbiol. Rev.* **2007**, *31*, 239–277. [CrossRef] [PubMed]
29. Nurnberger, T.; Brunner, F.; Kemmerling, B.; Piater, L. Innate immunity in plants and animals: Striking similarities and obvious differences. *Immunol. Rev.* **2004**, *198*, 249–266. [CrossRef] [PubMed]
30. Sexton, A.C.; Howlett, B.J. Parallels in fungal pathogenesis on plant and animal hosts. *Eukaryot. Cell* **2006**, *5*, 1941–1949. [CrossRef] [PubMed]
31. Ortoneda, M.; Guarro, J.; Madrid, M.P.; Caracuel, Z.; Roncero, M.I.; Mayayo, E.; Di, P.A. Fusarium oxysporum as a multihost model for the genetic dissection of fungal virulence in plants and mammals. *Infect. Immun.* **2004**, *72*, 1760–1766. [CrossRef]
32. Navarro-Velasco, G.Y.; Prados-Rosales, R.C.; Ortiz-Urquiza, A.; Quesada-Moraga, E.; Di Pietro, A. Galleria mellonella as model host for the trans-kingdom pathogen Fusarium oxysporum. *Fungal Genet. Biol.* **2011**, *48*, 1124–1129. [CrossRef]
33. Segorbe, D.; Di Pietro, A.; Perez-Nadales, E.; Turra, D. Three Fusarium oxysporum mitogen-activated protein kinases (MAPKs) have distinct and complementary roles in stress adaptation and cross-kingdom pathogenicity. *Mol. Plant Pathol.* **2017**, *18*, 912–924. [CrossRef]
34. Abdallah, M.F.; De Boevre, M.; Landschoot, S.; De Saeger, S.; Haesaert, G.; Audenaert, K. Fungal Endophytes Control Fusarium graminearum and Reduce Trichothecenes and Zearalenone in Maize. *Toxins* **2018**, *10*. [CrossRef]
35. Ausubel, F.M. Are innate immune signaling pathways in plants and animals conserved? *Nat. Immunol.* **2005**, *6*, 973–979. [CrossRef] [PubMed]
36. Neville, B.A.; Forster, S.C.; Lawley, T.D. Commensal Koch's postulates: Establishing causation in human microbiota research. *Curr. Opin. Microbiol.* **2018**, *42*, 47–52. [CrossRef] [PubMed]

MDPI
St. Alban-Anlage 66
4052 Basel
Switzerland
Tel. +41 61 683 77 34
Fax +41 61 302 89 18
www.mdpi.com

Microorganisms Editorial Office
E-mail: microorganisms@mdpi.com
www.mdpi.com/journal/microorganisms

www.ingramcontent.com/pod-product-compliance
Lightning Source LLC
LaVergne TN
LVHW070455100526
838202LV00014B/1725
* 9 7 8 3 0 3 9 4 3 4 0 8 4 *